W0037006

Earthquake Science and Seismic Risk Reduction

NATO Science Series

A Series presenting the results of scientific meetings supported under the NATO Science Programme.

The Series is published by IOS Press, Amsterdam, and Kluwer Academic Publishers in conjunction with the NATO Scientific Affairs Division

Sub-Series

I. Life and Behavioural Sciences	IOS Press
II. Mathematics, Physics and Chemistry	Kluwer Academic Publishers
III. Computer and Systems Science	IOS Press
IV. Earth and Environmental Sciences	Kluwer Academic Publishers
V. Science and Technology Policy	IOS Press

The NATO Science Series continues the series of books published formerly as the NATO ASI Series.

The NATO Science Programme offers support for collaboration in civil science between scientists of countries of the Euro-Atlantic Partnership Council. The types of scientific meeting generally supported are "Advanced Study Institutes" and "Advanced Research Workshops", although other types of meeting are supported from time to time. The NATO Science Series collects together the results of these meetings. The meetings are co-organized bij scientists from NATO countries and scientists from NATO's Partner countries – countries of the CIS and Central and Eastern Europe.

Advanced Study Institutes are high-level tutorial courses offering in-depth study of latest advances in a field.
Advanced Research Workshops are expert meetings aimed at critical assessment of a field, and identification of directions for future action.

As a consequence of the restructuring of the NATO Science Programme in 1999, the NATO Science Series has been re-organised and there are currently five sub-series as noted above. Please consult the following web sites for information on previous volumes published in the Series, as well as details of earlier sub-series.

http://www.nato.int/science
http://www.wkap.nl
http://www.iospress.nl
http://www.wtv-books.de/nato-pco.htm

Earthquake Science and Seismic Risk Reduction

edited by

Francesco Mulargia
Università degli Studi di Bologna,
Italy

and

Robert J. Geller
University of Tokyo,
Japan

Springer-Science+Business Media, B.V.

Not just a proceedings volume:
This book builds on the lively discussions that took place at the NATO Advanced
Research Workshop, on State of Scientific Knowledge Regarding Earthquake
Occurrence and Implications for Public Policy, held in Sardinia, Italy, but it is a uniformly
edited collection of contributions by leading authorities, designed to provide a
comprehensive one-volume summary of the state of the art in both the basic and applied
aspects of earthquake occurrence and seismic risk reduction.

A C.I.P. Catalogue record for this book is available from the Library of Congress.

Additional material to this book can be downloaded from http://extras.springer.com.

ISBN 978-1-4020-1778-0   ISBN 978-94-010-0041-3 (eBook)
DOI 10.1007/978-94-010-0041-3

Printed on acid-free paper

All Rights Reserved
© 2003 Springer Science+Business Media Dordrecht
Originally published by Kluwer Academic Publishers in 2003
Softcover reprint of the hardcover 1st edition 2003
No part of this work may be reproduced, stored in a retrieval system, or transmitted in any
form or by any means, electronic, mechanical, photocopying, microfilming, recording or
otherwise, without written permission from the Publisher, with the exception of any
material supplied specifically for the purpose of being entered and executed on a computer system, for exclusive use by the purchaser of the work.

Associate Editors

S. Castellaro (Bologna)   M. Ciccotti (Bologna)

Principal Contributors   **Co-Contributors**

D. Albarello (Siena)   H. Akman (Istanbul)
S. A. Anagnostopoulos (Patras)   M. Demircioglu (Istanbul)
S. Castellaro (Bologna)   G. Di Toro (Padova)
M. Ciccotti (Bologna)   E. Durukal (Istanbul)
M. Erdik (Istanbul)   Y. Fahjan (Istanbul)
R. J. Geller (Tokyo)   N. Field (Pasadena)
D. D. Jackson (Los Angeles)   A. Frankel (Lakewood)
Y. Y. Kagan (Los Angeles)   D. A. Freedman (Berkeley)
J. Kertész (Budapest)   J. Gomberg (Memphis)
I. Main (Edinburgh)   Y. F. Rong (Los Angeles)
A. Michael (Menlo Park)   F. Sansò (Milano)
M. Mucciarelli (Potenza)   K. Sesetyan (Istanbul)
F. Mulargia (Bologna)   K. Shedlock (Lakewood)
R. B. Olshansky (Urbana-Champaign)   B. Siyahi (Istanbul)
P. A. Pirazzoli (Paris)
F. Rocca (Milano)
P. B. Stark (Berkeley)

Contents

Preface

What is the first thing that ordinary people, for whom journalists are the proxy, ask when they meet a seismologist? It is certainly nothing technical like "What was the stress drop of the last earthquake in the Imperial Valley?" It is a simple question, which nevertheless summarizes the real demands that society has for seismology. This question is "Can you predict earthquakes?" Regrettably, notwithstanding the feeling of omnipotence induced by modern technology, the answer at present is the very opposite of "Yes, of course".

The primary motivation for the question "Can you predict earthquakes?" is practical. No other natural phenomenon has the tremendous destructive power of a large earthquake, a power which is rivaled only by a large scale war. An earthquake in a highly industrialized region is capable of adversely affecting the economy of the whole world for several years. But another motivation is cognitive. The aim of science is 'understanding' nature, and one of the best ways to show that we understand a phenomenon is the ability to make accurate predictions.

While it is unquestionable that our present understanding of earthquake physics is poor, leaving deterministic prediction of individual large earthquakes well beyond our reach at present and for the foreseeable future, it would be incorrect to state that earthquakes are totally unpredictable phenomena that are equally likely to strike anywhere and at any time. In fact, it is well known that seismogenesis is not completely random and that earthquakes tend to be more localized in space, primarily on plate boundaries, and more clustered in time and space than would be expected for a completely random process.

The scale-invariant nature of fault morphology, the frequency-magnitude distribution of earthquakes, the spatio-temporal clustering of earthquakes, their relatively constant dynamic stress drop, and the apparent ease with which they can be triggered by small perturbations in stress are all clues that can be used to achieve a semiempirical predictive power even without the capability to physically model the earthquake source process. However, our present predictive power falls far short of that envisioned by journalists and the public. Whether or not there are prospects for future improvements, and, if so, to what extent, is a topic of intense scientific discussion; some of the arguments will be introduced in this volume.

Notwithstanding the less than satisfactory state of our present scientific knowledge, earthquakes do occur, and recent earthquakes have caused substantial human casualties and economic losses in many countries. It is extremely rare for earthquakes to be the direct cause of casualties; almost all casualties are due to the failure of buildings or other structures, or to secondary effects such as fires. It is necessary to take all feasible steps to reduce seismic risk. The main arena for practical steps to reduce seismic risk is that of earthquake engineering. However, the measures taken by engineers must be soundly based on the best state of present scientific knowledge (including its uncertainties). Our goal for this volume is to provide a concise summary of the present state of earthquake science and to discuss how this can be reflected in practical measures for seismic risk reduction.

This volume is based on the lively discussions at the NATO Advanced Research Workshop (ARW) on "State of scientific knowledge regarding earthquake occurrence and implications for public policy" held in Arbus, Sardinia, from October 14 to October 19, 2000, under the Co-Direction of Stathis Stiros and M. Nafi Toksöz. The program and list of participants may be found at http://ibogeo.df.unibo.it/arw2000/index.html.

It is traditional for symposia and workshops to publish a proceedings volume, but in many cases these are just a collection of papers that have been stapled together. The participants in this ARW decided that it would be worth the extra effort to produce a work in which the various contributions would be more tightly integrated. To this end, an initial outline was agreed on, and authors were commissioned to write particular sections or chapters. The authors of each section or chapter are identified at its beginning. The contributions were then edited extensively to produce a more uniform and coherent work. Much time and effort on the part of the Associate Editors, Silvia Castellaro and Matteo Ciccotti, was required to perform this integration. The editors thank them for the great job they have done. They also thank Kenji Kawaii for his help in checking the references.

The contributors to this volume are listed on the inside cover page. Principal contributors are first authors or co-authors who also attended the ARW; co-contributors are co-authors who did not attend the ARW.

Why is the earthquake problem so difficult?

Our goal in this volume is to provide a coherent 'snapshot' of the state of the art in earthquake science and seismic risk reduction, summarizing both what is known and what is still the subject of research and controversy. Why is there still so much that we don't yet know?

The Earth is comprised of a core composed primarily of iron (with some nickel, and traces of other elements). The inner core is solid, while the outer core is liquid. Above the core, with a thickness of about 3000 km, is the mantle, composed mostly of silicate rocks. The mantle is viscoelastic, behaving as a solid on time scales of minutes or hours, but deforming viscously on a time scale of tens of thousands or millions of years. Above the mantle is the crust, with thickness ranging from a few km to about 100 km. The crust and the uppermost few tens of km of the mantle compose the lithosphere, which tends to release stress by by brittle failure (earthquakes) rather than by viscous deformation. Lithospheric slabs that subduct at oceanic trenches sink into the mantle and in some cases are still sufficiently cool to allow earthquakes at depths of up to about 700 km. However, with the exception of slabs, almost all earthquakes occur at shallow depth (< 30 km).

Excepting the brittle lithosphere, the mantle as a whole is thermally convecting. It acts as a giant engine, carrying thermal energy from the core-mantle boundary to the Earth's surface. Strain energy builds up in the lithosphere as it is dragged along by the convection in the mantle. Some of this strain energy is released from time to time by earthquakes. The lithosphere can be modeled as consisting of a small number (approximately ten) of rigid plates that move a constant velocity relative to one another. This model, plate tectonics, can explain the large scale geology of the Earth's surface, but it breaks down in several ways. For example, the plates are not perfectly rigid—earthquakes sometimes occur in the interior of plates, not just at plate boundaries. Also, in some regions around plate boundaries the deformation takes places over zones with a width of several hundred km or more, rather than just at a single sharp boundary. Some of the strain that is built up in the lithosphere is released by slow slip rather than by earthquakes. Over 90 per cent of the total energy of earthquakes is released at plate boundaries. Most of the earthquake energy release at plate boundaries occurs at subduction zones or at boundaries where two continents converge.

The mismatch of geological and human time scales is a fundamental barrier to our understanding of earthquakes. The characteristic time scale for mantle convection is about 10^8 yr, but we have only about 100 yr of instrumental seismic data. Basically weather runs on an annual cycle, so in 100 yr we would see 100 cycles. On the other hand, we have seen only about one millionth of one cycle of mantle convection in our 100 years of instrumental recording of earthquakes. By analogy to weather, this is like seeing one millionth of a year, i.e. less than one minute, of weather data and trying to extrapolate from that. This places a fundamental limitation on what earthquake scientists can accomplish by a brute force inductive approach.

How can we cope? First, both earthquake scientists and the users of our research (i.e., engineers, government officials, the public) must accept that even when we have done our best, estimates of future seismicity will inevitably be quite uncertain. Second, we can try to get more data in several ways. We can go backwards in time, using geological studies to gain information on earthquakes over the past, say, ten thousand years. We can use space-based observing techniques to study the ongoing deformation of the plates. We can also use laboratory studies and theoretical modeling to try to improve our understanding of the earthquake process. But while all of these can help, none can make an order of magnitude difference in eliminating the uncertainties we must live with. This may not be what people want to hear, but to do the best job possible of reducing seismic risk we must start with the way things are, not the way we would like them to be.

Many earthquake scientists, including some of the authors in this book, use the term 'earthquake cycle'. If this is understood as being a kind of characteristic time there is no problem, but if this term is mistakenly interpreted as implying periodicity in a rigorous sense—the same earthquake repeating on the same fault at regular intervals—this is a highly unfortunate misunderstanding. Almost every significant earthquake that has occurred in the past hundred years has had one or more aspects that differed markedly from what was expected. This pattern is so consistent that the only thing that should really surprise us is an earthquake that is not a surprise in any way. We must, as the noted seismologist Hiroo Kanamori[1] admonishes us, prepare for the unexpected.

Guide to this book

The first three chapters synthesize what is known—and what is not known—about earthquakes. Chapter 1 emphasizes methodological issues. The empirical approach is the starting point of any scientific research, but if we sift through large volumes of data until we find what we want we can get into trouble, as an apparently meaningful pattern might be merely a random fluctuation in a sea of noise. Proper use of statistical methods can save us from many pitfalls.

Chapter 2 is a summary of the classical approaches to studying earthquakes. Geological and seismological data are presented, and the basic earthquake source parameters are defined. This leads into a discussion of the classical elastic rebound model of earthquakes, and a discussion of whether earthquakes have a well-defined nucleation process which begins a long time before the event. Finally, we consider laboratory experiments and field data on slip and fracture, and their applicability to earthquakes. The material in this chapter is based on a classical continuum-mechanics based view of earthquakes. While this approach is intuitively appealing, it fails in many ways to satisfactorily explain observations associated with earthquakes. This suggests that a new paradigm is needed.

The most important developments in physics in the first half of the 20*th* century were in quantum electrodynamics, nuclear physics, and high energy particle physics. However, some of the most important developments in the second half of the 20*th* century came in the field of the physics of complex non-linear systems. It appears that in many cases such systems have common properties independent of the particular physical system being considered. Chapter 3 discusses this 'new physics' and its applicability to earthquakes.

We now shift gears and consider more applied topics. Chapter 4 presents a discussion of time-independent hazard estimation. Standard techniques and also some of their uncertainties are discussed. Chapter 5 discusses time-dependent hazard estimation and forecasting. If we use information on past earthquakes

---

[1] *Seism. Res. Lett.*, **66**(1), 7–8, 1995.

together with a model then we can perhaps obtain better estimates of hazards than by time-independent estimates. The first section of chapter 5 presents some approaches to this issue. The second section of chapter 5 presents a data-based approach to estimating earthquake occurrence probabilities based on the history of past seismic activity. The third section of chapter 5 looks at hazard estimation from a statistical point of view and strongly questions the significance of some of the approaches now being used. These disparate points of view reflect an ongoing controversy within the scientific community; the presentations in chapter 5 make it clear what the issues are. Note that all three sections of chapter 5 were written independently, and that none of the authors saw the other sections in the course of writing or editing their own contributions.

Chapter 6 presents discussions of two important sources of new data. One is space-based techniques for observing the ongoing deformation of the Earth's surface, and the second is paleo-seismology, which allows us to extend the instrumental catalogue of earthquake data (about 100 yr) by using geological data in coastal areas to study earthquakes which occurred in, say, the past 10,000 yr.

The last two chapters turn to issues that directly affect society. Chapter 7 discusses risk mitigation, based on recent earthquakes in Greece and Turkey. The conclusions drawn from the experiences in these two countries are remarkably similar. The main cause of loss of life was the collapse of substandard construction. Unfortunately once substandard structures exist, it is expensive to reinforce them, and this usually cannot be justified in purely economic terms. Government intervention is therefore required, but, because of the high costs, it is impossible to reinforce all unsafe structures. The best way to reduce risk is therefore to ensure that all new construction satisfies modern codes for earthquake resistance. The experience in both countries shows that it is essential to carry out strict checks to ensure that the code requirements are actually being rigorously followed.

Finally, in chapter 8, public policy issues are discussed. A review of research in social science and actual experience shows that short-term earthquake prediction could be useful to society only if it is highly accurate and reliable, but there are no scientific prospects for this at present. Many instances of social disruption caused by false earthquake predictions are discussed, and methods for dealing with such problems are suggested.

Following the NATO format, this book includes only black and white illustrations. However, in many cases the original illustrations were in color, and this is essential to fully understanding the information. You will find the original artwork in the CD which accompanies this book.

Francesco Mulargia, Bologna
Robert J. Geller, Tokyo
July 2003

Recommendations adopted by the ARW

At the conclusion of the NATO Advanced Research Workshop (ARW) 'State of scientific knowledge regarding earthquake occurrence and implications for public policy", held in Le Dune, Piscinas - Arbus, Sardinia, Italy, from October 15 to October 19, 2000, the participants adopted the following recommendations.

1. *Reduce seismic risk.* Measures should be taken to reduce seismic risk, with particular emphasis on urban areas. Since the costs of such measures can be high, it is necessary to identify critical facilities and buildings for immediate action, and also to give incentives to owners to strengthen their buildings. Governments should develop retrofitting guidelines. At least a minimum level of seismic resistance should be incorporated into building codes for zones of low seismic activity within seismically active regions; the resulting increase in construction costs will be relatively modest. All risk-reduction actions should take local site conditions into account.

2. *Encourage broad cooperation and public awareness.* Reducing earthquake risk requires the cooperation of scientists, engineers, statisticians, social scientists, government authorities, the news media, non-governmental organizations, and the general public. Communication and cooperation among these groups, efforts to educate the public about earthquake risks, and international cooperation should be strongly encouraged.

3. *Improve scientific understanding of earthquakes.* As our present scientific understanding of earthquakes is far from satisfactory, further theoretical, observational, experimental, computational and historical basic research should be strongly encouraged. The establishment of common standards, and the continuous and stable operation of networks of observational instruments, data centers, and systems for international data distribution are essential.

4. *Improve methods for making seismic hazard estimates.* It is not possible at present to make reliable and accurate warnings of imminent individual large earthquakes. Even if such warnings were possible, they would not be a substitute for efforts to reduce seismic risk, in which seismic hazard estimates are a key parameter. The limitations of the data should be accounted for by incorporating appropriate conservatism in estimates of seismic hazard. To reduce these uncertainties, research to identify faults and to measure their slip rates and earthquake histories and the deformation of the surrounding earth should be encouraged. Data from such studies and from other modern techniques, including satellite geodesy and advanced geological methods, should be incorporated into studies on hazard estimation.

5. *Enforce high scientific standards.* Research on making quantitative and

objectively testable statements regarding future seismic hazards should be encouraged. Such work must meet the highest standards of scientific and statistical rigor. Methodology in this area should be systematically validated and upgraded by comparison of forecasts to actual seismicity. Scientists and engineers should accurately represent the current state of the art when seeking funding for work in earthquake science. Funding agencies should evaluate proposals on the basis of well-established and rigorous rules of peer review.

6. *Establish policies for evaluating earthquake warnings.* Governments should establish policies and systems for evaluating earthquake warnings and deciding what actions, if any, should be taken. All persons are strongly encouraged to comply with these policies. The evaluators should work with the news media to discourage the dissemination of warnings that have not been approved and should inform the media on the scientific issues involved in making such warnings.

Addendum

The above recommendation are reproduced exactly as they were adopted by the Workshop. However, the editors would now propose that the following recommendation should also have been adopted:

7. *Enforcement.* Building codes and other regulations for seismic safety must be strictly enforced. Corruption and administrative laxity greatly increase the risk to the public.

Chapter 1

Modeling earthquakes

[By F. Mulargia and R. J. Geller. P. B. Stark's SticiGui[1] has been more than an inspiration for section 1.2 and S. Castellaro contributed to section 1.3]

1.1   Phenomenology

Geophysicists began studying earthquakes long before geophysics existed as a recognized field. Two centuries ago Montessus de Ballore identified crustal faulting as the basic phenomenon of earthquake physics. The picture remained essentially qualitative up to the late 1950s, when *Eshelby* (1957), *Keilis-Borok* (1959), *Maruyama* (1963), *Burridge and Knopoff* (1964) and *Haskell* (1964) realized that the faulting problem could be treated quantitatively using methods originally developed by applied mathematicians in the late 1800s to solve engineering problems in metals. The ensuing research progressively developed into the new discipline of 'quantitative seismology'. This discipline views the crust as a deterministic mechanical system with the following features: (1) the lithosphere is subject to stresses of tectonic and gravitational origin; (2) the lithosphere is brittle, and when these stresses exceed some fixed limit the crust ruptures (i.e., an earthquake occurs), thereby emitting elastic waves; (3) the ruptures are similar to plane dislocations.

1.1.1   The lack of a coherent phenomenology

A widely accepted view of the seismic mechanism is that it is strictly deterministic and can be viewed as frictional sliding (in Coulomb form) on a planar fault. Re-

---

[1]http://www.stat.berkeley.edu/~stark/SticiGui/

finements of this basic model incorporate more complicated friction laws (*Scholz*, 1998), 'asperities' (*Kanamori and Stewart*, 1978) or 'barriers' (*Aki*, 1979), or other phenomena observed in fracture experiments in the laboratory (*Dieterich*, 1994).

However, unless some stochastic elements are introduced on an *ad hoc* basis (*Ward*, 1991, 2000), the classical deterministic model is incapable of reproducing even the most simple dynamical features of seismicity, such as the irregularity of earthquake occurrence, the distribution of events in size, their tendency to cluster, etc. For this reason, as well as others discussed elsewhere in this volume, a new paradigm thus appears to be required.

The lack of a coherent phenomenology for earthquake occurrence is evident from the wide range of contradictory statements found in the seismological literature. We give some examples below.

A) The regularity (or lack thereof) of the seismic cycle

A.1) The large earthquakes of a given region take place in regular cycles (characteristic earthquakes – *Schwartz and Coppersmith*, 1984; Parkfield – *Bakun and Lindh*, 1985).

A.2) Better: cycle durations are distributed lognormally (*Nishenko and Buland*, 1987).

A.3) No! Large earthquakes occur at random (*Sieh et al.*, 1989).

A.4) No! Large earthquakes occur in clusters (*Kagan and Jackson*, 1991a).

B) The regularities (or lack thereof) within each seismic cycle

B.1) Large earthquakes at plate boundaries occur where there is a seismic gap (*McCann et al.*, 1979; *Sykes and Nishenko*, 1984).

B.2) No! This is not true (*Kagan and Jackson*, 1991b, 1995).

B.3) Earthquake recurrence is time-predictable (*Shimazaki and Nakata*, 1980).

B.4) No! This is not true (*Mulargia and Gasperini*, 1995).

B.5) The rate of occurrence of foreshocks increases according to a power law (*Bufe and Varnes*, 1993).

B.6) Better: it increases according to a power law with a lognormal oscillation (*Saleur et al.*, 1996).

B.7) No! More than 80 per cent of the earthquakes occur without foreshocks, and these can in any case be identified only *a posteriori* (*Abercrombie and Mori*, 1996).

B.8) According to the frictional stick-slip mechanism, earthquakes occur at high stress/strain conditions. This is implicit in the classical seismological view (see e.g., *Aki and Richards*, 1980; *Scholz*, 1990).

B.9) Not necessarily! *In situ* stress and heat flux measurements call for the opposite (*Lachenbruch and Sass*, 1980).

B.10) Strain release on the principal fault systems occurs primarily through earthquakes (*Scholz*, 1990).

B.11) No! Aseismic slip on the San Andreas fault is comparable to seismic slip (*Gladwin et al.*, 1994).

B.12) The level of tectonic strain is the fundamental variable (the classical view; e.g., *Aki and Richards*, 1980).

B.13) No! Fluid migration can be equally important (*Hickman et al.*, 1995; *Roeloffs et al.*, 1989) and it is also important to distinguish strain from the influence of pore pressure on the stress field, which is harder to observe directly.

B.14) Strain on the principal fault systems is mostly released by the largest earthquakes, as a corollary of B.10 and B.12, above and the Gutenberg-Richter law.

B.15) No! Strain is mostly released by faults too small to produce earthquakes (*Walsh et al.*, 1991; *Marrett and Allmendinger*, 1992).

C) On the earthquake mechanism

C.1) Fault slip is the basic mechanism of earthquakes, which can be approximated as a dislocation on a single plane fault (e.g., *Aki and Richards*, 1980).

C.2) Yes, but actually no! To fit geodetic data a very large number of small subfaults with variable slip is needed (e.g., *Hardebeck et al.*, 1998: 186 subfaults are necessary to fit the InSAR data for the Landers quake).

C.3) No! Faults are fractal objects, not planes, and epicenters are fractally distributed (*Kagan and Knopoff*, 1980). Real faults have a fractal geometry, and although their fractal dimension seems to approach 1.0 (i.e. Euclidean geometry) at a restricted length scale around 1 km, in no case can faults be taken to have a planar geometry (*Aviles et al.*, 1987; *Okubo and Aki*, 1987).

C.4) What controls fault rupture is the fracture of the strongest asperities (barriers), so that the governing mechanism is fracture mechanics in mode I loading configuration (*King and Sammis*, 1983).

C.5) Stick-slip dry friction sliding with well defined material thresholds rules the process (*Harris*, 1998).

C.6) Coulomb static friction law applies (*ibid.*).

C.7) No! A rate- and state-dependent friction law applies, as in the laboratory (*Ruina*, 1983; *Dieterich*, 1994; *Parsons et al.*, 2000).

C.8) Neither one is compatible with the data: a zero friction coefficient would be needed to fit the data (*Kagan and Jackson*, 1998) and no rotation in the stress axes is induced by large earthquakes (*Kagan*, 2000).

In addition to these, there is a plethora of minor but equally contradictory views of generic seismicity patterns (*Keilis-Borok et al.*, 1988; *Romanowicz*, 1993), or, more specifically, 'quiescence' (*Reasenberg and Matthews*, 1988; *Wyss and Martirosyan*, 1998), earthquake 'shadows' (*Harris and Simpson*, 1996), etc.

In summary, the phenomenology of earthquakes shows everything and its opposite, and each modeler feels free to pick the data fitting the theory he decides to propose. Let us attempt to understand the problem.

1.2   Retrospective selection bias

Statistics has developed very effective methods of sampling integrated with data analysis, which are the bases of modern agricultural sciences, pharmacology, production control, etc. (cf. *Cochran and Cox*, 1957). Such techniques, which may be categorized under the heading of experimental design, assume the ability to acquire new data at will and to control the experimental conditions. Both of these are often impossible in geophysical studies. Data availability induces major constraints and this is a crucial problem in many fields of geophysics. This lack of control on the experiment, and the scarcity of data, open the door to selection bias. In general, once identified, selection bias can be corrected or, at a minimum, its extent can be evaluated, allowing the accuracy of the analysis to be properly defined. The most dangerous type of selection bias is therefore that which is not identified.

In some cases, research is focused on phenomena where limited understanding is combined with difficulty in acquiring new data, rendering all analysis essentially empirical and retrospective. In this case, a particular type of selection bias, which goes under the name of retrospective selection bias (*Mulargia*, 1997, 2001), may be largely concealed in the process of data analysis itself. Earthquake physics, a field in which even the most basic processes are poorly known (*Ben-Menahem*, 1995), and for which only semi-qualitative models have been so far derived (*Main*, 1996), is a field in which this is a major problem.

1.2.1   Using statistics to find the 'truth'

Sir Harold Jeffreys, who helped to lay the foundations of both modern geophysics as well as modern statistics, said that "You don't need a statistician to find a real effect". While this is an extreme view, and counterexamples can be cited, the main role of statistics in earthquake science is to allow us to rule out hypotheses that are not significant beyond random chance. This particular branch of statistics is called 'hypothesis testing'.

Since a sound theory of earthquake physics is unavailable, great effort is devoted to the search for purely phenomenological features, such as apparent patterns, correlations, etc. Many artifacts can emerge from such data, and almost any pattern can be found if enough data are sifted. The mathematical theory that studies such selection artifacts is called Ramsey theory (*Graham and Spencer*, 1990).

While this problem is common to all sciences, it is aggravated in earthquake physics by the intrinsic impossibility of reproducing the experiments at will in forward time, and by the long time scale of geological phenomena. Also, a variety of models may all be able to 'explain' the data. These are mostly *ad hoc* quantitative descriptions of a single or at most a few cases involving a handful of data and a large number of free parameters.

1.2.2 Hypothesis testing

Statistical hypothesis testing provides the tools to *validate* a model, i.e. it gives a quantitative objective measure of the extent to which the model agrees with the data. Validation is a crucially important step, since models which are not practically effective should be discarded. In spite of its vital importance, the validation step is too often disregarded by modelers, who sometimes confound it with *verification*. The latter means proving that a model is right, an impossible task for both statistics and science. As will become clear in the following, statistics can only prove that a hypothesis is not wrong in the face of a given set of data. The rationale is that a model which cannot be shown to be wrong is acceptable (*Kirchherr et al.*, 1997).

Let us consider the simplest example of statistical hypothesis testing. In statistical jargon, the two competing hypotheses are called the 'null hypothesis' and the 'alternative hypothesis'. Typically, the null hypothesis is the absence of any effect, while the alternative hypothesis is that some specified effect is present. The two types of error that arise in hypothesis testing are a 'Type I error', rejecting a true null hypothesis, and a 'Type II error', failing to reject a false null hypothesis. A risk no higher than a given level α of committing a type I error, which is called the 'significance level' of the test, is selected. This value is the maximum acceptable risk of committing an type I error. Values of 0.01, or at most 0.05, are commonly adopted. These values represent a common compromise between Type I and Type II errors.

1.2.3   Data mining and fishing expeditions

When a field lacks a valid theoretical framework, research is necessarily phe-nomenological, in which 'discovering' a deterministic feature requires the ob-servation of evidence which is 'anomalous' with respect to a general situation of 'normality'. Normality, which constitutes the null hypothesis, is described in terms of a number of variables, the values of which will follow an assumed distri-bution. The observation of measured values in the tails of such a distribution is at the basis of the identification of possible anomalies.

The common procedure is therefore to sift suitable datasets retrospectively with the explicit aim of finding something 'anomalous'. This is called 'data min-ing' or a 'fishing expedition', since it essentially implies an endeavor to keep looking until something is found. Disregarding as unacceptable the studies (un-fortunately still common in geophysics) in which no quantitative validation at-tempt is made, a typical fishing expedition gauges its catch by hypothesis testing, under the null hypothesis that no effect is present, with the alternative hypothesis that some effect exists. There are two requirements for this procedure to be valid: (1) an appropriate statistical model must be known, and (2) the sample must be random.

As regards the first requirement, in many cases the sample space can be rea-sonably guessed and the central limit theorem allows one to use the Gaussian dis-tribution. Even if this is not the case, a wealth of data is usually available for the 'normal' situation, so that the first requirement is usually approximately satisfied. The real problem is the second requirement, because the retrospective researcher by definition is not dealing with a random sample, as he carefully sifts the data to identify subsets which exhibit a deviation from 'normal' behavior. His samples are thus as far as possible from being random. The consequence is that testing hypotheses on retrospective data using standard methods which assume random sampling produces a severe 'retrospective selection bias', which in turn opens the door to 'false discoveries' (*Scheffé*, 1959; *Eckhardt*, 1984; *Stark and Hengartner*, 1993; *Benjiamini and Hochberg*, 1995; *Mulargia*, 1997, 2001).

Since to first approximation the intervals of definition of the different param-eters can be assumed to be mutually independent, the total number of possible choices is given by the Cartesian product of the number of possible choices for each parameter. Formally, denoting by r_i the $i - th$ parameter (out of a total of N) which defines the working set, and assuming that its interval of definition has n_i values (typically $n_i = 2$, i.e. a lower and an upper bound) and that each of the latter is chosen among m_{ij} values, the total number of cases considered, N_T, is

$$N_T = \prod_{i=1}^{N} \prod_{j=1}^{n_i} \prod_{k=1}^{m_{ij}} r_{ijk} \, . \tag{1.1}$$

In other words, the retrospective researcher may tacitly consider a very large number of cases, and, disregarding the fact of having done so, may conclude that he has observed some 'unlikely' case that occurs 'surprisingly' far more often than expected by chance, and therefore may claim a 'discovery'.

For example, consider the following example (cf. section 2.3.2). The Gutenberg-Richter *b* value in a given region *A* seems to be exceedingly high, and according to the 'normal' distribution of the *b* values it seems to have a probability of random occurrence of 1×10^{-4}. Is this a random fluctuation or a real effect? While the *b* value appears low enough to comfortably reject the null hypothesis, consider the following. The very high *b* value was identified by retrospectively scrutinizing past seismicity after (1) selecting a lower magnitude threshold among the three values 3.0, 3.5, 4.0; (2) specifying region *A* as a cell of a square grid 50 km × 50 km; and (3) analyzing two different catalogues in three different time periods each.

Following equation (1.1), one might expect to find in one cell a value of *b* as unlikely as 3×10^{-5} by mere chance, and in several tens of cells *b* values as unlikely as 10^{-4}. The value of *b* in region *A* is therefore far from being anomalously high. Why then was it concluded that the *b* value was anomalous? Simply because the above retrospective selections were implicitly made, but this was not taken into account in the statistical analysis. Forgetting that such optimal choices were made is very common in the enthusiastic retrospective search for deterministic effects, since the choices are mainly made unconsciously.

1.2.4 *Post hoc* correction of optimal retrospective selections

If all retrospective selection choices were conscious and explicit, one could count *all* the cases considered, and obtain an unbiased estimate of the probability of occurrence by chance of any given deterministic effect. Statisticians call this a multiplicity problem, since it involves the simultaneous testing of not one but many hypotheses, each linked to each of the possible choices. If one tests each of these hypotheses, say *m* in number, at significance level α, the chance of making at least one type I error is at least α, and typically larger. The classical Bonferroni approach states that the probability is equal to *m*α, but this estimate is overly conservative. Improved Bonferroni methods have been proposed by *Simes* (1986), *Scheffé* (1959), *Benjiamini and Hochberg* (1995), *Hsu* (1996), and solutions to specific problems have been developed (*Eckhardt*, 1984; *Stark and Hengartner*, 1993; *Mulargia*, 1997).

The problem is that many of the options considered are implicit, often hidden in the procedure of analysis itself, or masked by weak arguments such as "California seismicity is considered". Sometimes tectonic arguments are advocated. Sound as they may appear, they are both unavoidable subjective and *post hoc* to some extent. They typically imply selecting events in a geographical region of complex geometry, the boundaries of which are identified according to geological interpretation. Since each vertex fixes 2 parameters, and since polygonal regions with more than 10 vertices are common, unverifiable arguments lead one to select a large number of parameters. Identical arguments apply to the selection of the other parameters of the operative set: the magnitude window and the time window; identifying each of the subjective choices can be particularly difficult.

1.2.5   The safest antidote to false discoveries: forward validation

Having realized that retrospective selection bias is both unavoidable and difficult to correct, one might consider the radical alternative of rejecting all apparent anomalies except those which have an extraordinarily low retrospective level of significance. The idea is that since retrospective optimal selections induce much lower apparent significance levels, rather than setting the threshold for significance at the usual 0.05 or 0.01, thresholds of 10^{-6} or lower (*Anderson*, 1990) should be required. This might appear to be a wise and practical option, but any attempt to lower the probability of a type I error, inevitably increases that of a type II error. In other words, any attempt to guard against false discoveries leads to the possibility of discarding genuine ones. Thus infeasibly large datasets would be required.

While large sample sizes can effectively control the problem, large samples are rare in geophysics. An alternative procedure requires both rigor and time but can provide definite answers. This procedure consists of two separate steps. The first is a retrospective analysis, and the second is a strictly forward validation, in which no parameter can be adjusted (*Mulargia and Gasperini*, 1996). In the first step, all available retrospective data are analyzed with the freedom to make all sensible optimizations. Standard tests and significance levels are then adopted, and all 'anomalies' which are apparently significant are the candidates which enter the validation step. This means that standard or marginally conservative significance levels, like 0.01 or 0.001, are used to identify the candidate 'anomalies'. It goes without saying that 'anomalies' which does not even reach a clear level of candidacy should be abandoned.

Keeping the unequivocal rules of the game which define an 'anomaly' strictly unchanged, validation is then performed on a separate, truly independent dataset. Standard significance tests can then be employed with no need to account for retrospective bias. The consequence is that much less stringent requirements are imposed on sample size than those for a retrospective validation. However, this also implies that no change in the rules of the game may be made, as otherwise a second candidacy step must be conducted. The main limitation of this approach is that the typically long time scales of earthquake recurrence require a long time to acquire sufficient forward data, the only data which are safe from retrospective selection bias.

One possible tool for identifying candidate hypotheses is pattern recognition. Modern computer science has developed powerful algorithms such as principal component analysis, discriminant analysis, cluster analysis and neural networks. Such techniques have the advantage of automatically extracting information from all of the various possible combinations of variables in the dataset. However, as noted above, complex patterns involving a large number of variables can easily lead to false discoveries. It therefore is essential to subject candidate hypothesis formulated on the basis of pattern recognition techniques to a rigorous forward validation using an independent dataset collected after the hypothesis was formulated.

1.3 Model building

We undoubtedly lack a valid earthquake model. Why? Possibly, some insight can be gathered by revisiting the basics of modeling.

Scientific understanding consists of two separate steps. The first is disassembling a phenomenon into a series of simple subphenomena, each of which can be described separately. The second is deriving a model (a set of assumptions and rules) that is capable of providing an effective simplified description of the subphenomena.

Not all models can be called scientific. One general definition of a model is 'a structure with a given representation'. The requirements for a model to be scientific are: (1) objectivity, (2) falsifiability, and (3) validation. Objectivity means independence from the observer. Falsifiability, as defined by *Popper* (1980), means that each statement must be accompanied by a precise set of rules which prescribe how it can be compared to facts by each observer. Validation means comparing the predictions of the model to real data, according to well defined statistical procedures. The data used in the validation step must be strictly independent of those used to formulate the model. If the validation gives satisfactory results, the model can be considered ready to use. If the validation results are unsatisfactory, the model must be changed and tested again *ab initio*. This cyclic procedure is called the 'abductive approach'. The abductive approach is often misused in earthquake physics, as some workers forget that (1) no model is valid before its capability to explain real data is proven, and (2) each change in the model requires a new validation.

Good models require strong phenomenological foundations. Since, as noted above, the phenomenology of earthquake occurrence is not yet well understood, this imposes limits on the extent to which modeling of seismogenesis can be useful. One important test of a model is its ability to make testable quantitative statements (predictions) about the future behavior of the system being modeled. Some level of reliable predictive power is normally a prerequisite for a model to be accepted by physicists. On the other hand, many models in geoscience make 'postdictions' but have little or no demonstrated predictive power for future events. This is quite understandable in view of the long time scale and lack of experimental control in geosciences. This key difference in the points of view of physics and geoscience is not necessarily fully appreciated by workers in the respective fields.

The advent of digital computers has brought a new class of models and further modified the meaning of 'understanding'. Models, which were traditionally tied to sets of partial differential equations simple enough to be solved at least approximately, were freed from the analytic formulation and could be cast directly as a set of rules sequentially executed as an algorithm. This makes it possible to take into account complex interactions beyond the pair-interaction or mean-field limits required by analytical treatments. Conversely, pattern recognition algorithms, and, in particular, those based on neural networks, provide sometimes surprisingly efficient models, which are *de facto* 'understood' by machines rather than by humans.

1.3.1   Choosing among models

There are often several hypotheses which are consistent with the experimental evidence. There may also be several parameters which are not constrained by the available data. As a consequence, several models will be able to 'explain' the data, which means that they are not in blatant disagreement with the data. In this case one speaks of 'multiple explanations' and of the 'principle of indifference', which states that "one should keep all the hypotheses which are consistent with the facts". The basic criterion of choice is then given by Ockham's razor: "among all the hypotheses consistent with the facts, choose the simplest". But why should one choose the most simple model? And what does 'simple' mean?

The key is the *Kolmogorov complexity* of a hypothesis or of a model s, which is expressed by the length in bits $K(s)$ required by the shortest computer algorithm necessary to describe it. Since these are binary bits, the probability of s is

$$P(s) = 2^{-K(s)} , \qquad (1.2)$$

which is called the 'universal distribution' (*Kirchherr et al.*, 1997). This gives a way to keep all competing hypotheses as required by the indifference principle, and assign them probabilities, as prescribed by a Bayesian approach. Note how in this framework a truly random sequence of 0 and 1 of length 10000 is a $K(s)$ which rewrites the string itself and has therefore a $P = 2^{-10000}$, while a string of alternating 010101010... would just require the bits to say "10000 alternating zeros and ones". Note that the first one provides a way to describe randomness, which is called the Martin-Löof definition.

Suppose now that we have two competing hypotheses, both of which appear to work. One has $K(s) = 100$, the other one has $K(s) = 200$. The relative likelihoods are given by their probabilities, i.e. the first has a probability of 8×10^{-31} and the second has a probability of 6×10^{-61}, so that the second is about 10^{30} times less likely than the first one. This is the theoretical justification for Ockham's razor.

Unfortunately, however, $K(s)$ is uncomputable, i.e. given a string s there is no way to compute the minimum length algorithm needed to describe it. However, $K(s)$ can be approximated. In practice, what is done is to adopt the simplest model which fits the data, simply measuring the fit with the squared residuals. Since it is obvious that a model with a larger number of parameters provides a better fit in the same way in which a higher degree polynomial provides a better fit to a set of generally distributed points, the performance of a model is weighted accounting for the number of parameters. There exists a variety of criteria to do this, like the Akaike Information Criterion AIC (*Akaike*, 1979), the Bayes Information Criterion BIC, etc. These are all variants of the original Gauss criterion, which gauges model performance with the function Ω of the residuals $(y_i - \hat{y}_i)$, where y_i indicates the measured values and $\hat{y}_i$ the corresponding values estimated by the model, defined as

$$\Omega = \frac{\Sigma(y_i - \hat{y}_i)^2}{n - m},$$ (1.3)

with n indicating the number of data and m the number of parameters. The best model is the one producing the minimum value of Ω. The AIC criterion looks for the minimum value of

$$AIC = -2\log L + 2(n - m)$$ (1.4)

where L is the likelihood, and the BIC criterion looks for the minimum value of

$$BIC = -2\log L - m\lambda \log(n)$$ (1.5)

where λ is a penalty function, usually equal to unity.

1.3.2 Deterministic, complex and stochastic cases

Consider the simple case of a process that is governed by a single variable x, with variability dx among the different realizations. Three cases are possible: (1) $dx \ll x$ always, (2) $dx \ll x$ in some particular cases, (3) $dx \ll x$ never. In the first case the process is called deterministic, in the second it is called complex (chaotic), and in the third stochastic. These correspond to the three main different ways of describing natural phenomena.

Deterministic processes can be described by a relatively simple set of mathematical equations. All variables and parameters are known with small uncertainties and are represented by narrow Gaussians which ideally tend to Dirac delta functions.

Stochastic processes occur in systems made up of a number of domains too large to be treated deterministically, or whose variables and parameters are known with uncertainties so large that they are comparable to the values of the variables themselves. A statistical approach, using average quantities, distributions and global properties, must be adopted. Solutions can only be derived in probabilistic terms.

Complex processes are different and, in some respect, intermediate between determinism and stochasticity, and yet cannot be successfully described by traditional techniques. While the description of deterministic and stochastic processes is well established, the picture is far less mature for complex systems. Complex systems appear to be potentially relevant to earthquake physics.

1.3.3   Complex systems

Complex systems, which will be discussed in chapter 3, are (cf. *Whitesides and Ismagilov*, 1999) either (1) systems which are sensitive to small perturbations or to initial conditions (so called 'deterministic chaos', cf. chapter 3), (2) open systems out of equilibrium, (3) systems in which there are multiple pathways possible for evolution, or (4) systems which are complicated by subjective judgments. In many fields, including Earth science, almost everything of interest is complex. Faced with the impossibility of handling such systems exactly, a series of approaches have been attempted for the treatment of complex systems, albeit limited to cases that are both complex enough to be interesting and simple enough to be tractable. These approaches are (1) reasoning by analogy, (2) averaging, (3) linearization, (4) drastic approximation, (5) pure empiricism, (6) detailed analytical solution. All of these have shown some success, but no general procedure has yet been developed.

The contrast between simple and complex systems is in terms of their descriptions. Reductionism describes natural systems in terms of fundamental processes, which are assumed to work like the bricks of a building. But systems which are not simple are dominated by the interactions between variables, and a stratified picture in terms of fundamental processes collapses. There are several hopes for making complex systems tractable. The first one is that drastic simplifications may occur due to the non-linear dynamics of the system itself. For example, self-organization (cf. chapter 3) can reduce the very large number of degrees of freedom. Its extreme view is 'universality', which attempts to describe systems with the simplest possible model which has apparently compatible features. A corresponding hope is found on the empirical side. The fact that complex behavior can be produced by 'deterministic chaos', i.e. by a few nonlinear deterministic equations with a handful of parameters, suggests the hope that such equations can be eventually discovered. Unfortunately, deterministic chaos has never been found in real data and appears therefore to be, at best, an asymptotic behavior. Self-organization and criticality presently show more promise and are being extensively studied, but so far no major breakthroughs have occurred.

In operational terms, a complex system is any system made up of parts interacting in a non-linear way. This results in behavior difficult to describe by a set of equations. The basic clues are therefore (1) non-linear interactions (effects are not directly proportional to causes), (2) non-equilibrium systems (undergoing changes), (3) non-uniform systems (containing structures in space and/or time), (4) systems with emergent properties (not explicable in terms of the single parts), (5) self-organizing systems (order not imposed from outside).

Problems involving complex systems arise in almost every branch of science. The behavior of the system as a whole can lead to properties which the single components did not individually have.

1.4 Prediction

As discussed above, predictive power is one of the criteria used by physicists to evaluate models. From the point of view of science, it is entirely natural to discuss what we can predict about future earthquakes. When we do so, we are using 'predict' in the sense of making quantitative and objectively testable statements about future seismicity. This ought not to be a particularly controversial or sensitive subject, as workers in every field of science rely on the cycle of hypothesis formulation followed by hypothesis testing to ensure that their theories are in accord with observed data.

Unfortunately, as noted above in the Preface, the term 'earthquake prediction' has taken on a life of its own with journalists and the general public. To these groups an 'earthquake prediction' is a warning with enough specificity in time, location, and magnitude that, for example, an evacuation could be conducted before a large earthquake strikes. A detailed discussion is beyond the scope of this volume, but it is easy to obtain a sample of public views on 'earthquake prediction' using a search engine such as Google (over 13,000 hits in English, and over 11,000 hits in Japanese as of February, 2002).

The fascination of journalists with the topic of earthquake prediction extends to the news and science news sections of *Nature* and *Science*, the two most prestigious general journals of science. These columns have reported extensively on earthquake prediction for many years (see *Geller*, 1997a for references), and *Nature* organized a debate on earthquake prediction on its web page[2] in 1999.

Nature's debate reached a consensus that the reliable and accurate prediction of imminent large and damaging earthquakes is not presently possible (this consensus was also shared by this ARW – see Recommendation #4.) However, there was a striking difference of opinion over whether large scale funding should be specifically allocated to earthquake prediction research. Supporters (also see *Wyss*, 2001; *Bernard*, 2001) cited the importance of the problem to society, while opponents argued that in light of past disappointments this was unwarranted in the absence of new, promising, and well substantiated research results. Perhaps the best way for funding agencies to deal with proposals for earthquake prediction research is to review them in competition with all other proposals in geosciences, using normal scientific review criteria. Good proposals in this area certainly warrant funding, but poor proposals should not be given any special treatment.

Extensive research on earthquake prediction (in the journalistic sense) has had no obvious successes (*Geller*, 1997a; *Kagan*, 1997). Claims of breakthroughs have failed to withstand scrutiny. Extensive searches have failed to find reliable precursors. Such research thus has no immediate relevance to practical efforts to reduce seismic risk. Furthermore, even if journalistic predictions were, hypothetically speaking, realizable to some extent, high accuracy and reliability would be required in order for the benefits to society to outweigh the costs (see chapter 8).

Notwithstanding the scientific consensus on the present impossibility of making predictions of imminent individual large earthquakes, claims to be able to do so are frequently made, some by amateurs and others by credentialed scientists (*Geller*, 1997a, b). Although none of these claims has withstood scrutiny, many have attracted extensive publicity in the media, and several have, at least for a short time, attracted funding from various national and international agencies. Much time and energy has been consumed in evaluating these claims, but these evaluation efforts have fortuitously led to the development of methodologies for objectively evaluating prediction claims.

1.4.1   Definitions of prediction

To discuss our predictive power we need a clear definition of 'prediction'. The absence of a clear and agreed-upon definition is one of the main reasons for the lack of consensus on earthquake predictability within the scientific community.

Prediction can be defined in four fundamental ways:

(1) The 'time-independent hazard' definition, which assumes that earthquakes

---

[2]http://www.nature.com/nature/debates/earthquake/

are a random (Poisson) process in time, and uses the past locations of earthquakes, active faults, geological recurrence times and/or fault slip rates from plate tectonic or satellite data to constrain the future long-term seismic hazard. These probabilistic estimates are the present standard for calculating the occurrence of ground-shaking from a combination of source magnitude probability with path and site effects. Such calculations are incorporated in building design and planning of land use, and for the estimation of earthquake insurance.

(2) The 'time-dependent hazard' definition, which assumes that earthquakes are the result of ruptures in the Earth's crust induced by a high level of tectonic strain. Since the latter can be measured geodetically as well as inferred from earthquake stress drop history, and since the rock mechanical properties are known at least on average, it would appear possible to calculate seismic hazard as a function of time. Central to this approach, which gained unexpected popularity after the 1999 Turkish earthquake, are: (a) the 'characteristic earthquake' paradigm, stating that large events occur with a relatively similar magnitude, location and repeat time, following a 'seismic cycle' in which a given portion of the crust accumulates strain and releases it according to a regular pattern; (b) the paradigm of the simple Coulomb failure stress criterion, which can be traced back to Reid's original elastic rebound model of the start of the 20*th* century, and of the more recent complex rate- and state-dependent friction laws, which seem to rule the stick-slip dynamics of sliding polished rock blocks in the laboratory. While each of these approaches has shown some mild promise in particular retrospective cases, none has survived objective validation analysis.

(3) The 'forecast' definition, which assumes that earthquakes are probabilistically predictable at medium and short term with better than random accuracy due to their tendency to cluster in time and space. In other words, since any small earthquake has a given, albeit very small, probability of being a foreshock to a large event, the likelihood of the latter increases whenever the former is observed. Curiously, the reverse will also be true, implying that the probability of occurrence of a large earthquake is lower in the absence of seismic activity. The predictive power in this case is unquestionable, but its practical utility is scarce since the relevant authorities would need extremely high, rather than very low, probabilities of occurrence to take any action to prepare for an impending event on a timescale of months to weeks.

(4) The 'deterministic prediction' definition, which coincides with the journalist's definition, seeks to make predictions with a high accuracy and reliability in size, location and time, so that drastic actions such as evacuations can take place. It is widely accepted that such predictions are presently impossible, since physical models of earthquakes are lacking and since reliable, unambiguous empirical earthquake precursors have never been identified.

1.5   References

Abercrombie, R. E., and Mori, J., 1996. Occurrence patterns of foreshocks to large earthquakes in the western United States, *Nature*, **381**, 303–307.

Akaike, H., 1979. A Bayesian extension of the minimum AIC procedure of autoregressive model fitting, *Biometrika*, **66**, 237–242.

Aki, K., 1979. Characterization of barriers on an earthquake fault, *J. Geophys. Res.*, **84**, 6140–6148.

Aki, K., and Richards, P. G., 1980. *Quantitative Seismology*, W. H. Freeman, San Francisco, 2 vols.

Anderson, P. W., 1990. On the nature of physical laws, *Phys. Today*, **43** (12), 9.

Aviles, C. A., Scholz, C. H., and Boatwright, J., 1987. Fractal analysis applied to characteristic segments of the San Andreas fault, *J. Geophys. Res.*, **92**, 331–344.

Bakun, W. H., and Lindh, A. G., 1985. The Parkfield, California, earthquake prediction experiment, *Science*, **229**, 619–624.

Benjamini, Y., and Hochberg, Y., 1995. Controlling the false discovery rate: a practical and powerful approach to multiple testing, *J. Roy. Stat. Soc.*, **57**, 289–300.

Ben Menahem, A., 1995. A concise history of mainstream seismology —Origins, legacy, and perspectives, *Bull. Seism. Soc. Am.*, **85**, 1202–1225.

Bernard, P., 2001. From the search of 'precursors' to the research on 'crustal transients', *Tectonophysics*, **338**, 225–232.

Bufe, C. G., and Varnes, D. J., 1993. Predictive modeling of the seismic cycle of the greater San Francisco Bay region, *J. Geophys. Res.*, **98**, 9871–9883.

Burridge, R., and Knopoff, L., 1964. Body force equivalents for seismic dislocations, *Bull. Seism. Soc. Am.*, **54**, 1875–1888.

Cochran, W. G., and Cox, G. M., 1957. *Experimental Designs*, 2nd ed., Wiley, New York.

Dieterich, J. H., 1994. A constitutive law for rate of earthquake production and its application to earthquake clustering, *J. Geophys. Res.*, **99**, 2601–2618.

Eckhardt, D. H., 1984. Correlations between global features of terrestrial fields, *Math. Geol.*, **16**, 155–171.

Geller, R. J., 1997a. Earthquake prediction: a critical review, *Geophys. J. Int.*, **131**, 425–450.

Geller, R. J., 1997b. Predictable publicity, *Astron. & Geophys.*, **38**(1), 16–18, (reprinted 1997, *Seism. Res. Lett.*, **68**, 477–480).

Gladwin, M. T., Gwyther, R. L., Hart, R. H. G., and Breckenridge, K. S., 1994. Measurements of the strain field associated with episodic creep events on the San Andreas fault at San Juan Bautista, California, *J. Geophys. Res.*, **99**, 4559–4565.

Graham, R. L., and Spencer, J. H., 1990. Ramsey theory, *Sci. Am.*, **263** (1), 112–117.

Hardebeck, J. L., Nazareth, J. J., and Hauksson, E., 1998. The static stress change triggering model: constraints from two southern California aftershock sequences, *J. Geophys. Res.*, **103**, 24427–24437.

Harris, R. A., 1998. Introduction to special section: stress triggers, stress shadows and implications for seismic hazard, *J. Geophys. Res.*, **103**, 24347–24358.

Harris, R. A., and Simpson, R. W., 1996. In the shadow of 1857 —the effect of the great Ft. Tejon earthquake on subsequent earthquakes in southern California, *Geophys. Res. Lett.*, **23**, 229–232.

Haskell, N. A., 1964. Total energy and energy spectral density of elastic wave radiation from propagating faults, *Bull. Seism. Soc. Am.*, **54**, 1811–1841.

Hickman, S., Sibson, R., and Bruhn, R., 1995. Introduction to special section: mechanical involvement of fluids in faulting, *J. Geophys. Res.*, **100**, 12831–12840.

Hsu, J., 1996. *Multiple Comparisons: Theory and Methods*, Chapman & Hall, London.

Kagan, Y. Y., 1997. Are earthquakes predictable?, *Geophys. J. Int.*, **131**, 505–525.

Kagan, Y. Y., 2000. Temporal correlations of earthquake focal mechanism, *Geophys. J. Int.*, **143**, 881–897.

Kagan, Y. Y., and Jackson, D. D., 1991a. Long-term earthquake clustering, *Geophys. J. Int.*, **104**, 117–133.

Kagan, Y. Y., and Jackson, D. D., 1991b. Seismic gap hypothesis: ten years after *J. Geophys. Res.*, **96**, 21419–21431.

Kagan, Y. Y., and Jackson, D. D., 1995. New seismic gap hypothesis - 5 years after , *J. Geophys. Res.*, **100**, 3943–3959.

Kagan, Y. Y., and Jackson, D. D., 1998. Spatial aftershock distribution, *J. Geophys. Res.*, **103**, 24453–24467.

Kagan, Y. Y., and Knopoff, L., 1980. Spatial distribution of earthquakes: the two-point correlation function, *Geophys. J. Roy. Astron. Soc.*, **62**, 303–320.

Kanamori, H., and Stewart, G. S., 1978. Seismological aspects of the Guatemala earthquake, *J. Geophys. Res.*, **83**, 3427–3434.

Keilis-Borok, V. I., Knopoff, L., Rotwain, I. M., and Allen, C. R., 1988. Intermediate-term prediction of occurrence times of strong earthquakes, *Nature*, **335**, 690–694.

King, G. P., and Sammis, C. G., 1983. The mechanism of finite brittle strain, *Pure Appl. Geophys.*, **138**, 611–640.

Kirchherr, W., Li, M., and Vitany, P., 1997. The miraculous universal distribution, *Math. Intelligencer*, **19**, 7–15.

Lachenbruch, A. H., and Sass, J. H., 1980. Heat-flow and energetics of the San Andreas fault zone, *J. Geophys. Res.*, **85**, 6185–6222.

Main, I., 1996. Statistical physics, seismogenesis and seismic hazard, *Rev. Geophys. Space Phys.*, **34**, 433–462.

Marrett, R., and Allmendinger, R. W., 1992. Amount of extension on 'small" faults: an example from the Viking graben, *Geology*, **20**, 47–50.

McCann, W. R., Nishenko, S. P., Sykes, L. R., and Krause, J., 1979. Seismic gaps and plate tectonics: seismic potential for major boundaries, *Pure Appl. Geophys.*, **117**, 1082–1147.

Mulargia, F., 1997. Retrospective validation of the time association of precursors, *Geophys. J. Int.*, **131**, 500–504.

Mulargia, F., 2001. Retrospective selection bias (or the benefit of hindsight), *Geophys. J. Int.*, **146**, 489–496.

Mulargia, F., and Gasperini, P., 1995. Evaluation of the applicability of the time– and slip–predictable earthquake recurrence models to Italian seismicity, *Geophys. J. Int.*, **120**, 453–473.

Mulargia F., and Gasperini, P., 1996. Precursor candidacy and validation: the VAN case so far, *Geophys. Res. Lett.*, **23**, 1323–1326.

Nishenko, S. P., and Buland, R., 1987. A generic recurrence interval distribution for earthquake forecasting, *Bull. Seism. Soc. Am.*, **77**, 1382–1399.

Okubo, P. G., and Aki, K., 1987. Fractal geometry in the San Andreas fault system, *J. Geophys. Res.*, **92**, 345–355.

Parsons, T., Toda, S., Stein, R. S., Barka, A., and Dieterich, J. H., 2000. Heightened odds of large earthquakes near Istanbul: an interaction-based probability calculation, *Science*, **288**, 661–665.

Popper, K. R., 1980. *Logic of Scientifi c Discovery*, Hutchinson, London.

Reasenberg, P. A., and Matthews, M. V. 1988. Precursory seismic quiescence: a preliminary assessment of the hypothesis, *Pure Appl. Geophys.*, **126**, 373–406.

Roeloffs, E. A., Burford, S. S., Riley, F. S., and Records, A. W., 1989. Hydrological effects on water level changes associated with episodic fault creep near Parkfield, California, *J. Geophys. Res.*, **94**, 12387–12402.

Romanowicz, B., 1993. Spatiotemporal patterns in the energy release of great earthquakes, *Science*, **260**, 1923–1926.

Ruina, A. L., 1983. Slip instability and variable friction laws, *J. Geophys. Res.*, **88**, 10359–10370.

Saleur, H., Sammis, C. G., and Sornette, D., 1996. Discrete scale invariance, complex fractal dimensions, and log-periodic fluctuations in seismicity, *J. Geophys. Res.*, **101**, 17661–17677.

Scheffé, H., 1959. *The Analysis of Variance*, Wiley, New York.

Scholz, C. H., 1990. *The Mechanics of Earthquakes and Faulting*, Cambridge University Press, Cambridge.

Scholz, C. H., 1998. Earthquake and friction laws, *Nature*, **391**, 37–41.

Schwartz, D. P., and Coppersmith, K. J., 1994. Fault behaviour and characteristic earthquakes: examples from Wasatch and San Andreas fault zones, *J. Geophys. Res.*, **89**, 5681–5698.

Shimazaki, K., and Nakata, T., 1980. Time-predictable recurrence model for large earthquakes, *Geophys. Res. Lett.*, **7**, 279–283.

Sieh, K., Stuiver, M., and Brillinger, D., 1989. A more precise chronology of earthquakes produced by the San Andreas fault in Southern California, *J. Geophys. Res.*, **94**, 603–623.

Simes, R. J., 1986. An improved Bonferroni procedure for multiple tests of significance, *Biometrika*, **73**, 751–754.

Stark, P. B., and Hengarter, N. W., 1993. Reproducing Earth's kernel: uncertainty of the shape of the core-mantle boundary from PKP and PcP travel times, *J. Geophys. Res.*, **98**, 1957–1971.

Sykes, L. R., and Nishenko, S. P., 1984. Probabilities of occurrence of large plate ruptur-ing earthquakes for the San Andreas, San Jacinto, and Imperial faults, California, *J. Geophys. Res.*, **89**, 5905–5927.

Walsh, J., Watterson, J., and Yielding, G., 1991. The importance of small-scale faulting in regional extension, *Nature*, **351**, 391–393.

Ward, S. N., 1991. A synthetic seismicity model for the Middle America Trench, *J. Geophys. Res.*, **96**, 21433–21442.

Ward, S. N., 2000. San Francisco Bay Area earthquake simulations: a step toward a standard physical earthquake model, *Bull. Seism. Soc. Am.*, **90**, 370–386.

Whitesides, G. M., and Ismagilov, R. F., 1999. Complexity in chemistry, *Science*, **284**, 89–92.

Wyss, M., 2001. Why is earthquake prediction research not progressing faster?, *Tectono-physics*, **338**, 217–223.

Wyss, M., and Martirosyan, A. H., 1998. Seismic quiescence before the M7, 1998, Spitak earthquake, Armenia, *Geophys. J. Int.*, **134**, 329–340.

Chapter 2

The classical view of earthquakes

Editors' introduction. What do we know about earthquakes? Either a lot or very little, depending on one's point of view and on one's definition of earthquake. If we define an earthquake to be the slip on a fault (or faults) that produces the observed seismic wavefield, then, as explained below, we have a good understanding of earthquakes. We call this the *kinematics* of earthquakes. On the other hand, it is indisputable that we have not yet attained a satisfactory understanding of the physical processes in the lithosphere that cause slip on faults, which we call the *dynamics* of earthquakes.

In this chapter we review the basics of the kinematics of earthquakes and summarize the state of knowledge and problems of the dynamics of earthquakes. We refer to these topics collectively as the classical view of earthquakes, as everything covered in this chapter is in the framework of the classical viewpoint of continuum mechanics. In contrast, the next chapter discusses earthquakes from the standpoint of the *new physics*.

We begin our discussion of the kinematics of earthquakes from the geologist's viewpoint, followed by the seismologist's viewpoint. We then turn our attention to laboratory experiments. As we shall see, the relation between actual earthquake faulting and laboratory experiments is problematical, owing to great differences in ambient conditions, time scales, and strain rates, among other parameters. Finally we discuss classical models of the so-called earthquake cycle. We shall see that these models, while simple and perhaps intuitively appealing, suffer from the fatal flaw that they do not agree with the data. In summary, the classical (continuum mechanics-based) view of faulting affords us an excellent understanding of earthquake kinematics, but comparatively little understanding of the dynamics of faulting.

2.1 A geologist's view of earthquakes

[By S. Castellaro, except section 2.1.3 which is by G. Di Toro and S. Castellaro]

The *in situ* recognition of the evidence of past earthquakes is a task which involves all the branches of geology. Signs of a past earthquake, in fact, can be identified on the landscape morphology and in the rocks generated by fault movements on the fault zone.

While for historical times, documents and the archaeological examination of ruins contribute to the reconstruction of the seismic history of a region, for pre-historic times the geologist alone can make estimates not only of the age of occurrence of earthquakes but also of their mechanisms. Detailed and systematic field studies performed on faults allow the reconstruction of the directions of stresses of the seismic mechanisms for many event series, which is essential for fixing the tectonic history of the Earth.

Furthermore, data from geology can help in assessing the environmental consequences of earthquakes and on this basis engineers can design structures suited to each seismic region.

The geologist, therefore, can study the past to capture in it the signs of catastrophic events and to apply this knowledge to present and future times, in order to minimize risks to the landscape, lives and structures.

Although earthquakes have always been present in man's life, it was only recently that their relation to faulting was understood. In the second half of the 18*th* century, the main school of scientific thought tried to explain the Earth's morphology and its origin in a 'catastrophic' way. Catastrophists were firmly convinced that the strangest landscape configurations, such as high mountains with rocks bent in remarkable shapes, deep gullies and valleys, could not have been created by anything other than a supernatural force. It appeared clear to them that the water contained in the rivers was insufficient to have eroded such valleys in the time available.

It was thanks to James Hutton (1726–1797) that this belief in the dichotomy between past and present was defeated. He supposed that the mechanisms of Nature are uniform and constant and on this basis he argued that a sufficient period of time must have passed to produce the events, the effects of which we observe. *Hutton* (1788) wrote:

"[...] For they are the facts from whence we have reasoned, in discovering the nature and constitution of this earth: therefore, there is no occasion for having recourse to any unnatural supposition of evil, to

any destructive accident in nature, or to the agency of any preternatu-
ral cause, in explaining that which actually exists".

For having formulated this theory, according to which we do not need to imagine causes different from those acting at present on our planet to explain its land-
forms, Hutton is considered to be the founder of modern geology.

Before Hutton, the continents were considered to be in a state of perfection
and it was only after him that the concept of "a time so long that things can develop without humans noticing it" arose and it was realized that the Earth's history
could be explained in a different way. The unifying theory arrived near the beginning of the 20*th* century, when Alfred Wegener (1880–1930) proposed a theory
of continental drift. Despite the initial skepticism of geologists, much evidence
was found in favor of this theory[1]. Such evidence ranges from the presence of
glacial deposits, morphologies and fossils in areas in which the climate does not
nowadays allow their formation, to the discovery of the mid-ocean ridges, fissures
in the oceanic floor in which oceanic basaltic crust is formed. The fundamental
role of earthquake occurrence in plate tectonics became clear when the map of the
world's largest earthquakes was available. The continuous band of shallow earthquakes under the mid-ocean ridges and transform faults, together with the concentration of earthquakes at the plate boundaries, suggested that the Earth's crust
is a dynamically evolving system, in which new crust is formed at the oceanic
ridges and other crust is consumed at the slabs in the subduction zones at the plate
boundaries.

We will now discuss the geological evidence of past earthquakes and the information that can be inferred regarding future seismicity.

2.1.1   Geology, geomorphology and earthquakes

A fault is a surface or a narrow zone along which one side has moved relative to
the other in a direction parallel to the surface or zone (*Twiss and Moores*, 1992).
Most faults are brittle shear fractures or zones of closely spaced shear fractures extending over distances of meters or larger, the production and movement of which
can be associated with seismic activity. The location of a fault plane, its geometric description and the direction and magnitude of the relative displacement
between its two sides represent the fundamentals of the earthquake description
from a kinematic point of view.

Faults can be recognized (1) for their intrinsic features, (2) for the effects on
the stratigraphic units and (3) for the effects on the landscape morphology (*Twiss
and Moores*, 1992). Concerning the intrinsic features, faults often present typical

---

[1]Wegener's original theory considered only the motion of continents but during the 1960s it
was superceded by the theory of plate tectonics.

			Random fabric		Foliated	
Incohesive			Fault breccia visible fragments > 30% of rock mass			
			Fault gouge visible fragments < 30% of rock mass		Foliated gouge	
Cohesive	**Nature of matrix**	Glass, devitrified glass	Pseudotachylyte			
		Granular: tectonic reduction in grain size dominates grain growth by recrystallization and neomineralization	Crush breccia fragments > 0.5 cm Fine crush breccia 0.1 < fragments < 0.5 cm Crush microbreccia fragments < 0.1 cm		0–10	**Percent of matrix**
			Cataclasite series — Proto-cataclasite	**Mylonite series** — Protomylonite	10–50	
			Cataclasite	Mylonite	50–90	
			Ultra-cataclasite	**Phyllonite series** — Ultra-mylonite	90–100	
		Pronounced grain growth		Blastomylonite		

Figure 2.1: Table of the terminology used for fault rocks. [Redrawn from table 3.1 in *Scholz* (1990)].

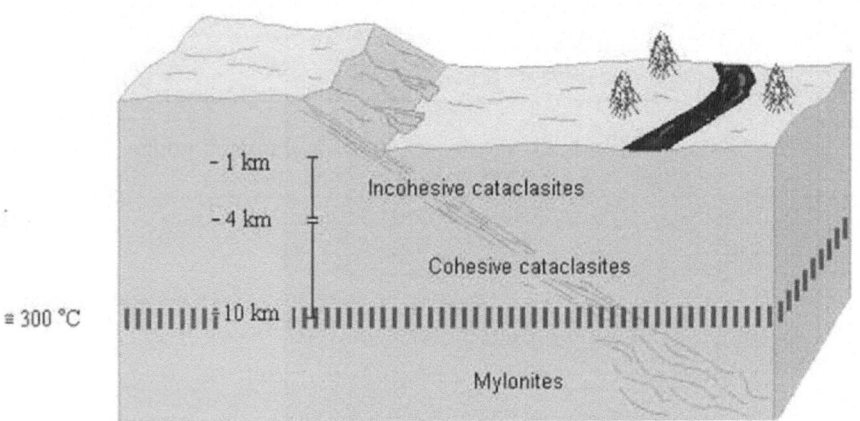

Figure 2.2: Schematic drawing of a fault scarp with associated landslides. The transition with depth of the fault rock types around the fault zone is also illustrated. Incoherent cataclasites are characteristic of the first 4 km of crust; coherent cataclasites can be found from about 4 km depth up to 10–15 km. At greater depths, when temperature reaches 250–350 $^{\circ}$C, mylonites are present.

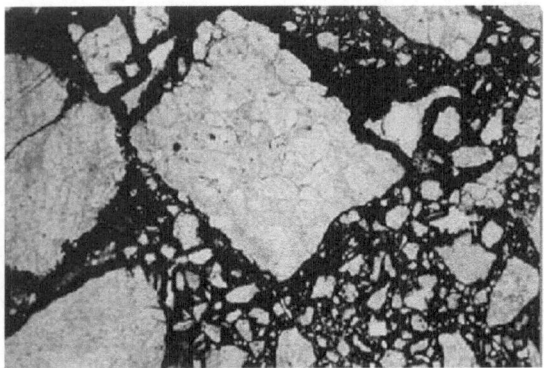

Figure 2.3: Photomicrograph of fault breccia in the Antietam Formation, Blue Ridge province (Virginia). Breccias form when rocks are extensively fractured in fault zones and are cemented together when minerals precipitate in the cracks and fractures. Note the angular fragments of quartz sandstone in a matrix of fine-grained iron oxide cement. Field of view 4 × 2.7 mm, Cross Polarized Light. Photo by Christopher M. Bailey, College of William & Mary.

structures resulting from shearing. Such structures differ depending on the depth and the environment (that is, temperature and pressure) in which they developed. As can be seen in figures 2.1 and 2.2, going downward along a fault zone we move from a brittle to a ductile domain, the boundary being generally at 10–15 km. Rocks characteristic of the brittle domain are generally called cataclasites, a term which indicates that they have been fragmented into clasts (breccia) or powder () during the deformation. A peculiarity of these clasts is their angular shape, which suggests that they did not undergo significant transport. At depths greater than about 4 km, cataclasites can present a certain degree of cementation. Breccia, gouge and cataclasite, the so called cataclastic rocks, are usually distinguished on the basis of clast size and the percentage of matrix: from breccia to cataclasite the percentage of matrix increases (usually up to more than 30 per cent) while the clast size reaches the minimum for gouge (less than 0.1 mm) and the maximum for megabreccia which can have clasts larger than 0.5 m.

Pseudotachylyte is a further rock which can be encountered along the fault zone, mainly in the brittle domain but sometimes also in the ductile domain. A paragraph is devoted to its description in section 2.1.3.

Mylonitic rocks are typical of fault zones in the ductile environment, which is generally met when temperature reaches at least 250 ^oC. Mylonitic rocks underwent recrystallization during rapid deformation, and due to the strong differential pressures acting at that time; they show strong planar and linear structures (foliation and lineation) parallel to the fault zone (*Bell and Etheridge*, 1973). Also in mylonitic rocks a matrix of very fine grains is present together with variable amounts of relict minerals, called porphyroclasts, which did not undergo recrystallization. The distinction among the different end-members illustrated in the bottom part of figure 2.1 is once again performed on the basis of the porphyroclasts and on the percentage of matrix.

On the fault's surface many other details of the movement that occurred can be found. Typically they are striations and slickenlines (in the forms of ridges or grooves generally parallel to the slip) from which the direction of displacement can be inferred. Simulations of the brittle and ductile mechanisms during fault shearing which try to model the evolution of the grain size distribution can be found in *Ozkan and Ortoleva* (2000), *Mair and Marone* (2000), *Morgan and Böttcher* (1999), *Place and Mora* (2000), and *Sleep et al.* (2000). *Cladouhos* (1999a, b) studied instead the textural fabric during the formation of intrafault structures (breccias, gouges, mylonites). This research helps to determine the hydrologic and mechanical properties of faults, which are needed to constrain large scale fault models.

Faults displace the stratigraphic series laterally and vertically, resulting in a discontinuity. Not all the discontinuities are due to faults: they can derive, e.g., from the intrusion of magmatic bodies, from erosion, or from other causes. How-

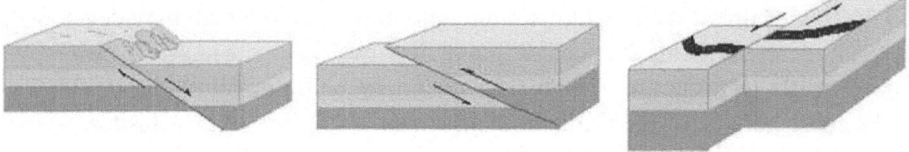

Figure 2.4: From left to right: a normal, reverse and strike-slip fault. In the normal
fault the top block moves down relative to the bottom block; in a reverse fault the top
block moves up relative to the bottom block; in a strike-slip fault the two blocks displace
laterally. Natural faults are usually derived by a combination of these end-members.

ever, once it is recognized that a discontinuity is a fault, if the local stratigraphic
series is known or other markers can be found on the fault's surface from which
the relative movement of the blocks can be inferred, then the fault can be classified
into one of the three main types: normal, reverse or strike-slip (figure 2.4).

Contrary to the schematic drawings in figure 2.4, natural faults are never planar. They change their orientation almost every time they pass through layers of
different lithology (*Ramsey and Huber*, 1983). Flats and ramps on a fault surface
are linked to rock competence. Ramps occur when a competent layer is met (massive limestones, sandstones...), while flats are found when incompetent layers are
met (shales, marls, evaporites...) (*Twiss and Moores*, 1992).

Although not all earthquakes produce visible faults on the Earth's surface,
faults remain the preferred structural elements to perform orientation analysis at
a certain scale, in the *brittle domain*. The same role in the ductile domain is
performed by folds and shear zones.

The total *displacement* of a fault is a vector which can be determined by geologists by matching points which were once in contact and which are now separated
by the fault plane. As already stated, on the fault surfaces, fault grooves and striae
in the direction of displacement are formed and by observing them it is then possible to reconstruct the last movements of the fault. On a smaller scale, as can be
seen in figure 2.5 and 2.6, indicators of the displacement direction in a shear zone
are the sense of curvature of the foliation defined by the mineral orientation or the
sense of asymmetry in the recrystallized porphyroclasts. The problem with the
small scale structure shown in figure 2.5 and 2.6 is that the magnitude of the displacement cannot be determined. All these features must be used with care since
they can record only the most recent fault slip, which is not necessarily parallel to
the average slip vector.

Topographic and morphological features assist in recognizing faults. Since

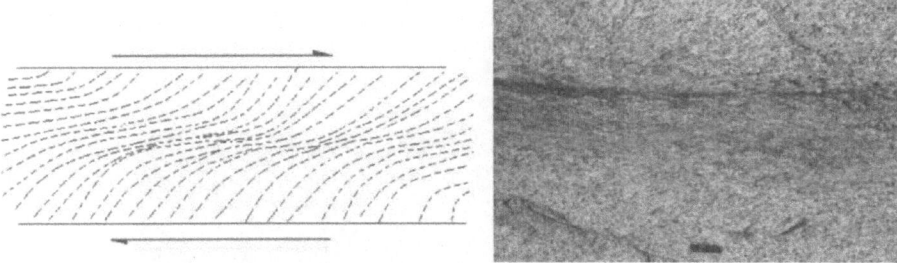

Figure 2.5: In a ductile shear zone the shear sense (arrows) can be inferred from the curvature of the foliation determined by the parallel alignment of the platy minerals. On the right, a mylonitic shear zone (amphibolitic facies) in a granitoid rock from the Adamello Massif (Italy) is shown. The sense of shear is dextral. Photo by G. Di Toro.

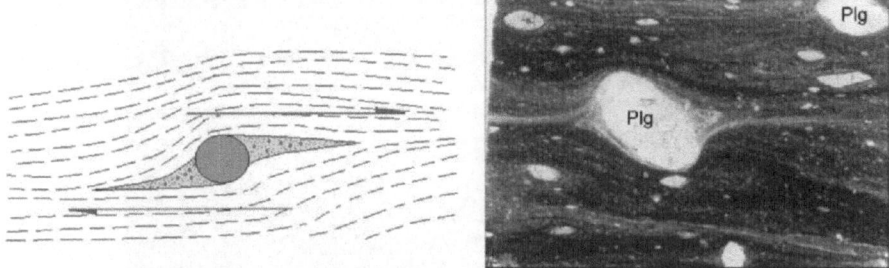

Figure 2.6: In a ductile shear zone, the sense of shearing can be inferred from the asymmetry of the tails of the recrystallized porphyroclastic material around the porphyroclast. The photo shows thin tails fed by a plagioclase porphyroclast. The orientation of the tails indicates a dextral sense of shear. The horizontal size of the picture is 1 mm. Photo by G. Di Toro.

faults are associated with the breakage of rocks, their surfaces are usually subject to more rapid erosion than the surrounding rock, and fault traces can be identified as local topographic lows, depressions or river valleys. At times a fault scarp (figure 2.7) is also present. Thanks to such features, faults can on some occasions be easily located on air photographs or satellite images, since they appear as lineaments on the Earth's surface. Note, however, that not all lineaments on the Earth's surface are fault traces.

For past earthquakes, for which seismic recordings do not exist, the structural geologists' work is the only possible source of information. But the direct investigation of faults is still needed for recent earthquakes to help seismologists remove the ambiguity left by the focal mechanism solutions (section 2.2). The ambiguity lies in determining which of the two nodal planes of the radiation pattern was the

Figure 2.7: A very clear normal fault scarp in Carnia (North-Eastern Italy). Photo by G. B. Vai.

fault plane that generated the earthquake.

Landslides

Intrafault structures are not the only markers of past earthquakes. There is in sedimentology and seismology a debate on a kind of deposit of not always certain origin, that is on submarine landslide deposits (*Prior and Coleman*, 1979; *Schwab et al.*, 1993; *Hampton et al.*, 1996; *McAdoo et al.*, 2000). Such deposits generally cover a much larger area than those produced under subaerial conditions and their slopes are more gentle. The larger extent of the deposits is due to the lower shear resistance of sediments under water. Under normal conditions, sediment under water is drained and the interstitial pressure has no mechanical effect since water can exit only under lithostatic load. It would then be difficult to find landslides under water since they would activate only on slopes higher than 15-20°, which are very rare on the sea floors and are, when they occur, very localized and usually depleted in sediments (fault scarps, canyons flanks). One could then infer that submarine landslide deposits can be triggered by seismic activity or fast subsidence. It is a matter of fact that earthquakes can trigger dozens of both subaerial and submarine landslides (*Fernandez et al.*, 2000) and in this second case, a submarine slump can generate a tsunami. As a consequence there is often a debate on the source of a given tsunami.

Many submarine landslides occur in the deep ocean basins and they do not interfere with human life and activities while they are partly responsible for the evolution of the submarine landscape (*O'Leary*, 1993). On the other hand, many cases are documented of landslides with significant impact on human activities. For example, the 1929 Grand Banks (Canada) landslides and turbidity current broke trans-Atlantic cables on the sea floor and produced a destructive tsunami on the sea surface (*Hasegawa and Kanamori*, 1987).

Those who support the idea of landslide tsunamis base this on some data which could not be explained if the tsunami source were an earthquake, for example, the unusually large wave amplitudes relative to the moment-magnitude, the very localized region of tsunami run-up, the extraordinary long duration of the earthquake and the 1:1 aspect ratio of the aftershock region (*Watts*, 2001; *Trifunac and Todorovska*, 2002). To better put the case for the landslide tsunami hypothesis, landslide tsunamis have amplitude proportional to their vertical center of mass displacement (*Murty*, 1979; *Watts*, 1998, 2000). Underwater landslides can have vertical displacements of up to several kilometers, contrary to coseismic displacement during earthquake which rarely surpasses 5 m (*Geist*, 1998). The maximum tsunami amplitude from the largest possible landslide on earth would then be dictated solely by the depth of the oceans. Many models have been proposed primarily based on the study of expected wave spectra tsunami amplitudes as a discriminant between earthquake or landslide origin (*Walker and Bernard*, 1993; *Watts*, 1998; *Jiang and LeBlond*, 1994; *Iwasaki*, 1987, 1997; *Imamura*, 1996).

Among the most recent tsunamis attributed, not without controversy, to submarine landslides is the 1998 Papua New Guinea event (*Tappin et al.*, 2001).

Paleoseismology

Geology is traditionally related to the study of the most ancient rock formation,
while geomorphology, and in particular its branch called *neotectonics*, deals with
recent and present (Pliocenic - from 2 million to 5 million years ago - and Quaternary - the last 2 million years) tectonic movements. The interest of geomorphology in earthquakes is thus mainly centered on their direct effects: (1) deformations and dislocations of layers; (2) deformation of fixed elements such as
marine, river and lake terraces; (3) modifications of the hydrographic net: deviations or inversions of the river flows, creation of lakes or of steps in fluvial valleys,
changes of the shorelines and in the circulation of underground water; or on the
indirect activation of other processes such as: (4) landslides, creation of fissures
and fault slopes which can be new or, usually, the reactivation of older morphologies, tsunamis and their consequences on the shorelines, collapse of the caves; (5)
subsidence.

A further sub-branch of geology and geomorphology is paleoseismology (*Mc-
Calpin*, 1996; *Michetti and Hancock*, 1997; *Yeats et al.*, 1997), which deals with
the study of large or moderate-sized earthquakes as recorded in geological units
during the Pleistocene and Holocene (the most recent 100,000 years). This field
received strong support in the USA and Japan starting in the 1970s, as a consequence of the need to evaluate the seismic hazard in areas of quickly increasing
urbanization and with large engineering structures. In regions for which the historical records of past earthquakes do not exist (for example because they were
not populated), this discipline is essential for documenting historical seismicity,
but it also has become an important source of data to supplement questionable or
incomplete historical seismicity catalogues.

The primary evidence of paleoearthquakes (*Meghraoui and Crone*, 2001;
http://www.astro.oma.be/PALEOSIS/) is the deformation directly related to
the coseismic displacement along the fault (tectonic features). The secondary evidence corresponds to the structures induced by the land-shaking (liquefaction –
see section 2.1.4, landslide, flooding etc.). The most typical way to study a recent
or an ancient earthquake fault is that of investigating the deposits on the two sides
of the fault surface by digging a trench (*Sieh*, 1978, 1984; *Sieh et al.*, 1989). In
this way it is possible to make direct measurements of the fault geometry and displacement due to the last earthquake. In some cases it is also possible to recognize
and date the deformations caused by more ancient seismic events.

The tectonic signature of an earthquake depends on the rate of deformation
and is generally expressed as a landform and scarp, the relief of which depends on

the ratio between tectonic and erosional processes. Most major fault scarps with prominent vertical offsets result from a series of repeated faulting and coseismic displacements affecting superficial and thus soft sedimentary units. These scarps naturally degrade with time as a result of climatic effects. If the rate of local erosional transport is known, a good estimate of the fault scarp age (*Andrews and Hanks*, 1985) can be made.

Under favorable circumstances old layers will be preserved under the post-seismic sediments. Applying the basic criteria of geology, if it is possible to date the layers rich in organic material which lie immediately above and immediately below the fault surface, then the age of the fault movement can be assumed to be included in this time interval. This information about the date of the earthquake can be combined with information on the earthquake's magnitude, approximately derived from the scarp extension and height.

To study geometry and displacements, trenches with a length of 10–100 m and a depth of about 4 m are excavated along the fault exposures. Samples collected from each significant sedimentary unit can later be analyzed for both the mechanical tests and ^{14}C radiometric dating. The latter is appropriate for samples which are rich in carbon and younger than 40×10^3–50×10^4 yr. Older samples must be dated using other methods such as U-Th or K-Ar, providing that their chemical composition allows this[2].

In addition to the study of fault structures, liquefaction, flooding and landslides are secondary evidence which can be used to infer coseismic displacements. For example, magnitudes ranging from 7.8 to 8.3 have been attributed to the 1811–1812 earthquake sequence of New Madrid (central USA), partly based on the detailed study of liquefaction features (see also section 2.1.4 and *McCalpin*, 1996).

In karst regions, ruptures and differential growth of speleothems[3] are also

---

[2]Radiometric dating is a method to obtain the absolute age of a material. Naturally occurring radioactive elements decay into other elements at known rates so that if the amount of daughter elements can be measured, one can calculate the age of the parent inclusion in the material. In igneous rocks one can date the time the magma cooled and the decay of radioactive elements such as U, Th, Rb, Sr and K began. In sedimentary rocks or fossils these methods cannot usually be applied (both because these elements are generally not present and because even if they were present, it would not be clear which type of age is computed). Thus in order to obtain the absolute age of sedimentary rocks or fossils, the ^{14}C method must be used. All living organisms contain a constant ratio of ^{14}C which is in equilibrium with the atmosphere. At death, ^{14}C exchange ceases and the ^{14}C contained in the tissues begins to decay to ^{14}N and then to ^{12}C. The ratio ^{14}C/^{12}C is the basis for dating. Radiometric methods can often be affected by large errors which we do not discuss here; nonetheless they remain a powerful tool for dating. For an introduction to geochronology, see *Poty et al.* (1990), or *Faure* (1986).

[3]A speleothem is any secondary mineral deposit that is formed in a cave. The best known forms of speleothems are stalactites, which develop downward from the ceiling of caves, and stalagmites, which develop upward from the floor of caves by the action of dripping water. They are usually composed of calcite although different mineralizations are possible (*Jackson*, 1987).

sources of information on the occurrence of past earthquakes. *Postpischl et al.* (1991) were able to date the paleoearthquakes of central Italy since the calcitic composition of stalactites and stalagmites allowed precise absolute dating using the U-Th method.

Another classical method, which can be used in regions rich in archeological artifacts is the evaluation of faulted archeologic sites. Although some uncertainty may exist about the seismic or non-seismic origin of such damage, *Stiros and Jones* (1996) applied this method to some deformed ancient graves in Greece, and *Levret et al.* (1996) applied it to the Nimes Roman aquaduct in France.

When faults are not clearly visible at the Earth's surface, classical geological methods must be accompanied by geophysical exploration (e.g. ground penetrating radar, electrical prospecting and seismic tomography, high resolution seismic profiles). The use of digitized topographic maps combined with Quaternary geology and geomorphology (distribution of alluvial terraces and young deposits) is the most appropriate methodology to identify the best fault sites for paleoseismic investigations.

Several studies of past and recent earthquakes have been performed on significant faults. The neotectonics of the huge Sumatran fault, for example, is described by *Sieh and Natawidjaja* (2000), while that of the Japanese fault lines have been investigated by, e.g., *Ohtani et al.* (2000) and *Okumura* (2001) through deep boreholes. Structural and paleoseismic studies on the faults of Taiwan can be found, for example, in *Lee et al.* (2001); studies on the California fault systems have been performed by, e.g., *Dolan et al.* (2000), *Lee et al.* (2001), and *Oskin et al.* (2000). Also for the low seismicity European areas surveys have been performed to identify active faults and to measure their activity in terms of historical and prehistorical large earthquakes (*Camelbeeck and Meghraoui*, 1996, 1998; *Valensise and Pantosti*, 2001). These studies, combining the geological, structural, textural and mineralogical information from the recent fault zones (sometimes investigated in depth, through boreholes, as in the Japanese cases), provide interesting constraints on the physico-chemical conditions in which faults acted. For instance, they provide information on the presence or absence of fluids, about the rheology, about the amount of dislocation recognized for each deformative event and about the probable stress axes, thus giving useful constraints on physical models of the faults.

2.1.2   Paleontology and earthquakes

As discussed in section 6.2, paleontology helps geologists in their reconstruction of past earthquakes. Fossils (ranging from the smallest animal and plant to the 'footprints' left by living beings on the ground and to pollens) are among the best indicators of past environments and climates. Due to the sensitivity of certain animals to environmental conditions, precise indications about these two factors can sometimes be obtained.

When an earthquake produces a change in the level of the shoreline, many consequences arise for the animals and plants living there. A relative change in the sea level, in fact, implies that some animals living at some specific depths must escape from their position to an upper or lower level. These escapes can be witnessed by the sediments as fossil traces or by changes in the fossil associations. The study of such associations, together with the sedimentological study of the shorelines, can help geologists in better characterizing past tectonic events.

For example, studying the Sumatran subduction zone, *Sieh and Ward* (1999) and *Zachariasen et al.* (2000) used the coral rings to describe the crustal deformation of that area. Corals reefs are known to live at a level close to the annual lowest low tide. In the absence of relative sea level variations, coral reefs can only grow outward, horizontally. If a tectonic movement raises the sea level, corals can grow upward. In the opposite case, the growth of the reef is obviously inhibited. Since each coral ring can be dated, the relative sea level change can be inferred.

2.1.3 Petrology and earthquakes

Each mineral association crystallizes only in specific ranges of temperature and pressure and, outside such intervals, minerals recrystallize or alter. Under favorable circumstances, geologists can therefore reconstruct the temperature and pressure acting at the time of crystallization and infer the orientation of paleo-stresses from the fabric of crystals.

Structural geologists commonly also study the rocks involved in seismic rupturing processes[4]. We describe now in more detail an interesting type of fault rock and the inferences which can be derived from its petrology.

Pseudotachylytes

Geological evidence. Shand (1916) classified as pseudotachylyte (figure 2.8) a fault rock made of clasts of the host rock immersed in a dark ultra-fine glassy matrix. Similar rocks had already been described by *Holland* (1900) in India and called trap-shotten charnokite gneiss.

Pseudotachylytes usually outcrop as fault vein or injection veins associations (*Sibson*, 1975) in inactive fault zones now exhumed by erosion (figure 2.8). Fault veins fill planar fault shear fractures, and rarely exceed 10^{-1} m in thickness,

---

[4]An example of what can be learned by such analyses is provided by a survey of the stratigraphy of a borehole (750 m deep) across the Nojima fault responsible for the 1995 Kobe earthquake. In that case, *Ohtani et al.* (2000) were able to determine the temperature of the fault core zone at the time of the earthquake by studying the hydrothermal alteration of minerals and fluid inclusions.

Figure 2.8: Example of pseudotachylyte fault vein and injection veins from the Adamello Massif (Southern Alps, Italy). The pseudotachylyte is the black vein with whitish clasts on the granodioritic host rock and it is associated to cataclasite (green bands). Photo by G. Di Toro.

whereas injection veins depart at about 80–110° from fault veins and intrude into the host rock. Several features are indicative of pseudotachylyte formation both at shallow (2–10 km) and mid-crustal levels (10–20 km). Some pseudotachylytes are clearly linked to brittle regimes, since they overprint and are reworked by cataclasites (e.g. *Magloughlin*, 1992; *Fabbri et al.*, 2000). Other pseudotachylytes, instead, appear closely related to the plastic-ductile transition regime since they overprint and are overprinted by ductile shear zones (*Sibson*, 1980; *Passchier*, 1982; *White*, 1996). Some authors consider them to be the result of the downward propagation of faults initiated at shallower levels, while others think they formed as ductile instabilities during plastic flow (*Hobbs et al.*, 1986).

The genesis of shallow level pseudotachylytes (2–10 km) as the product of frictional melting during earthquakes or ultracomminution without melting was debated for a century (*Holland*, 1900; *Philpotts*, 1964; *McKenzie and Brune*, 1972; *Wenk*, 1978; *Allen*, 1979; *Maddock*, 1983; *Spray*, 1987). *Spray* (1995) demonstrated experimentally that grain size reduction during frictional sliding is a precursor of melting during coseismic slip. Thus comminution and frictional melting are two related processes as already suggested in the so-called 'abrasive wear' model by *Swanson* (1992). Today, pseudotachylyte textures are unambiguously recognized as the product of solidification of frictional melts and indicative of seismic activity on exhumed faults (*Cowan*, 1999).

Among the structural, textural and chemical evidence for a melt origin of pseudotachylytes, let us first recall the intrusive habit of injection veins and the presence of flow structures, indicative of a fluid (i.e. melt) intruded in a host rock and of hydrofracturing (*Sibson*, 1975). Second, pseudotachylyte veins often exhibit microlitic textures, chilled margins, and/or glass, indicating the rapid chilling of a melt (figure 2.9, see also *Maddock*, 1983; *Lin*, 1994). However, the presence of glass is not a necessary feature of these rocks, since the environmental conditions under which pseudotachylytes form (usually between 3–15 km in depth) are unfavorable for glass preservation so that devitrification textures are commonly found. Third, analysis of size distribution of clasts in pseudotachylytes shows that small grains (5 μm) are much fewer than expected according to the number of larger grains (*Shimamoto and Nagahama*, 1992). This lack of small grains is interpreted as a consequence of preferential melting of the finer part of the clast distribution

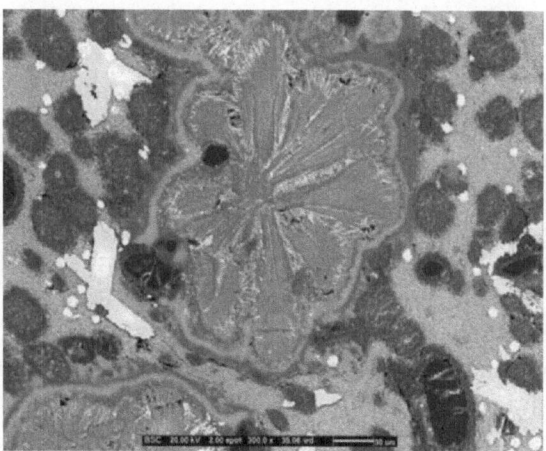

Figure 2.9: Cooling structures visible in a Scanning Electron Microscope (SEM) image of a pseudotachylyte from the Adamello Massif (Southern Alps, Italy). Spherulitic structures consist of a plagioclase core rimmed by plagioclase microlites (middle gray) and very small acicular biotite (white). Spherulites are externally rimmed by devitrification haloes of K-feldspar (middle-bright gray) and silica (dark gray). Quartz grains are black and rounded, suggesting interaction with the melt and dissolution. Bright small dots (5 μm) are clusters of titanite and biotite, probably crystallized in the residual melt. Late overgrowth of epidote (large bright rods), suggests host rock temperature of 250–300 ^{o}C. These microstructures are indicative of very rapid energy release with instantaneous melting and rapid cooling in the upper crust, providing additional evidence for their episodic (earthquake) origin. Photo by G. Di Toro.

(*Shimamoto and Nagahama*, 1992). Lastly, frictional melting is a non-equilibrium process and this implies that, compared to the host rock or to the cataclastic precursor, pseudotachylytes usually contain survivor clasts of quartz and feldspar and a matrix enriched in Fe, Mg, Al, Ca, H_2O (*Sibson*, 1975). This mineral selection and chemical differentiation is due to the preferential breakdown and selective melting of minerals which have lower yield stress, fracture toughness and melting point (i.e. biotite, chlorite, amphibole, *Spray*, 1992).

Pseudotachylyte production mechanisms. A possible mechanism for producing pseudotachylyte textures at shallow crustal levels is the conversion of elastic strain energy into heat during coseismic slip (*McKenzie and Brune*, 1972; *Spray*, 1992). Given the low thermal diffusivity of the host rock (10^{-6} m^2/s), the high effective stress in the seismic source area (2–15 km in depth) and the particle velocities (0.1–2 m/s) achieved during the rapid (1–5 s) coseismic slip, heat can be supposed to be produced in large amounts and to remain *in situ* (section 2.8). The process is thus adiabatic and fault rock would be instantaneously and locally heated, producing a thin layer of melt on fault planes.

Two different wear mechanisms have been suggested for the production of frictional melt depending on the mechanical behavior of the asperities, or protrusions between opposite surfaces which interact during sliding and determine friction in rocks (e.g. *Scholz*, 1990). In abrasive wear (pseudotachylyte-cataclasite association), after rupture propagation, initial slip is impeded by brittle breaking of the asperities and by surface refinement due to clast rotation and fracturing of the initiation breccia. The opposite sliding surfaces are then uncoupled and a cushion of cataclasite develops in between during sliding. Frictional melting occurs where cataclasites are highly comminuted and the strain rate is higher (*Swanson*, 1992). In adhesive wear (pseudotachylyte-mylonite association) rupture propagates along a preexisting planar anisotropy (i.e. mylonitic foliation) and slip is retarded during welding of asperities by plastic deformation. Surface refinement of the opposing slip surfaces lead to total area adhesion of the fault and plastic flow without the production of a lubricating cushion of cataclasite. The adhesion of the opposite surfaces abruptly increases the frictional resistance and promotes the onset of frictional melting (*Swanson*, 1992).

The role of water during frictional melting is still debated (*Sibson*, 1975; *Magloughlin*, 1992; *O'Hara et al.*, 2001). When fault rocks are depleted in intergranular fluid, pseudotachylyte production seems to be favored, since the absence of fluid pressure along a fault increases the dynamic frictional shear resistance and the heat release during coseismic slip (*Allen*, 1979; *Sibson*, 1975). On the other hand, microstructural evidence such as the presence of fluid inclusions, vesicles and amygdalae (*Maddock et al.*, 1987; *Magloughlin*, 1992; *Boullier et al.*, 2001) in the pseudotachylyte matrix suggests that sometimes water must be present at the time of pseudotachylyte generation. In these cases, water acts to lower the melting temperature of minerals, thus helping fusion (*Allen*, 1979). Moreover, the dramatic rise in temperature during coseismic slip expands the superheated water present in minerals as fluid inclusions and this favors rock fragmentation and subsequent melting due to the reduced grain size of the rock (*Sibson*, 1975).

Pseudotachylytes are not only geological markers of past seismic activity of a fault, but can also help in constraining physical parameters during earthquakes. On the basis of the pseudotachylyte volumes and through energy balance calculations, *Sibson* (1975) estimated the dynamic shear stress resistance during coseismic slip and *Wenk et al.* (2000) compared past earthquakes in exhumed paleofaults with actual microseismic events at Parkfield, along the San Andreas fault. The presence of amygdales and vesicles in pseudotachylytes has been used to estimate the depth of formation of frictional melts (*Maddock et al.*, 1987). Recently *O'Hara* (2001) proposed a geothermometer based on the clast/matrix ratio in pseudotachylytes to estimate the temperature attained by the frictional melt or the temperature of the host rock during seismic faulting.

Pseudotachylytes remain, up to now, the least abundant representatives of the fault rock family. Their occurrence seems to be much lower than expected according to energy balance calculations from *McKenzie and Brune* (1972) and the widespread seismic activity in the upper crust (*Sibson*, 1973; *Spray*, 1987). One reason for this depletion could be that the conversion of work to heat is a low efficiency process in the upper crust (*O'Hara*, 2001). Moreover different processes may weaken frictional resistance during coseismic slip before the onset of frictional melting such as fluid pressurization (*Lachenbruch*, 1980), acoustic fluidization (*Melosh*, 1996) or hydrodynamic lubrication (*Brodsky and Kanamori*, 2001; *Goldsby and Tullis*, 2002). Finally, devitrification might obliterate pseudotachylyte textures from the geological record.

Although a landmark for the use of pseudotachylytes as kinematic and dynamic indicators of past earthquakes already exists in the form of the work of *Sibson* (1975), much work - both in the field and in the laboratory - is still needed to better understand the phenomenon of frictional melting. This task is complicated by the variety of host lithologies and crustal depths in which pseudotachylytes are met (e.g., pseudotachylytes produced at 60 km depth, *Austrheim and Blundy*, 1994).

2.1.4 Applied geology and seismic hazard

The identification of faults is essential in applied geology since such discontinuities may have severe consequences especially for dams and artificial reservoirs, where an infiltration of water along faults can compromise the stability of the system.

Particular attention also deserves to be given to the faults which are crossed during tunnel construction, since the surrounding rocks are often finely fragmented, and therefore unstable and sometimes permeable.

In these terms, the interest of applied geologists in earthquakes is primarily devoted to the static aspects or to post-seismic effects. For the dynamic (co-seismic) aspect, the main concern is the identification of sites at which liquefaction can occur. Liquefaction is a technical term used to indicate the loss of resistance in saturated earth under dynamic or static stresses, the consequence of which is that the soil loses its resistance to shear stresses. Liquefaction is one of the main causes of damage to buildings, which generally remain standing but are strongly inclined to one side and can sink deeply into the surface sediments.

This thixotropic phenomenon is typical of saturated fine sand deposits. When a load or a force is applied to the soil, the pressure of the water contained in the pores progressively increases until the effective pressure (and the shear resistance) falls to zero.

All unconsolidated deposits can in principle undergo liquefaction. The Coulomb relation states that $\tau_f = \sigma \tan\phi$, where τ_f is the shear resistance of the unconsolidated material along a possible sliding plane, σ is the normal stress acting on that plane and ϕ the shear resistance angle. When the material is saturated with water, the Coulomb law must be rewritten as $\tau_f = (\sigma - u)\tan\phi'$, where u is the interstitial pressure. When u increases up to the total pressure σ, the effective pressure $\sigma' = (\sigma - u)$ goes to zero as does the shear resistance. However, $\sigma' = 0$ does not always lead to liquefaction, since the initial compactness, i.e. the initial pore volume, can play an important role.

The liquefaction of natural and artificial unconsolidated deposits during an earthquake is due to the progressive increase of the interstitial pressure produced by the upward propagation of the S-waves. Pressure is transferred from particles to water, since water cannot instantly leave the sediment under pressure. If the original material is not very compact and if the earthquake has a sufficient strength and duration, then, after a certain number of cycles, the interstitial pressure can be so high that the sediment particles float on the water they previously included. After that, water is expelled to the surface and the solid particles are redeposited in a denser structure.

The granulometry of the sediment is a controlling parameter in the liquefaction process. Both the dimension of the pores in the sediment and the forces acting among particles depend on it. Sediments which can undergo liquefaction have a particle diameter in the range 50 μm–1.5 mm. Below this dimension (even more so below 4 μm), the clay-regime is characterized by coherent earths, the cohesion of which is due to the crystallographic structure of the clay minerals (phyllosilicates) which crystallize into very thin leaves, with the negative charges toward the external surfaces. This favors, at very small distances, an electrostatic attraction giving cohesion. For larger particles (typically sands and gravels) cohesion can only be ensured by the friction among grains.

Fine sands have a pore size ideal to initiate liquefaction, since it does not allow a dissipation of interstitial water overpressure as fast as in gravels, which have very large inter-granular pores. On the other hand, although clays can have a large total porosity, the connected pore size is so small that permeability is inhibited and water overpressure has no time to spread out coherently inside the sediment.

Liquefaction can occur for the cyclical application of loads (as happens with earthquakes) as well as for monotonically increasing loads, if the interstitial pressure cannot be freed (that is in undrained conditions), and for a water level change on earth banks and dams.

Liquefaction has been assumed by some (*Morgensten*, 1967) as the cause of submarine landslides (see section 2.1.1).

Geologists have observed that the sites most prone to liquefaction are young (and therefore not consolidated) soils: for example, recent soils at the bottom of valleys, deltas, swamps, river meanders or soft loess deposits.

In relatively flat-lying areas, earthquakes of magnitude higher than 6 and duration longer than 15 s are required in order to produce liquefaction. These phenomena can occur also in the far field, thanks to the longer duration of the shaking. Liquefaction generally does not involve sediments deeper than 15–20 m and the presence of clay or silt, providing cohesion to the deposit, highly decreases the possibilities that the deposit will undergo liquefaction. The same site can be involved in liquefaction more than once. In any case, the duration of the earthquake has a dominant role since it contributes to the growth of the interstitial pressure and to the progressive expansion of fluidification (*Crespellani et al.*, 1988).

In sloping areas, conditions for liquefaction can be even more common since local amplification of seismic waves can occur. The problem is aggravated because, in sloping areas, liquefaction easily leads to landslides. In many cases, the presence of gravel layers is insufficient to inhibit the liquefaction process and, in contrast to flat-lying areas, illiquefiable layers on the top of sands can contribute to the liquefaction of the underlying layers since they hamper the upward dissipation of the pore pressure (*Crespellani et al.*, 1988). Artificial dams or terraces built with earth material obviously belong to this last category. Among the historical cases of almost total collapse of artificial earth dams we recall one which occurred in 1925 at Sheffield, and one in 1971 at San Fernando (California).

Soil liquefaction can lead to the sinking or overturning of buildings. A typical example of severe damage induced by liquefaction is that of the Niigata earthquake (June 16, 1964) in Japan (*Olson and Stark*, 2002).

Some criteria, based on laboratory experiments and on the study of the granulometry and the density factor, the age of the deposit, its origin, the stratigraphy and the groundwater level, have been developed for an empirical forecast of the zones prone to liquefaction. These criteria are those taken into account in seismic microzonation.

At the same time, some techniques can be applied to reduce the liquefaction potential of soils. These are based on the improvement of the soil compaction (generally through the employment of vibrations), on increasing the effective pressure (lowering the underground water table), on the consolidation through injection of cements or resins and on the improvement of the earth permeability with natural or artificial drains. Of course these methods can be applied only to restricted areas.

2.2 Seismology and geodesy

[By M. Ciccotti]

2.2.1 Introduction

Since the origin of seismology, seismologists have made instrumental recordings of the ground motion at observatories on the Earth's surface in order to estimate the location and origin time of earthquakes as well as the mechanism of the source. Several magnitude scales have been developed for estimating the size of the source. The 'hypocenter' (coordinates of the onset of rupture), and origin time were the first source parameters to be routinely determined for all significant events (note that the 'epicenter' is the point on the Earth's surface directly over the hypocenter). By the 1950s the magnitude was added to the standard set of source parameters. Because definitions of magnitude varied greatly and were not always unambiguous, care should be exercised when using older magnitude values (*Geller and Kanamori*, 1977).

As instrumental networks and methods of data analysis improved, the set of source parameters that are routinely determined also steadily improved. Current practice is summarized below.

In order to determine accurate values of earthquake source parameters it was necessary for seismologists to also determine increasingly accurate models of

Earth structure (the spatial distribution of seismic velocities and density). This work contributed greatly to our understanding of geodynamics (e.g., mantle convection), and is one of the major successes of the past century of seismological research. However, as this is peripheral to the present topic, further discussion will be omitted.

One of the first kinematic models of the physical source of seismic waves was that of *Haskell* (1964), which consists of a simple planar rectangular dislocation taking place in an elastic medium. While this is a very simple model, it provides a satisfactory description of the asymptotic radiation field at low frequency. In particular, it explains the polarity and the angular radiation pattern of the radiated P and S waves. For the case of an earthquake which is a slip on a fault plane, an analysis of far-field seismological data cannot uniquely determine the fault plane and slip direction. What can be determined are two mutually perpendicular 'nodal planes', either of which could be the fault plane, and the earthquake slip vector corresponding to each nodal plane (if that nodal plane were the fault plane). The slip vector for each of the nodal planes is a vector normal to the other nodal plane. Determining which of the two nodal planes was the fault plane requires additional evidence, such as geological observations, the spatial distribution of aftershock hypocenters, or more sophisticated seismological analyses.

The development of broad-band digital networks in the 1970s and later led to further developments in seismological data analysis. The analysis of the initial data to determine the two nodal planes and slip vectors was supplanted by an analysis of the waveforms of long period surface waves to determine the centroid and moment tensor (abbreviated as CMT) of each large earthquake.

A more detailed description is provided below. However, we note here that the hypocenter, origin time, nodal planes and slip vectors all pertain to the onset of rupture, while the CMT and moment tensor characterize an average over the entire space-time rupture process for the particular earthquake being studied. The overall size of each earthquake is now characterized by the scalar moment[5]

$$M = \mu A \Delta s, \tag{2.1}$$

where μ is the rigidity of the rocks surrounding the fault, A is the area of the fault, and Δs is the average slip on the fault plane. Since not only seismologists but also related professions (engineers, government officials, the news media) were accustomed to characterizing the size of an earthquake by the magnitude, *Kanamori* (1977) devised an empirical formula (equation 2.15 on page 48), sometimes referred to as the moment-magnitude relation, for converting the scalar moment

---

[5]Editors' note: throughout this book we use M (upper case) to denote the scalar seismic moment, and m (lower case) to denote the magnitude. Where a particular magnitude scale (surface wave, moment-magnitude, etc.) is being discussed we add a subscript to m.

back to a magnitude-like quantity, m_w. m_w is a more robust measure of the size of the largest events, since unlike earlier magnitude scales it does not saturate.

2.2.2 Inversion for the Centroid and Moment Tensor (CMT)

Inversion for the CMT (*Dziewonski et al.*, 1981) is a mildly non-linear least squares procedure that determines the centroid coordinates (in space and time) and the six components of the moment tensor. CMT inversion was first performed using long period body and surface waves from global broad-band world-wide digital networks (e.g., IRIS, GEOSCOPE). However, in many cases CMT inversion is now also performed using broadband digital data from regional or local seismic networks. As noted above, the centroid is the optimal point-source location for the moment release for an earthquake and should not be confused with the hypocenter and origin time, which specify the starting point of dynamic rupture.

If the centroid were fixed, inversion for the moment tensor[6] would be a linear problem, but as inversion for the centroid is (mildly) non-linear, CMT inversion must be conducted by iterative linearization. CMT inversion is now a relatively routine process and except in cases of insufficient data or pathological distribution of the stations, the process converges after two or three iterations.

Global catalogues of CMT solutions are routinely available on the web[7]. These catalogues are complete down to about magnitude $m_w = 5.5$. The completeness threshold is gradually decreasing as station coverage improves.

For events smaller than magnitude 5.5 or so CMT solutions are also available from national or regional seismological organizations in most developed countries and some underdeveloped countries. In some areas however only hypocentral

---

[6]The *seismic moment tensor* M_{ij} is a second rank tensor that describes the body forces acting in a seismic source as the superposition of three fundamental dipoles, plus three double couples. Since M_{ij} is a symmetric second rank tensor, it has six independent components. When actual seismic data are analyzed the resulting moment tensor is never a perfect double couple, but the best fitting double couple can usually be determined from the moment tensor solution. For further details see standard textbooks such as *Lay and Wallace* (1995).

The moment tensor for a shear dislocation in an isotropic medium (i.e., the usual model of an earthquake source) is given by the following:

$$M_{ij} = \mu(\overline{\Delta s_i}n_j + \overline{\Delta s_j}n_i)A,$$

where μ is the shear modulus, $\overline{\Delta s_i}$ is the average slip vector, A and n_j are respectively the fault area and its normal vector.

The above equation shows that the moment tensor for a shear dislocation is a double couple without net moment. Alternatively, this moment tensor can be rotated 45° to decompose it into the sum of mutually perpendicular dipoles of compression and dilatation, each with moment $M = \mu\overline{\Delta s}A$.

[7]Harvard $\Rightarrow$ http://www.seismology.harvard.edu/CMTsearch.html or USGS $\Rightarrow$ http://neic.usgs.gov/neis/FM.

locations, origin time and (for some events) nodal planes from first motions are available. Of course for events smaller than about 5.5 under the oceans in many cases only the hypocentral locations and origin times are available. Progress is rapid, so if a more detailed description were presented here it would quickly be outdated.

The ability to routinely determine CMT solutions is evidence of the great progress that seismologists have made in understanding the kinematics of faulting. In the following subsections we address other aspects of the kinematic problem, before turning to the dynamic problem.

2.2.3 Geodetic constraints

Geodetic data are also important information for inferring the focal mechanism of an earthquake. However, the geodetic displacements are part of the near-field displacement, which means they decay inversely as the square of the distance from the source. In contrast, far field displacements decay with distance, rather than distance squared, so they are easier to use in trying to infer details of the earthquake source.

A key result for inferring the displacement on a fault plane from the surface displacement was derived by *Okada* (1985, 1992), who obtained an analytical expression for the surface displacement field produced by the slip of a rectangular fault in an elastic half space for general orientations of the fault and slip vector.

Since the observed displacement field of sizable earthquakes appears more complicated than that of a single rectangular fault, these results are generally used by modeling the seismic source as a superposition of rectangular patches, each with its own slip vector. The surface displacement field produced by such a source may be modeled by the superposition of the contributions of each segment estimated through the above analytical solutions.

In order to use the above approach it is necessary to know the coseismic displacement, i.e., the displacement that occurred at the time of the earthquake. However, a geodetic survey measures the position of benchmarks on the Earth's surface with respect to a reference frame also on the Earth's surface. In order to obtain the displacement it is necessary to take the difference of surveys that were conducted before and after the earthquake. If the former are not available then the displacement cannot be inferred by the geodetic approach.

The reliability of the inversion of the source parameters is clearly also related to the accuracy of the models of deformation. Although the equations of Okada are formulated in an elegant analytic form, the model is quite simple and does not account for the heterogeneity of the Earth's crust and the topographic effects. The influence of these effects is currently being investigated by the use of finite element three dimensional modeling (*Armigliato*, 2001), but no general procedure is available.

In spite of its age, one of the best documented events is the 1906 San Francisco earthquake ($m = 7.7$). *Thatcher et al.* (1997) have recently revised the geodetic data of several triangulation networks evaluating the trade-off between slip resolution and uncertainty. This work is particularly valuable, because it evaluates which information can be effectively constrained by the available data. In particular the fault was initially divided into 48 segments, but these were then grouped in order to obtain a stable inversion and segments of the fault for which no significant information could be drawn from the data were excluded.

The recent development of new satellite geodetic techniques such as InSAR and GPS has led to the possibility of measuring surface displacements with a very high precision, thus allowing more detailed measurements of moderate earthquakes. GPS (see section 6.1) measures provide good precision (some mm), and the possibility of continuous monitoring, but their pointlike character heavily limits their ability to constrain source parameters.

SAR interferometry (see section 6.1) shows more promise since large portions of the earth are imaged at least once a month and the interferogram between two successive images allows one to obtain millimetric precision in surface elevation on a grid of cells of 100 m side, which provides a good constraint on the displacements.

However, the efficiency of the method is limited by several problems. Since the nature of the terrain does not always allow the highest potential accuracy, the most significant measures are often limited to some special points called permanent scatterers. Moreover, if the displacement gradient is too large, the coherence conditions for the reconstruction of the displacement field are lost, with the consequence that large earthquakes cannot always be analyzed with this technique. For more moderate earthquakes the near fault displacements are generally unresolvable. A further constraint is that a digital elevation model of the area prior to the earthquake is necessary for analyzing the image. This is often available for regions with high human density or elevated seismic risk, but may not be available uniformly for all regions.

Some examples of InSAR results are given by *Massonnet et al.* (1993) and *Massonnet and Feigl* (1998).

2.2.4   Space-time history of faulting and physical implications

CMT inversion (section 2.2.2) obtains the best-fitting point source for each earthquake; thus it is a long wavelength approximation. On the other hand, we know that earthquake fault planes can have lengths of hundreds of kilometers for the largest events, and temporal durations of tens (or in a few cases) even hundreds of seconds. Rupture along fault planes is well known not to be a smooth process in space and time, and many efforts have been made to infer the space-time history of rupture from recorded seismograms. One notable example is the work of *Kikuchi and Kanamori* (1991), who use the far-field waveforms of P-waves as primary data. They invert for a sequence of subevents positioned on a grid on the ruptured fault. Their inversion method consists of an adaptive chi-square method that determines a moment tensor for each subevent. Unfortunately, this procedure is unstable, since there are multiple sequences that fit the data, and to obtain a stable solution additional constraints must be imposed on the source parameters (such as that the main axis of the focal mechanism should not change significantly among the events in the sequence). However, this approach in effect is shifting the problem to the choice of the constraints, which requires either a subjective choice or a comparison with some other source of data on the event, such as geologic or geodetic data.

This problem is common to many inverse studies for source parameters. *Matsu'ura and Hirata* (1982) discussed several quasi-linear inversion techniques based on generalized least squares methods in a stochastic framework. In particular, they discussed the possibility of using *a priori* information to constrain the model parameters when the inversion results unstable. However, although these methods have a sound statistical foundation, the significance of the *a priori* information is often questionable.

Up to this point we have considered only the kinematics of earthquakes. Much less is known on the dynamics of the source mechanism, i.e., on the causes of earthquake initiation and on the effective physical laws that govern the faulting process. The process is certainly very complex, involving an elaborate interplay between fracture dynamics, friction, plasticity, melting, fluid migration, and chemical activity. Many of these phenomena have been approached on their own in laboratory experiments, but whether, and if so how, these results can be applied to the Earth, is still an open question. Moreover, when these phenomena act together in the Earth's crust they probably produce emergent properties, i.e. a complex dynamics with properties that do not appear in any one single phenomenon.

While the preparatory process of the earthquake is possibly an inseparable mixture of the above phenomena, the rapid unstable stage associated with the onset of the earthquake probably can be approached with a combination of fracture mechanics and an effective friction law. Although some constitutive friction relations have been investigated in laboratory experiment such as slip-weakening (*Byerlee*, 1970), velocity weakening (*Rabinowicz*, 1965), or rate- and state-dependent friction law (*Dieterich*, 1979a, b), their scaling relation and effective applicability to faulting in the Earth's crust is still unknown.

An investigation on the constitutive relations of fault slip was made by *Ide and Takeo* (1997) using an inversion of strong motion seismograms. Although the rupture process involves a volume with a very irregular shape surrounding the fault, its geometry can be considered planar if the wavelength of the seismic waves considered is larger than the thickness of the damaged volume. In this context, the dynamics of faulting results from the interaction of the constitutive friction relations on the fault plane with the elastic behavior of the surrounding body. The method proposed by *Ide and Takeo* (1997) consists of first determining the spatiotemporal distribution of slip, then using it to determine the evolution of stress on the fault by the finite difference method. This makes it possible to evaluate the local constitutive relations between stress and slip or slip-rate on the fault. Their results are consistent with slip weakening behavior, while it is not possible to observe a relation with the slip-rate. The weakening rate appears to be stronger at depth, along with a shorter rise time. However, an evaluation of the resolution limits inherent in their analysis suggests that these results are at the threshold of significance.

2.3   Scaling laws for earthquakes

[By Y. Y. Kagan, except section 2.3.2 which is by F. Mulargia]

2.3.1   The Gutenberg-Richter law

Formulas for quantifying the size of earthquakes by a logarithmic measure, the magnitude, were first developed in the 1930s by two seismologists at the California Institute of Technology, Gutenberg and Richter. Within a few years they had begun to make empirical measurements of the relation between frequency (the number of earthquakes) and magnitude (*Gutenberg and Richter*, 1941, 1944, 1956). These empirical relations had the form:

$$\log_{10} N(m) = a_t - b(m - m_t) \qquad \text{for} \quad m_t \leq m, \qquad (2.2)$$

where $N(m)$ is the number of earthquakes with magnitude $\geq m$, m_t is a catalogue completeness threshold (observational cutoff), and a_t and b are distribution parameters: a_t is the logarithm of the number of earthquakes with $m \geq m_t$ and b characterizes the dependence of the number of earthquakes on the magnitude ($b \approx 1$ is usually observed).

Equation (2.2) has become widely known as the Gutenberg-Richter (G-R) equation. For extensive recent reviews of the G-R distribution see *Utsu* (1999) and *Wiemer and Wyss* (2002). Its importance is not limited just to seismology, as it is the prototype for similar power laws that have become ubiquitous in the study of non-linear systems.

2.3.2 Empirical roots of the Gutenberg-Richter law

The Gutenberg-Richter relation (and its predecessor, *Ishimoto and Iida*, 1939), extending over at least 5 orders of magnitude (cf. figure 3.8), is one of the major features of earthquake phenomenology. What are its bases? The empirical rationale, shared by most papers, is very simple (see e.g. the appendix of *Rundle*, 1989). Starting from a typical expression

$$\log N = \text{const} - b\,m \tag{2.3}$$

for magnitude m, assuming that the scalar seismic moment M can be approximated by

$$\log M = c\,m + \text{const} \tag{2.4}$$

and combining them, one obtains a similar law for moment

$$\log N = -\frac{b}{c}\log M + \text{const.} \tag{2.5}$$

Now, we recall from equation (2.1) that the scalar moment is $M = \mu A\,\Delta s$. Since in the crust $\mu \simeq \text{const}$, and taking the average stress drop $\Delta\sigma$ as approximately constant, Hooke's law yields that

$$\Delta\sigma \simeq \text{const}\,\Delta\varepsilon, \tag{2.6}$$

where ε is the strain, from which it follows that also the strain drop is constant

$$\Delta\varepsilon = \text{const.} \tag{2.7}$$

Consider then the definition of the strain drop, which is the strain induced by fault slip:

$$\Delta\varepsilon = \frac{\Delta s}{s}, \tag{2.8}$$

where s is the linear dimension of the fault. Since the latter is constant then $\Delta s \propto s$ so that

$$M \propto \mu A\,s. \tag{2.9}$$

Now, if the shape of the fault remains the same, i.e. if it is *self-similar*, then $w/s = \text{const}$, where w is width and $A \propto s^2$, consequently

$$M \propto s^3. \tag{2.10}$$

For large earthquakes, which cut the whole crust (w fixed at about 30 km), the constancy in shape is not possible and larger earthquakes necessarily imply longer faults. For them

$$M \propto s^2. \tag{2.11}$$

Substituting the latter two equations in the Gutenberg-Richter relation yields for the 'small' events:

$$\log N = -3\frac{b}{c}\log s + \text{const} \tag{2.12}$$

and for the 'large' events

$$\log N = -2\frac{b}{c}\log s + \text{const}. \tag{2.13}$$

A law similar to Gutenberg-Richter exists in the time-domain for aftershocks. This is the Omori law, which describes the rate of decay of the number of aftershocks in a given sequence as

$$(t - t_0)^{-p}, \tag{2.14}$$

where t_0 is the time of the mainshock and p is a positive constant.

2.3.3   Moment-frequency relation

Although the original G-R equation was an empirical relation between magnitude and frequency, seismologists now use the scalar seismic moment as the parameter for specifying the size of earthquakes. *Kanamori* (1977) presented the following empirical relation for converting the moment M to the moment-magnitude m_W:

$$1.5m_w + 9.1 = \log_{10}M, \tag{2.15}$$

where m_w is the moment-magnitude and M is the scalar moment in Nm.

As noted in the previous section, routine CMT determinations have been published since the mid-1970s; about 25 years of data are available in the Harvard catalogue. Since moment data provide a more reliable indication of the size of an earthquake than older magnitude scales, it is appropriate to restate the G-R relation as a frequency-moment relation, which is the Pareto distribution for scalar seismic moment M (*Kagan*, 2002a)

$$\phi(M) = \beta M_t^\beta M^{-1-\beta} \qquad \text{for} \quad M_t \leq M, \tag{2.16}$$

where β is the index parameter of the distribution, $\beta = \frac{2}{3}b$ and $\phi(M)$ is the probability density function.

The finiteness of the seismic moment flux or of the potential energy available for an earthquake generation (*Knopoff and Kagan*, 1977), requires equation (2.16) to be modified at the high end of the moment scale and it to have a decay stronger than $M^{-1-\beta}$ (with $\beta > 1$). An additional parameter called the *maximum* or *corner* moment (M_x or M_c) is therefore introduced into the distribution. The resulting tapered G-R (TGR) relation has an exponential taper applied to the normalized cumulative number of events with seismic moment larger than M (*Shen and Mansinha*, 1983; *Main and Burton*, 1984; *Vere-Jones et al.*, 2001; *Kagan*, 2002a)

$$\Phi(M) = (M_t/M)^\beta \exp\left(\frac{M_t - M}{M_c}\right) \qquad \text{for} \quad M_t \leq M < \infty, \tag{2.17}$$

where M_c is the parameter that controls the distribution in the upper ranges of M (the corner moment). *Leonard et al.* (2001) formally showed that the incremental form of the distribution (2.17) fitted global earthquake data better than alternatives such as (2.16) (a single-slope model) or a double-slope model.

Figure 2.10 displays cumulative histograms for the scalar seismic moment of earthquakes in the 1977-2001 Harvard catalogue for three depth ranges. The curves display a scale-invariant segment (linear in the log-log plot) for small and moderate values of the seismic moment. At large M, the curve for shallow events is bent downward. It is more difficult to see the bending of curves for deep earthquakes, since the number of earthquakes is relatively small. Two exceptionally large deep earthquakes (the 1994 Bolivia and Tonga events) control the shape of the curve for large moment values.

By determining the maximum likelihood estimates of β and the corner moment parameters (*Kagan*, 2002a), we show that the former has a universal value of about $\beta = 0.63$ for all earthquakes in the Harvard catalogue. The apparent variation of this parameter, noted by many earlier studies, is caused by various factors, among which the most influential is variation of the corner moment. If a variable corner moment is introduced in the distribution, mid-ocean events appear to have the same value of β as the rest of the global set of earthquakes. The β-value for the shallow events in figure 2.10 is slightly higher than for deeper earthquakes; this is an effect of shallow mid-ocean earthquakes (see *Kagan*, 2002a for further details of the estimates of the uncertainties of the TGR distribution).

We found that the corner moment has approximately the same value $M_c = 10^{21.0} - 10^{22.0}$ Nm (corner moment-magnitude $m_c = 8.0 - 8.7$) for all

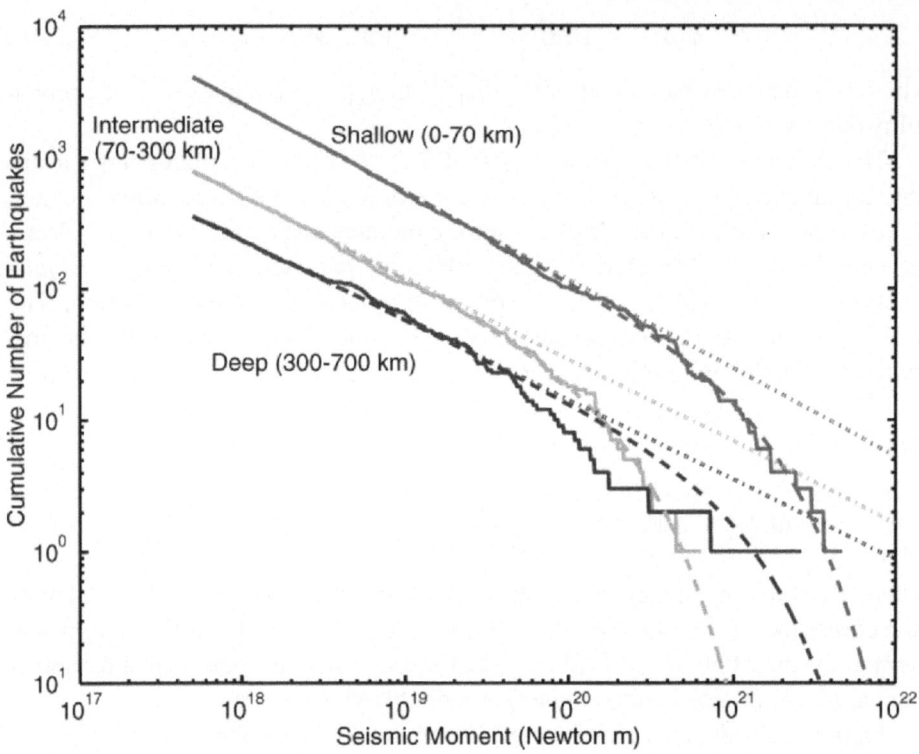

Figure 2.10: Cumulative number of earthquakes *versus* seismic moment for the global earthquake distribution in the January 1, 1977 – December 31, 2001 Harvard catalogue. The solid curves show the numbers of events with moment greater than or equal to M for deep, intermediate and shallow earthquakes. The dotted (straight) lines are the classic G-R distribution (equation 2.2) while the dashed curves which taper off exponentially are the tapered G-R distribution (equation 2.17). The latter fi t the data whereas the former do not. The slopes of the linear part of the curves correspond to β equal to 0.672 ± 0.011, 0.623 ± 0.024, 0.608 ± 0.033, and the corner moment $M_c = 1.6 \times 10^{21}$, 2.1×10^{20}, 1.2×10^{21} Nm, for shallow, intermediate, and deep earthquakes, respectively.

continental and near-continental areas. For mid-ocean earthquakes, m_c is significantly smaller; for spreading ridges, the corner magnitude is about 5.8, and for strike-slip earthquakes on transform faults m_c decreases from 7.2 to about 6.5 as the relative slip velocity of faults increases (*Bird et al.*, 2002).

The upper limit of the corner moment can be reliably estimated only for large catalogues. *Kagan* (2002a) estimates that at least 2–3 earthquakes exceeding M_c need to be in a catalogue in order for the corner moment estimates to be marginally reliable. Given that the Harvard catalogue for 1977–2001 contains only 14 shallow earthquakes $m \geq 8$, it is clear that with available data one cannot distinguish a M_c variation in continental plate boundaries and their interiors.

By integrating equation (2.17) we may estimate the total seismic moment rate. Comparing the seismic rate with a tectonic deformation rate, $\dot{M}_T$, and assuming the universality of the β-value, we estimate the corner moment (*Kagan*, 2002b)

$$ M_c \simeq \left[\frac{\chi \dot{M}_T (1 - \beta)}{\alpha_t M_t^\beta \Gamma(2 - \beta)} \right]^{1/(1-\beta)} , \qquad (2.18) $$

where χ is the seismic coupling (or seismic efficiency) coefficient, α_t is the seismic activity level (occurrence rate) for earthquakes with moment M_t and greater, and Γ is a gamma function. By analyzing the tectonic moment rate and the earthquake moment distribution for several types of tectonic environments, including subduction zones, plate bounding transform faults, and deforming continental regions, we confirm (*Kagan*, 2002b) that the corner magnitude values for plate boundary zones and continental areas are 8.3–8.8, i.e., similar to those values obtained by statistical analysis (*Kagan*, 2002a).

The results of statistical analyses of earthquake catalogues present convincing evidence that the standard G-R relation needs to be re-evaluated as a tool for description of the earthquake size distribution. Physical and geometrical considerations – conservation of energy and finiteness of seismogenic regions – require the introduction of an upper seismic moment/magnitude bound into the frequency-moment distribution. The empirical evidence reported above also supports the need for modification of the G-R law.

Introduction of the distribution upper bound parameter also makes it necessary to re-interpret the moment-frequency or magnitude-frequency relations. A mix of earthquake populations with varying corner moment/magnitude may lead to an appearance of a more-or-less linear G-R relation with a substantially higher *b*-slope (*Vere-Jones et al.*, 2001; *Kagan*, 2002a). *Kagan* (2003) shows that the uncertainty of conventional magnitudes is 2–3 times greater than those of moment-magnitude errors, and these uncertainties can strongly influence the empirical estimates of the *b*- and m_c-values (*Kagan*, 2002a). Thus, we propose a new paradigm for the earthquake size distribution: most reported variations of the *b*-value (in the magnitude-frequency relation) are artifacts due to magnitude deficiencies and defects of analysis. Some apparent changes in the β-value (e.g., in volcanic areas, for very shallow earthquakes, etc.) may be genuine effects due to the mixing of earthquake populations with different corner magnitudes, m_c (*Kagan*, 2002a).

2.4   The elastic rebound model and its successors

[By I. Main and F. Mulargia]

In this section several models are discussed that are based primarily on direct observation of deformation associated with faulting and actual earthquake occurrence. The models are all linear in the sense that the failure time is determined essentially by a linear increase in the remote strain. They are phenomenological in the sense that they were developed on the basis of observations rather than quantitative dynamical modeling. Non-linear interaction and feedback are generally neglected by these models. Large regions of the Earth are treated as structurally homogeneous and isolated from stress perturbations due to nearby faults. These are clearly approximations. The question is whether they are so drastic that they will render the conclusions reached from the models inconsistent with observed data.

After the 1906 San Francisco earthquake, *Reid* (1910) proposed that energy had been stored in the region around the fault before the earthquake, in the form of elastic strain energy built up in response to a linear increase in strain due to movements of the earth very remote from the fault surface. The assumption of constant remote strain rate is also a key assumption of the hypothesis of plate tectonics, now validated to first order by numerous direct and independent observations of surface deformation rates. For example *DeMets* (1995) documents a surprisingly good correlation between current plate motion velocities from 10 yr or so of satellite data and 3 Myr or so of palaeomagnetic data, confirming just how regular and constant the remote driving velocities and strains are for earthquake generation. Reid then proposed that, once the accumulated strain reached a critical value, dynamic failure would occur, with some of the energy built up in the surrounding strained rocks being released as radiated energy. He termed this process 'elastic rebound'.

The elastic rebound model is a very simple model for earthquake occurrence. Reid's original approach views the earth's crust as a mechanical specimen which, when subjected to stresses, deforms elastically up to a given threshold and releases the accumulated strain in a sudden jerk which produces large amplitude seismic waves. Once this happens, a new cycle of the process takes place and the specimen is slowly loaded again up to its mechanical threshold and so on. The constancy of the load externally imposed by tectonics might seem to support Reid's view, but, as appealing as it may look, it also appears immediately too simple to be applied to earthquakes. Some of its assumptions have been validated on other data sets, but others have proven less robust. For example, in repeated laboratory experiments on composite materials, the 'critical breaking strain' has been shown to be highly variable, and certainly cannot be regarded as a material parameter (e.g. *Gathercole et al.*, 1994). Similarly, analysis of earthquake recurrence on the San Andreas fault, based on trenching of quaternary sediment laid down in the last two millennia, has shown that the occurrence of similar-sized events has in fact been highly variable. For example in Southern California the inter-event time for the largest earthquakes has fluctuated between 50 and 350 years (*Sieh*, 1978; *Sieh et al.*, 1989). This confirms that the critical breaking strain is not a material constant in a region near the area used to derive Reid's original model. Thus the concept of 'average repeat time' is not well defined.

2.4.1 The time- and slip-predictable models

Reid's original model was modified in two important ways by *Shimazaki and Nakata* (1980), who proposed the time-predictable and slip-predictable variants of the elastic rebound model, and by *McCann et al.* (1979) who made a testable long term forecast of seismicity at plate boundaries based on the seismic gap model (*Kelleher et al.*, 1973, 1974).

The phenomenology of the time- and slip-predictable earthquake recurrence models (abbreviated as TP and SP for the remainder of this section) can be summarized as follows (*Bufe et al.*, 1977; *Shimazaki and Nakata*, 1980): in the TP model the time between two occurrences in a seismic region is proportional to the 'size' of the first one, while in the SP model the 'size' of an event is proportional to the time elapsed since the previous one (see figure 2.11). It is implicit that the models are applicable to the 'largest' earthquakes in each region. If either of these models could be shown to be valid, this would provide an invaluable tool for estimating the time-dependent seismic hazard of a given region.

From a theoretical point of view, the above models are clearly oversimplified (see for example *Scholz*, 1990), but the lack of a detailed understanding of the earthquake source mechanisms does not allow us to exclude *a priori* the possibility that the many contributing effects might combine to yield a simple phenomenology. On the other hand, in the absence of a theoretical underpinning, the practical applicability of such models requires empirical evidence. This would preferably be in the form of objective testing.

The TP and SP models were widely used to make forecasts of time-dependent hazard in the 1980s (e.g., *Sykes and Quittmeyer*, 1981; *Kiremidjian and Anagnos*, 1984; *Anagnos and Kiremidjian*, 1984; *Nishenko*, 1985; *Papazachos*, 1989). Although it is easy to make such forecasts, considerable time is required before validation exercises can be carried prospectively rather than retrospectively. However, by the end of the decade doubts were arising. *Thatcher* (1989) pointed out that several unresolved issues limited the confidence that could be placed in such forecasts, including the facts that some 'gap filling' events (see also next section)

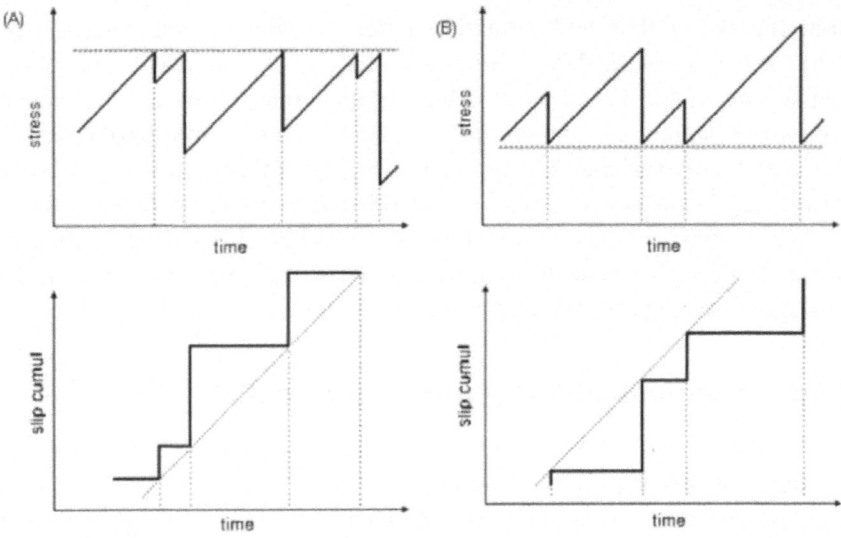

Figure 2.11: The slip-time behavior according to the (A) time- and (B) slip-predictable models (*Shimazaki and Nakata*, 1980). The two upper graphs show the non cumulative curves while the two lower ones show the two cumulative curves.

had differed from those that were expected, that some observations suggested irregular recurrence times, and that many recurrence time estimates were based on plausible but idealized models.

The TP and SP models were originally proposed to describe the behavior of earthquake faults constituting 'patches' of a single megafault (*Bufe et al.*, 1977; *Shimazaki and Nakata*, 1980) but the megafault concept applies only to simple plate margins. Virtually all other tectonic settings are not amenable to such a simple scheme. Italy, as well as the whole central Mediterranean basin, is a typical example. It appears straightforward to rewrite the TP and SP models in three rather than in two dimensions, reconciling them, at least in principle, with a generally complex non-megafault tectonic setting, which is applicable in most zones of continental deformation. However, *Mulargia and Gasperini* (1995) applied the modified TP and SP models to Italian seismicity using rigorous statistical procedures, but they found that these models fit the data satisfactorily in just two regions and one region respectively, out of the 19 that were studied, which is about what would be expected by random chance. They also conducted rigorous reanalyses of data from other regions that had earlier been cited by positive studies, and obtained similar negative results. They therefore concluded that the TP and SP models were not tools of practical utility for time-dependent hazard estimates. The reason for these negative results is probably that the simplifying assumptions made by the TP and SP models are too crude. In particular, it is probably essential to consider non-linear interactions between different earthquakes and different fault segments.

There is also a statistical problem in assessing empirical models such as TP or SP, especially when only a few instances of large events are available in the earthquake record for each location. In the general case, with a given random component in the distribution of breaking stress and residual stress, we would also be likely to see a few examples of apparently time- and slip-predictable behavior by random chance. This highlights the general point that broad and comprehensive testing of hypotheses is required in earthquake physics, not just a few case studies. This principle has not always been followed in the history of examining the predictability of earthquakes. In the original paper by *Shimazaki and Nakata* (1980) three examples of time-predictable earthquake recurrence were shown. There was no attempt to evaluate the model statistically in a systematic way based on all of the data then available. Since then, neither TP nor SP has been shown to outperform the more general assumption of a variable breaking and residual stresses.

2.4.2 The seismic gap hypothesis

The models discussed above are deterministic in the sense that they apply to the precise occurrence of a specified individual event. *McCann et al.* (1979) accepted the broad tenets of the elastic rebound model, but relaxed this requirement by proposing that the time of occurrence could be estimated only in a probabilistic way. A key tenet of their model, amended slightly from Reid, is that the probability of occurrence of an earthquake increases systematically with time. Thus, in their model, when an earthquake has just happened, it is likely to be several years until sufficient strain can be accumulated to produce the next earthquake, even if the breaking strain and residual strain have a random component. This introduced the notion of a temporal 'stress shadow' and a spatial 'seismic gap' that could be defined if an earthquake had not happened in a region in the last n years. Although n should in principle depend on the subsequent slip and on the local rate of strain accumulation, they initially proposed $n = 30$ years as a broad average for the circum-Pacific belt. Their model was posed in the scientific tradition of being objectively testable, and thus being falsifiable and was applied to a large region that dominated global earthquake occurrence, thereby minimizing selection bias. On the other hand, it also had the disadvantage of requiring a large number of degrees of freedom, mainly from the definition of specific fault segments in a way that required a significant subjective component. The choice of $n = 30$ years in 1979 meant that an objective statistical test of the original hypothesis was possible after a decade or so. When tested, it did not predict the ensuing seismicity better than the alternative hypothesis of random temporal occurrence at the same spatial positions (*Kagan and Jackson*, 1991).

The original seismic gap hypothesis of *McCann et al.* (1979) was subsequently tuned by *Nishenko* (1991) to account for the variabilities in the recurrence interval (assumed to be log-normal) and to allow for local triggering or sub-events within a 'gap'. Thus, the original hypothesis was modified to include many more degrees of freedom, and to include the possibility of triggering from neighboring large fault segments. This revised hypothesis also failed to outperform the null hypothesis of a Poisson process (*Kagan and Jackson*, 1995).

In summary, the seismic gap model does not describe the data temporally better than a Poisson process. Why might this be? First, it is hard to distinguish a log-normal recurrence from a Poisson distribution with only a few data points. Thus the probability gain of a component of quasi-periodic recurrence in the log-normal distribution is actually quite subtle, even if the inferred physical process could in principle be confirmed. Second, if triggering by neighboring segments and sub-segment failure is allowed, and the gaps themselves become segmented, then where should this cascade of new model parameters at ever higher resolution stop? The above considerations suggest that perhaps a new paradigm be adopted instead.

2.4.3   The characteristic earthquake model

The characteristic earthquake model, which to some extent is a corollary of the seismic gap model, is based on the notion that similar-sized earthquakes repeat on the same fault segment over several 'earthquake cycles' (see figure 2.12), with an elevated probability of occurrence of these events compared to a linear extrapolation of the log-linear incremental frequency- magnitude relation (*Schwartz and Coppersmith*, 1984). This model was used as the basis for a long-term prediction (see section 8.3.3) of the next magnitude 6 or so earthquake on the Parkfield segment of the San Andreas fault within a specific time window (95 per cent probability by 1993), now passed (*Roeloffs and Langbein*, 1994). Several authors have pointed out scientific fallacies of the original prediction that could have been foreseen in advance (e.g. *Savage*, 1991, 1993). For example *Kagan* (1997) has pointed out that, in an area the size of California with one magnitude 6 or greater earthquake per year with events randomly distributed around the mapped neo-tectonic faults, it is likely that a sequence of six apparently regular repeated events could occur at one place by chance. This illustrates once more that a Poisson distribution is quite capable of producing clustering and patterns if data are treated selectively.

At the time of writing, there is no evidence for the characteristic earthquake model based on the most reliable instrumental seismic data. First of all, characteristic earthquakes would imply a bulge in the earthquake frequency-magnitude distribution which has never been observed in instrumental data (see figure 2.10

and 3.8) although it is not precluded by physical models, such as the supercritical ones (see section 3.1.1). Second, the characteristic model did not pass the test of formal validation (*Kagan*, 1993; *Kagan and Jackson*, 1996; but also see the opposing viewpoint of *Wesnousky*, 1996). The problem is also illustrated by *Thatcher*'s (1990) comprehensive study of the circum-Pacific belt, which showed that characteristic earthquakes were the exception rather than the rule.

Characteristic earthquake models are often derived from geological data that suffer from three significant sources of bias in assessing earthquake physics. The first is that only the largest events $m > 6$ or so actually produce slip that can be observed at the surface. Since such events produce slip on the order of 1 m, and the number of events of larger magnitude decreases rapidly thereafter, it is not surprising that event sizes often appear to be similar. Second, frequency data are often plotted assuming that events have exactly the same magnitude (within 0.1 magnitude units), whereas in reality the 'repeated' events would be expected to have some statistical variability of size between 'cycles'. This variability would tend to reduce and broaden the 'characteristic' peak. Third, the area of study is often focused very close to a particular fault, thereby excluding seismicity in the volume of the lithosphere responsible for the build-up of strain energy. If instead we examine source rupture areas from recent earthquakes, we are forced to look over time periods which are small compared with the average repeat time of the largest events, and replace a true temporal evolution with a probabilistic estimate based on a spatial average (*Thatcher*, 1990). Thus the characteristic earthquake model is unproven as a physical hypothesis, and there is much evidence against it.

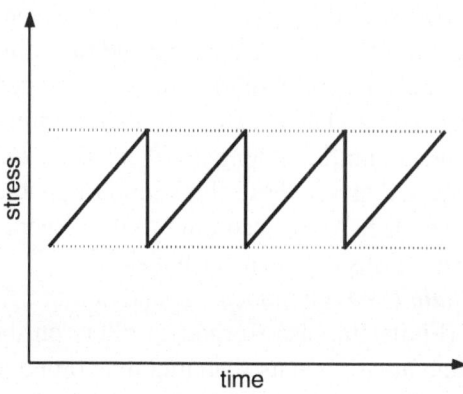

Figure 2.12: The stress vs. time cyclic and regular behavior according to the characteristic earthquake model.

2.5   Nucleation or not?

[By S. Castellaro]

The long-standing problem of discovering whether any phenomena exist that can be reliably identified in advance as precursors of specific large earthquake, has led scientists to study 'source nucleation' in the field, in the laboratory and through computer simulations. The fundamental questions are: (1) is there a nucleation stage which can be distinct from the critical rupture?, (2) if so, can the nucleation processes tell us anything about the forthcoming earthquake size?, and (3) how is the nucleation process (if it exists) linked to the rheology of the rocks in the source region?

2.5.1   Is there any evidence for a nucleation phase?

Ellsworth and Beroza (1995) analyzed data from 30 earthquakes with magnitude ranging from 2.6 to 8.1, searching for evidence of a nucleation phase. They concluded both that a nucleation phase exists and that the size of the eventual earthquake depends on the size of the nucleation region and on the duration of the nucleation process. For the events they investigated, they found that a period of weak ground motion precedes the strong ground motion induced by the mainshock. They quantified the temporal evolution of the process and reported, as a general feature, that the duration, seismic moment, source dimension and average slip of the process generating the seismic nucleation phase all scale with the moment of the eventual shock. However, two explanations are possible. Earthquakes could start in the same way, independent of their size. In this case (the 'cascade model') the eventual size would be determined by the chance accumulation of small events occurring from the beginning to the end of rupture. Alternatively, in the 'preslip model', it could be the case that the initial stages of small and large earthquakes differ. Failure would then start with stable sliding over a small region which gradually accelerates until reaching a critical size. *Ellsworth and Beroza* (1995), as noted above, are favorable to this second hypothesis, indicating that earthquakes can nucleate differently according to their eventual size. *Iio* (1995) arrived at the same conclusions for microearthquakes.

Scherbaum and Bouin (1997) conducted a study on the artifacts produced by the widely used FIR (Finite Impulse Response) filters on the recordings of the modern seismic instruments for events scanning nine orders of magnitude, from 10^{10} Nm to 10^{19} Nm. They demonstrated that, although such filters have the desirable property of not distorting the phase spectrum of the input signal, they can generate precursory artifact signals to impulsive seismic arrivals. Such artifacts are often undistinguishable from a visual inspection. To avoid interpreting an artifact as a nucleation phase, extreme care should be taken to remove any FIR-filter acausal effect.

Diametrically opposite evidence on the initial stages of earthquakes rupture was presented by *Mori and Kanamori* (1996). Examining 49 earthquakes in the range $1.5 \leq m \leq 4.2$, they found no differences in the shape of the initiation of the P waves, concluding therefore that it is not possible to predict the eventual size of an earthquake from the initial portion of the waveform. The final size of the event would then be explained by the dynamical properties of rupture (the cascade model) rather than by its initiation. They ascribe the curvature observed at the beginning of the velocity waveforms as an effect of anelastic attenuation. This view is shared also by *Brune* (1979), *Abercrombie and Mori* (1994) and *Anderson and Chen* (1995) who found no difference in the initiation of the Landers (California) $m = 7.3$ and a $m = 4.4$ aftershock and the $3 \leq m \leq 8$ events in Michoacan (Mexico).

2.5.2 Models of a hypothetical preparatory process

Ohnaka (1992) observes that foreshocks, when they exist, are located in the vicinity of the epicenter of the pending mainshock (*Ohnaka and Kuwahara*, 1990; *Ohnaka*, 1990). He considers the nucleation process a short-term (or immediate) precursor occurring in very localized zones and develops a model which takes into account the parameters indicative of the rupture growth resistance, namely the breakdown strength, the breakdown stress drop, the critical slip displacement and their variation with depth, that is with temperature and with the normal stress, estimated from laboratory data for granitic rocks.

Previous laboratory work by *Ohnaka and Kuwahara* (1990) led to the conclusion that the nucleation phase can be divided into two phases. The first one is a quasistatic steady progression of the tip of the breakdown zone; the second is a stable but accelerating progression of the tip of the breakdown zone up to a critical state beyond which the earthquake instability occurs. Whether or not foreshocks occur during the mainshock nucleation depends on how the rupture growth resistance varies inhomogeneously at local scale on the fault. The nucleation, in fact, begins where the rupture growth resistance is at a minimum but a barrier, that is a local patch of greater rupture growth resistance, can stop it. In this state the criterion for quasistatic or quasidynamic instability depends on how the parameters included in the model vary on the fault.

Ohnaka's model leads to the conclusion that when an earthquake nucleates within the brittle seismogenic layer and its hypocenter is located next to the base of the seismogenic layer, immediate foreshocks for this mainshock are necessarily restricted to lie within a localized region, shallower than the hypocentral depth of the mainshock. On the other hand, when the nucleation process starts at the base of the seismogenic layer, which is aseismic in nature, no foreshocks will occur. These conclusions seem to fit at least approximately the observable data. However we remind (see section 1.1) that *Abercrombie and Mori* (1996) found that more than 80 per cent of earthquakes occur without foreshocks and that foreshocks, when present, cannot be recognized in advance.

While *Ohnaka and Kuwahara* (1990) studied how the crack grows to a critical size in laboratory stick-slip experiments, *Matsu'ura et al.* (1992) did the same from a theoretical point of view. They developed a mathematical model for frictional sliding, describing the slip-strengthening and slip-weakening processes occurring in materials nucleating ruptures. The model incorporates the constitutive relation between frictional slip and fault slip into the equation of motion for an elastic body, obtaining in this way a non-linear system (previously solved by *Andrews*, 1976a, b, 1985) which describes the process of fracture.

In this model the seismogenic zone is represented by a weak area with a locally strong part (asperity) on a fault plane which is statistically self-similar in topography within a broad but bounded wavelength range. The nucleation process proceeds quasi-statically at the weak portion of the fault with the increase in the external shear stress. If the constitutive relation at the asperity has a high and narrow peak of the stress value, brittle dynamic rupture occurs in a limited area around the asperity, but otherwise it does not. This process is aseismic. Only when the stress peak is high and broad, then a catastrophic rupture can develop involving the whole region. In summary, with the progress of fault slip, frictional stress increases up to a peak value and then decreases to a residual value when the critical rupture occurs. In the same way, the slip-weakening process derives from a decrease of frictional resistance due to the abrasion of surface asperities. The rate of abrasion is in turn proportional to normal stress and the normal stress acting on the asperities will be proportional to their wavenumber and amplitude.

As the fault slip is accompanied by the abrasion of asperities, the state of fault surfaces changes with the progress of slip. To explain multiple occurrence of earthquakes on the same fault, some healing mechanism is required. The authors conclude that diversity in the nucleation process must be ascribed to differences in the constitutive behavior along the fault. However, the above models still lack validation for real cases.

2.5.3 Theoretical models

From the above discussion, it is not yet clear what governs earthquake size although it seems reasonable to think that small and large events start in the same way. The problem has been studied through cellular automata models by *Steacy and McCloskey* (1998) who found that large events occur when stress is highly correlated with strength over the entire fault but that the magnitude of the correlation has no predictive power since events of all magnitudes occur during times of high stress/strength correlation. Rather, the size of any particular event depends on the local stress heterogeneity encountered by the growing rupture.

Castellaro and Mulargia (2001, 2002) found from a series of cellular automata models that the general behavior is a progressive increase in foreshock size and 'rate of occurrence' when approaching the mainshock, independently of the mainshock size.

The dynamics of rupture propagation has been studied theoretically by a number of authors. *Cochard and Madariaga* (1994), using a rate- and state-dependent friction simulation, proposed a model of seismic rupture in which earthquakes are due to the catastrophic development of a frictional instability along a pre-existing fault surface. The model is based on a boundary integral equation and considers faults with single and twin asperities (where asperities are regions with higher prestress values) in the presence of nonlinear rate-dependent friction. They found that slip velocity weakening friction leads to the propagation of healing phases and to the spontaneous arrest of fracture if the stress around the asperities is low enough. A complex distribution of stress after the rupture, depending on the details of the initial distribution of asperities and on the details of the friction laws, is observed.

Similar work, differing in the frictional law used, was performed by *Okubo* (1989) but without leading to any complex and heterogeneous final stress distribution on the fault. Okubo used the friction law proposed by *Dieterich* (1972), and Cochard and Madariaga used the one suggested by *Carlson and Langer* (1989).

Other models to reconstruct the spatio-temporal slip on faults have been developed by *Rice* (1993). These models are based on a rate- and state-dependent friction law, but with constitutive properties (velocity weakening and strengthening) varying with depth. The governing equations are solved on a grid of cells and the emerging complexity of slip is ascribed to an undersampling of the friction law or, equivalently, to an oversizing of the cells. The spatio-temporal complexity vanishes as the cell size is reduced and cells larger than a certain threshold can fail independently while those much smaller can slip only as part of a cooperating group of cells.

While Rice considered an homogeneous elastic half-space, a model with rheologically layered strata has been considered by *Lyakhovsky et al.* (2001), who studied the evolution of the coupled system consisting of a seismogenic elastic upper crust over a viscoelastic substrate. When rupture occurs, the ongoing deformation modifies the elastic properties of the crustal seismogenic layer. The physical properties in the model must be derived from geophysical, geodetic, rock mechanics and seismological data. In the simplest case of damage, initially randomly distributed, the authors found that a long healing timescale leads to geometrically regular fault systems, while a short healing timescale leads to a network of disordered faults, with Gutenberg-Richter earthquake statistics.

2.6   What is an earthquake? Fracture, slip or both?

[By I. Main and F. Mulargia]

One of the several persisting 'grey areas' in the description of earthquakes is the fact that the physical process governing its occurrence is assumed to be either (a) fracture or (b) stick-slip sliding. The latter process is assumed following the additional hypothesis that earthquakes repeat on the same faults, which can be taken as plane surfaces on which no substantial chemical healing occurs, and this is essentially all that is required by seismology. Conversely, fracture is assumed to be the basic process when chemical healing is thought to be important. Both processes obviously derive from what is known from laboratory experiments.

2.6.1   Laboratory-based hypotheses

The linear, phenomenological models described in section 2.4 depend on making fundamental observations on the Earth, making some simplifying assumptions to predict future occurrence, and then testing these heuristic predictions against actual future occurrence. Some of the assumptions, notably the very nearly constant plate driving velocities, have been confirmed unequivocally by independent observation. In contrast, many apparently 'common–sense' assumptions, such as the notion that the most likely place for future events in the circum-Pacific should be in places where earthquakes had not occurred recently, turn out not to outperform random chance in prospective mode. So, if this 'top-down' approach is unsuccessful, can the more traditional atomistic or 'bottom-up' scaling process often adopted in the recent history of science be more successful? In this sense our minimum scale is not usually atomistic, due to the inherently granular nature of crystalline, metamorphic and sedimentary rocks. This implies that material properties are more dependent on intergranular coupling (cementation, annealing etc.) than on the intrinsic properties of the relevant crystal grains (see, e.g., *Scholz*, 1990, chapter 1). Porous granular media in fact show such distinctive behavior that they have been called a 'fourth phase' of matter.

A major advantage of the laboratory approach is that macroscopic thermodynamic variables, such as stress, strain, pore pressure and pore volume, can all be independently measured under controlled conditions. Laboratory tests on small rock samples inherently include the effects of material heterogeneity in the form of their granular structure. The major disadvantage is the spatial and temporal scales involved. Typical strain rates for tectonic processes are on the order of 10^{-15} s^{-1}, whereas a typical laboratory test that can be completed in one day's work is of the order 10^{-5} s^{-1}. This means that the effect of any time-dependent, chemically assisted, processes may be severely underestimated. Constraints on spatial scale in the laboratory also introduce fixed sample boundaries that have no direct analogue in the natural Earth. The boundary conditions are therefore applied in the intermediate rather than the true far-field, as in plate tectonics, with step changes in elastic properties between the loading system and the rock. Typically rock samples are chosen to be homogeneous on the scale of a few cm, so the larger-scale heterogeneities, such as bed thickness, are also not generally taken into account.

This section reviews some of the laboratory-based approaches, including: fracture mechanics, based on the properties of a single growing crack; damage mechanics, based on the ensemble response of a population of microcracks; and rate- and state-friction theory, based on the pre-existence of faults with sliding surface.

2.6.2 Stick-slip friction

The detailed understanding of earthquake nucleation underwent a step change in the late 1970s with the advent of a rate- and state-frictional constitutive law that also predicted the development of instability. Based on laboratory experiments (*Dieterich*, 1979a) and theoretical work (*Dieterich*, 1979b; *Ruina*, 1983) a full theory for the sliding properties of faults was developed that included a mechanism for either dynamic stress drop or quasi-static stable slip, depending on the model parameters. There are two classical approaches to modeling fault rupture as a frictional sliding episode based on the basis of laboratory evidence on the friction sliding of two specimens in contact through a plane surface. These are the *Coulomb Failure Stress* theory (from now on CFS) and the *rate- and state-dependent friction* approach.

The CFS theory is based on the dry friction theory in its simplest form. In a material, stick-slip occurs when the combined effect of shear stress τ and normal stress σ is larger than the cohesion c, that is when

$$\tau \geq c + \mu\sigma, \tag{2.19}$$

where μ is the internal friction coefficient and compression is assumed as positive. Extending the concept to a fault rupture in the Earth's crust, one calculates the static stress changes generated by a given earthquake on the fault planes of future earthquakes. Equation (2.19) is used after slight modifications. The cohesion force c is ignored while the fluid pressure P is taken into account. The variation on CFS is thus defined as

$$\Delta \text{CFS} = \Delta\tau - \mu(\Delta\sigma + \Delta P), \qquad (2.20)$$

where ΔP is the fluid pore pressure change, $\Delta\tau$ the shear stress change along the slip direction of the fault plane under consideration and $\Delta\sigma$ is the normal stress change on the fault plane. Equation (2.20) is generally rewritten as:

$$\Delta \text{CFS} = \Delta\tau - \mu'\Delta\sigma, \qquad (2.21)$$

including in the term μ' both the frictional properties of the material and the effect of fluid pore pressure. This step makes major assumptions concerning the hydraulic gradient. The modifications in ΔCFS are thus computed primarily from the redistribution of the static stresses $\Delta\tau$ and $\Delta\sigma$ induced by the earthquake. The most widely-used method is presently the one developed by *Okada* (1992), which calculates the static stress variations induced by normal, reverse and transcurrent faulting described as a uniform plane dislocation in an elastic half-space. Such a formulation has been shown to be capable of providing a satisfactory match of geodetic and seismological evidence, provided that the slip and stress drop of the source faults are known in detail. Since each sizable earthquake is composed of a large number of subfaults - for example 186 subfaults are necessary to fit the coseismic slip of Landers earthquake (*Hardebeck et al.*, 1998) - this detail can obviously be known only in retrospect. In the *far-field* (that is, far enough from the epicenter to assume that the fault is point-like), the details of the geometry and slip have less influence on the results of the calculation.

The actual geographic distribution of earthquakes can be easily compared with the Coulomb Failure Stress variation ΔCFS. Positive values of ΔCFS suggest that the fault plane has been moved closer to rupture by an increase in τ or a decrease in σ. Negative ΔCFS values suggest instead that the fault plane under consideration has been moved away from rupture and that it is in a *stress shadow* zone.

The absolute values of static stress are generally unknown while its changes can be inferred from geometric considerations on the fault plane and on the slip direction and magnitude. The main limitation of this method as a time-dependent hazard criterion is that, even if it were proved to be right, it cannot say anything about *when* a future event will take place. It can simply indicate whether a fault has been moved in time 'closer' to or 'further' from rupture (see figure 2.13) but the absolute time is unknown. It should also be noted that the CFS model is founded on pure elasticity assumptions and that it completely ignores the load-unload history of rocks while it is known from geomechanics that material behavior is strongly dependent on previous history.

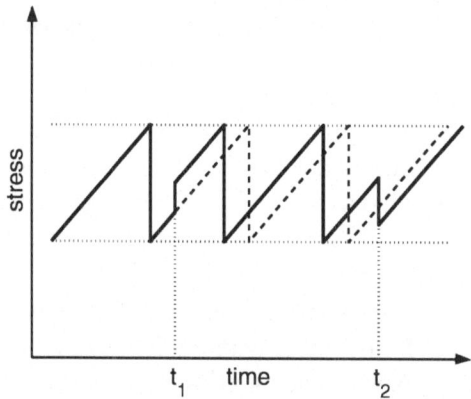

Figure 2.13: According to the CFS model, an earthquake occurring at time t_1 would take a hypothetical nearby fault closer, in time, to rupture. Another earthquake occurring at time t_2 would, instead, take the fault further in time from rupture. The characteristic model has also been reported to show time 'advance' and time 'delay'. Note that this is just speculative since it would imply a lower and upper threshold which have not been found to be compatible with the data (see text).

A typical result of the CFS method is shown in figure 2.14. Several authors seem to have found relations between the ΔCFS increase and the occurrence of small aftershocks (*Das and Scholz*, 1981b; *Stein and Lisowski*, 1983; *Reasenberg and Simpson*, 1992; *Stein et al.*, 1992, 1994; *Gross and Kisslinger*, 1994) or medium to large aftershocks (*Hudnut et al.*, 1989; *Harris and Simpson*, 1992; *Jaumè and Sykes*, 1992, *Stein et al.*, 1992, 1994; *Du and Aydin*, 1993; *Simpson and Reasenberg*, 1994; *King et al.*, 1994; *Harris et al.*, 1995; *Harris and Simpson*, 1996; *Deng and Sykes*, 1996).

These positive results have granted the CFS approach some popularity. However, they hardly support general conclusions since they all relate to retrospective case histories and thus suffer at various degrees of the retrospective selection bias problem (cf. section 1.2). The main drawbacks appear to be (1) a presentation of positive case histories rather than systematic validation tests and (2) an *ad hoc* selection of the value of the friction coefficient.

From a general point of view the static Coulomb approach has several problems. First, the calculated 'triggering' stresses are very small (at most of the order of 0.1 MPa), and difficult to reconcile with the solely static earthquake models (cf. *Huc and Main*, 2003). Second, the CFS model is based on the existence of fixed upper and lower rupture thresholds, and the existence of two such thresholds implies the characteristic earthquake model, which did not pass validation (*Kagan*, 1993). Third, even relaxing the assumption of two thresholds and assuming only one, would produce either a time-predictable model (upper threshold alone) or a slip-predictable model (lower threshold alone), and none of these models passed validation (*Mulargia and Gasperini*, 1995). Last, but certainly not least, *Kagan and Jackson* (1998) studied directly the spatial distribution of aftershocks with respect to the focal mechanism of mainshocks, which, according to the CFS model, should be asymmetric, with aftershocks concentrated in the dilatancy quadrant. Such an asymmetry was found to be absent, which suggests that the effective frictional coefficient μ must be close to zero, with the normal stress exerting little influence on the aftershock location.

A parallel approach to CFS is that of rate- and state-dependent friction, which has an equally empirical basis, but it is more detailed, since its equations were derived to describe the time dependent stick-slip behavior of the same sliding at the surface of contact of two bodies. The basic rate- and state-constitutive

Figure 2.14: Map showing the calculated CFS change after the 1999 $m_w = 7.4$ Izmit (Turkey) earthquake. Figure after http://www.ingv.it.

equation is

$$\tau = \sigma \left[\mu + A \ln \left(\frac{\delta'}{\delta'^*} \right) + B \ln \left(\frac{\theta}{\theta^*} \right) \right] \qquad (2.22)$$

where δ' is the velocity of sliding, θ is a state variable representing the time of contact of the two surfaces, δ'^* and θ^* are normalization constants and A, B are empirical coefficients governing the sliding dynamics. $A - B > 0$ gives stable sliding, while $A - B < 0$ gives stick-slip.

According to this approach failure occurs as a catastrophically fast slip episode, i.e. a catastrophic increase in δ'. Time to failure t can then be expressed as (*Dieterich*, 1994)

$$t = \frac{A\sigma}{\tau'} \ln \left(1 + \frac{\tau'}{H\sigma\delta'} \right) \qquad (2.23)$$

where τ' is the long term stress rate and H a material constant. If some external factor, like a variation in the loads, imposes a change in the sliding velocity δ', the time to failure is consequently modified according to

$$\frac{\delta'}{\delta'_0} = \left(\frac{\sigma}{\sigma_0} \right)^{\sigma/A} \exp \left(\frac{\tau}{A\sigma} - \frac{\tau_0}{A\sigma_0} \right). \qquad (2.24)$$

Favored also by a degree of optimism in the interpretation of the data, some apparently successful applications to real cases have been presented, like the ones regarding the North-Anatolian fault (*Nalbant et al.*, 1998; *Parsons et al.*, 2000), and the Izu islands earthquakes (cf. the regression in figure 4 of *Toda et al.*, 2002). But the latter are mere case histories and it is impossible to state whether this success is due to the genuine merits of the model or to retrospective selection bias (cf. section 1.2) combined with the fact that using a larger number of parameters than the CFS approach a better fit is perforce expected. As a consequence, lacking an appropriate validation procedure, it is impossible to draw a definite conclusion about the general applicability of the rate- and state-approach to earthquakes.

Many of the predictions of earthquake phenomenology based on the rate- and state-friction theory are consistent with observations (*Scholz*, 1998). These include the clustering properties and temporal statistics of aftershocks (*Dieterich*, 1994), the triggering of individual events (*Stein*, 1999), the typical depth for the nucleation of large earthquakes (*Scholz*, 1990), and the deformation style in exhumed fault zones (*Scholz*, 1990). As in the crack nucleation model, the rate- and state-theory implies accelerating precursory slip on a nucleating fault patch that should generate precursors (*Rudnicki*, 1988). However, the nucleation patch may be very local, as in the dilatancy problem, and hence precursors, if they exist, may also in practice be hard to detect.

As a final note it is important to remind the reader that friction theory is not necessarily mutually exclusive with fracture mechanics. For example *Scholz* (1990) cites several examples of constitutive rules that can be derived in either framework. If friction occurs due to the contact properties of a few asperities, then their fracture properties on a smaller scale may determine the nucleation properties seen on a larger scale in the frictional constitutive rules. Similarly, the nucleation of a slipping patch can be incorporated in the friction theory in a way that unifies the two competing hypotheses, in the alternative formulation of *Ohnaka and Yamashita* (1989). The important common elements are non-linearity, time dependence, and a clear instability criterion.

2.6.3   Fracture mechanics

This subject began with the classic experiments of *Griffith* (1920, 1922), who showed, around a decade after Reid's work was published, that the breaking strength of composite materials was not in fact a material constant. In an elegant series of experiments, he cut notches of variable length into glass rods held under tension by weights. The rods with larger notches broke more easily. He concluded that fractures nucleated from microscopic flaws in the material, such as microcracks or interstitial pores inherent in the formation of composite materials. He also explained the quantitative dependence of rock strength on the length of the initial flaw using linear elasticity combined with Gibbs' nucleation theory, originally derived for the nucleation of raindrops from a vapor. Thus Griffith's theory is cast in terms of fundamental thermodynamic concepts, by considering an internal energy supply U held in the volume and an energy demand Γ on the crack surface of area A. (In the raindrop problem the energy supply from the vapor is similarly distributed in the cloud volume, whereas the energy demand for condensation is due to the surface tension). Both theories are based not so much on a simple energy 'balance' of supply exceeding demand ($U + \Gamma > 0$), but on a *nucleation theory* where the Gibbs free energy F is at a maximum with respect to growth of the object

$$\frac{\partial F}{\partial A} = \frac{\partial U}{\partial A} + \frac{\partial \Gamma}{\partial A} = 0. \tag{2.25}$$

If γ is the specific surface energy, then for a two-dimensional elliptical crack of semilength c under a remote tensile stress σ then equation (2.25) holds when

$$G = \frac{\partial U}{\partial A} = \frac{\pi \sigma^2 c}{E} = 2\gamma, \tag{2.26}$$

where E is the Young's modulus and G is the energy release rate per unit surface area, sometimes known as the crack extension force. The Griffith theory advanced our physical understanding in several ways. First, it specified a criterion for instability – once nucleated even a small initial flaw can grow arbitrarily large at rates limited only by inertia. The criterion for instability could be shown to be a result of the localization of energy demand on a two-dimensional fracture surface, embedded in a volume containing the stored energy, similar to the assumption of *Reid* (1910). However, it showed that strength was not a material property, hence introducing a random component to strength in composite materials dependent on microscopic variables that would be hard to measure directly in an extensive and opaque material. On the other hand it cannot explain why earthquakes stop – this requires additional sources of energy loss such as local stress relaxation, or loss of energy to radiation, heat, or the creation of a damage zone containing multiple fractures and fault gouge. This problem has recently been solved by recasting the Griffith theory with a stochastic critical energy release rate that varies spatially as a random walk by *Rundle et al.* (1998).

The Griffith theory is a state theory in the sense that it contains no time-dependent elements. Laboratory experiments have shown though, that cracks in rocks can grow under sub-critical conditions, $G < 2\gamma$, by thermally-enhanced chemical processes attacking weakened strained bonds ahead of the crack tip (*Lawn*, 1993). Thus many of the parameters in the Griffith formulation are actually time-dependent. The time-dependence of such sub-critical crack growth is found empirically to depend very non-linearly on G via

$$\frac{dc}{dt} = V_0 \left(\frac{G}{G_0} \right)^{\alpha}, \tag{2.27}$$

where V_0 is the initial crack velocity and G_0 is the initial crack extension force at time $t=0$ (e.g. *Atkinson and Meredith*, 1987). For $\alpha > 1$, the crack grows unstably, with $\alpha = 15$ being typical for crystalline rocks, in the form

$$c = c_0 \left(1 - \frac{t}{t_f} \right)^{-\nu}, \tag{2.28}$$

where $\nu = 1/(\alpha - 1) > 0$, so that the crack size becomes singular at the failure time t_f (*Das and Scholz*, 1981a; *Main*, 1999). In principle, if we knew the size and growth rate of the nucleation patch at a given time, then we could predict the time of occurrence of a future earthquake (but not its magnitude) using this time-dependent approach.

2.6.4   Damage mechanics

In this approach the mechanical properties of the medium depend on a population of fractures (e.g. *Scholz*, 1968; *Horii and Nemat-Nasser*, 1995; *Ashby and Hal-*

lam, 1986; *Costin*, 1987; *Ashby and Sammis*, 1990). One of the first hypotheses to be developed that used the laboratory properties of a population of fractures was the dilatancy-diffusion hypothesis of *Scholz et al.* (1973). The physical principles were based on well-validated laboratory observation of geophysical precursors, calibrated against natural 'precursor' data, provided by the literature available at the time. Dilatancy may be due either to microcracking (*Scholz*, 1968) or the propping property of two rough surfaces sliding past one another (*Rudnicki and Chen*, 1988).

The dilatancy-diffusion hypothesis postulated that a 'source area' of extent S would require large-scale dilatancy by microcracking in a volume of the same dimensions as the mainshock. If this were true, then the magnitude of an impending earthquake could be predicted. Having dilated, the system would have moved away from failure due to the hardening effect of a drop in pore pressure p and a consequent increase in the effective normal stress $\sigma_n^e = \sigma_n - p$ across the incipient fault. The time for the pore pressure to recover to its initial value would then depend on the pore pressure drop, the hydraulic diffusivity, and the linear dimension of the source zone. *Scholz et al.* (1973) postulated that the earthquake would be triggered by the pore pressure recovery at this time in a way that could be predicted by the hydraulic diffusion equation.

The dilatancy-diffusion hypothesis was grounded in clear laboratory observation. The notion of earthquake triggering by fluid pressure change is still fundamental to much of earthquake physics, notably to seismicity induced due to mining or dam impoundment. However, the hypothesis failed in the sense that the predicted precursors failed to materialize (*Scholz*, 1990; *Turcotte*, 1991). The notion of dilatancy remains an important part of rock and earthquake physics, but so far it has not proven of practical utility in terms of earthquake predictability.

So, again, why did a plausible physical model fail in prospective mode? The main reason is the general problem of the difficulty of identifying any precursors in an unambiguous way. First, the precursors cited in *Scholz et al.* (1973), notably the observation of compressional to shear wave velocity anomalies, simply did not stand up to close inspection of the primary data (*Geller*, 1997). Second, the notion of a localized preparation volume, clear enough in a laboratory context, is less obvious in the three-dimensional continuum of the Earth, where there is a also significant stochastic effect from nearby events that may also load or unload the postulated preparation zone of the fault. The preparation volume may then be much larger than the volume in which any precursory signal may occur, perhaps, for the largest events, even including a component of plastic deformation in the lower crust and mantle. In contrast, the constitutive rules for dilatancy were based on the rather large volume dilatancy found for small laboratory samples with large step changes in material properties on the sample boundaries that are not easily scalable to conditions in the seismogenic zone. Thirdly it has recently been shown

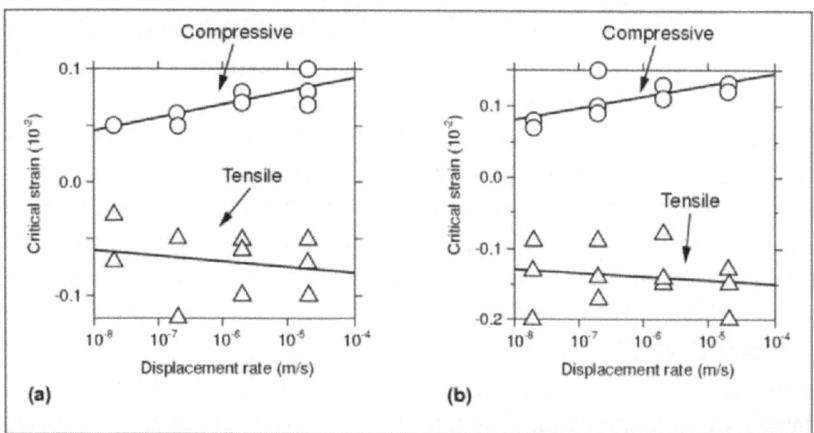

Figure 2.15: Critical strain in the axial (compressive) and radial (tensile) directions at the moment of dynamic failure as a function of strain rate, for (a) Inada Granite and (b) Noboribetsu welded tuff, after *Fujii et al.* (1998). Dilatancy in the radial direction (the tensile component marked) decreases systematically with respect to strain rate.

that there is a systematic negative correlation between the magnitude of volume dilatancy with respect both to increasing sample size and strain rate (see figure 2.15 in *Fujii et al.*, 1998). The size scaling of dilatancy in porous rocks is most probably due to the easier strain accommodation problem in a granular material with a large size to grain scale ratio (*Main et al.*, 2001). On both counts, extrapolations predict finite but much smaller volume dilatancy at crustal scales and strain rates. Modern surface and satellite-based geodetic data at well-instrumented sites typically show no clear precursory signals in the surface strain prior to earthquakes where ground deformation has been measured in the lead-up to the earthquake, in contrast to the accelerating strain of laboratory tests (figure 2.16). In conclusion, precursory dilatancy is a physical requirement for localized unstable slip, but its magnitude may be small, very local, and very hard to measure within current error bars (e.g. figure 2.16b), compared to the original hypothesis.

The time-dependence in the dilatancy-diffusion hypothesis comes from the hydraulic diffusion time. In the fracture mechanics formulation it comes from the chemical effect of hydrolytic weakening of chemical bonds ahead of the crack tip (*Lawn*, 1993). In the damage mechanics of rock samples in compression, this time dependence is modified by the interaction of a population of fractures. The constitutive behavior turns out to be similar to that exhibited in equation (2.28), but with modified model parameters, notably the non-linear exponent v (*Main et al.*, 1993). *Bufe and Varnes* (1993) have applied this equation to several earthquake populations, using seismic event rate, seismic moment rate and cumulative Benioff strain (defined as the square root of the seismic moment) in order to test the hypothesis that natural seismicity follows this equation. To achieve tractable results the point process of individual earthquakes is integrated with respect to time, assuming *a priori* a zero random component, to produce a smoother signal which is then fitted to the integral form of equation (2.28). They discovered that event predictability of the failure time (in retrospect) was very low when using event rate or seismic moment, mainly since the resulting constitutive rules were extremely non-linear. As a result the concept of a Benioff strain was invoked, since it appeared to predict event time better in retrospect, on empirical rather than physical grounds, at the expense of introducing an extra free parameter (the exponent of moment of 1/2) in the parametric search for precursors.

The Benioff 'strain' is not an actual strain. Its evolution cannot be checked, or the model validated, in comparison with any independent geodetic measure of strain. In fact the actual strain has been demonstrated in a rigorous way to be proportional to the linear sum of seismic moment tensors (*Kostrov*, 1974), and a formulation has since been validated by a welter of seismotectonic studies (e.g. *Jackson and McKenzie*, 1988). In prospective mode the predictions of magnitude are very uncertain due to the interdependence of predicted magnitude and time, given the strong non-linearity of equation (2.28). As a consequence, the method is commonly applied in retrospective mode, fitting curves to data that commonly include the magnitude and time of the mainshock.

The treatment of Benioff strain for a population also commonly fails to formally reject the null hypothesis of random occurrence (*Main*, 1999). In fact, when global catalogues are examined the signal is dominated by the random component

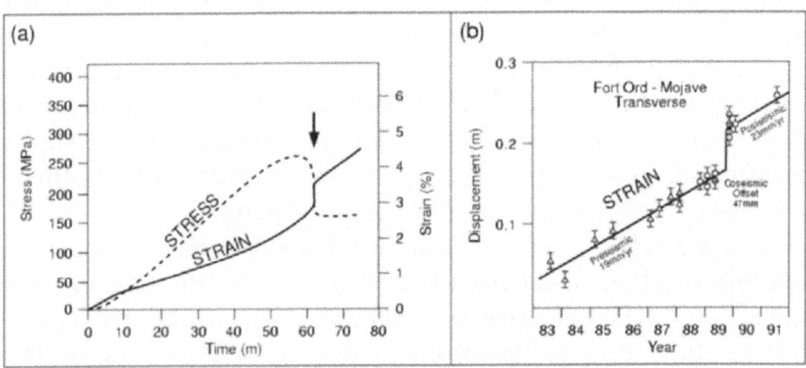

Figure 2.16: Comparison of (a) laboratory and (b) field measurements of precursory strain (solid lines). The accelerating strain seen in the laboratory sample (after *Main et al.*, 1993) is not observable within the resolution of the geodetic method used to measure crustal strain around the time of the earthquake Loma Prieta, California, in 1989 by *Argus and Lyzenga* (1994). It is only in the laboratory that we can directly monitor the stress, which may decrease as well as increase before dynamic failure (dashed line).

(*Gross and Rundle*, 1998). Similarly, model parameters in equation (2.28) can be systematically in error if no account is taken of a background random component to the seismicity (*Main*, 1999). Further studies on a global level are required as new data become available to test this hypothesis in a systematic way. From a physical and geophysical perspective, a strain based on cumulative moment is preferable to the Benioff 'strain'.

2.7   Stress: the basic yet unknown quantity

[By F. Mulargia]

In spite of its vital importance, crustal stress, which is the motor of the earthquake process, is broadly unknown. Almost all the direct *in situ* measurements sample only depths above 2 km, which is a very small fraction of the crust (cf. *Wyss*, 1977). The data are scanty and, for some methodologies of measurement, imprecise. Most of the available data are relative to measurements of stress relief in boreholes in various regions of the world. These can obviously be taken as typical of the crust *not* close to an earthquake state. All of these measurements suggest a state of tectonic stress which is highly dependent on tectonic setting, with horizontal stresses always compressive, reaching values up to twice of the vertical stress (*Hast*, 1969; *McGarr and Gay*, 1978). Small errors are often claimed, but these are likely to be very optimistic. The few measurements available in earthquake focal regions were all related to areas of modest seismicity, and were all based on less accurate hydrofracture methods (*Baumgartner and Zoback*, 1989; *Tsukahara et al.*, 1996). They indicated the presence of large horizontal compressive stress values, up to 4 times the overburden. Note that such a picture is not consistent with the widespread issue that the state of stress is everywhere close to rupture (cf. *Grasso and Sornette*, 1998) since seismicity is easily induced by fluid injection or withdrawal (*Zoback and Harjes*, 1997; *Segall*, 1992). In fact, while the above discussion has highlighted the crucial role of interstitial fluids in earthquake generation, there are sites in which the construction of dams, which induces large variations in fluid pressures, produced no variation in seismicity, and also in the cases in which such triggering occurred, it was a localized property, since only small events and shallow events were induced (for refs. see for example http://www.cadvision.com/retom/). The few apparent exceptions to this, like the December 11, 1967 Koyna earthquake, India, which had $m \simeq 7$ and was preceded by a cluster of obvious foreshocks, occurred in the case of dams built in active tectonic regions which were known sites of large earthquakes in the past. In the case of Koyna an earthquake most likely in excess of $m = 6$ occurred in

August 1764 (see *Seismotectonic Atlas of India and its Environs*, 2000).

In conclusion, the increase in pore pressure immediately below a reservoir always favors induced seismicity, although a true model is lacking since hydrological properties at depth are poorly known. Regardless of tectonic setting, stresses of the order of 1 MPa appear sufficient to trigger quakes. This compares with the calculated stresses of 0.1–1 MPa induced by large earthquakes inferred by the CFS approach (see *Harris*, 1998). This same level of stress, 0.1 MPa, in reservoirs often induces the sustaining of or an increase in seismic activity, once this has been triggered (*Rydelek et al.*, 1988). In both cases it should be noted that there is *not* a one to one correspondence: they must act on a situation close to failure, and often stresses larger than these have not triggered any seismicity in both cases. The stress rate seems also to be important. In this respect, the analysis of the tidal stresses and stress rates on the fault planes of more than 13,000 earthquakes in California has shown that the tidal triggering is not significant (*Vidale et al.*, 1998). This means that preseismic stress rates must be larger than tidal stress rates (10^{-2} MPa/h) and the tectonic stress rates (10^{-5} MPa/h). Let us then attempt to draw an estimate of the state of stress in the Earth's crust, based on the assumption that maximum overall deviations of about a factor of two in the horizontal stresses are possible. We will derive these values in terms of the eigenvalues (the *principal stress values*) and eigenvectors (the *principal stress axes*) of the stress tensor for the three main faulting types: normal, inverse and strike-slip (cf. *Mulargia*, 2000).

2.7.1 Stress in the Earth's crust

The Earth's interior is dominated by a compressive regime originated by gravitational forces. There would be no other forces and the stress field would be isotropic if the Earth were a liquid in equilibrium. This obviously is not the case, and in the lithosphere, the only part which is relevant for earthquakes, there are deviatoric *tectonic* stresses superimposed on the gravitational component. These tectonic stresses are primarily originated by the convective motions in the *asthenosphere*.

In light of the presence of a generalized compressive stress field - no tensional stresses exist in the crust at a macroscopic scale - rupture can only occur in shear and therefore fresh faulting will occur on the plane of maximum shear, which is always bisecting the planes of maximum and minimum stress, the normal vectors to which are also called the *P* (compressional) and *T* (tensile) axes. Note that tensile is an improper term since there is no tension, but the tensile axis just coincides with the direction of minimum compression.

Let us now examine the three different faulting cases.

Normal faulting

In absolute value the maximum stress is vertical and is due to the overburden. The minimum stress is horizontal and orthogonal to strike (figure 2.17). The intermediate eigenvector is along the strike direction. Taking the density of crustal rocks to be $\rho = 2700$ kg/m^3 then the overburden σ_Z at depth z, measured in km, is always (in physical notation, i.e. with negative stresses applied to the system)

$$\rho g z = -2.7 \cdot 10^4 \, z \qquad \text{Pa} \qquad (2.29)$$

Taking a range of approximately a factor of 2 between the maximum and minimum eigenvalues gives a stress tensor equal to

$$\begin{vmatrix} -1.4 & 0 & 0 \\ 0 & -2.0 & 0 \\ 0 & 0 & -2.7 \end{vmatrix} \cdot 10^4 \, z \qquad \text{Pa} \qquad (2.30)$$

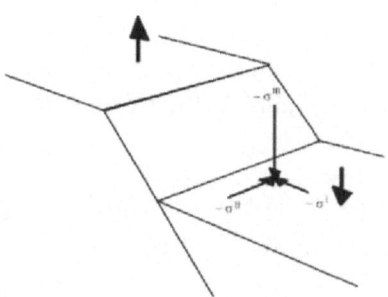

Figure 2.17: The principal stress values and axes for a normal fault.

Reverse (thrust) faulting

In this case the absolute value of the vertical stress, due to the overburden, is minimum. The maximum stress is horizontal and orthogonal to strike (figure 2.18). The intermediate eigenvector is along the strike direction. Taking a range of approximately a factor of 2 between the maximum and minimum eigenvalues gives a stress tensor equal to

$$\begin{vmatrix} -2.7 & 0 & 0 \\ 0 & -4.0 & 0 \\ 0 & 0 & -5.4 \end{vmatrix} \cdot 10^4 \, z \qquad \text{Pa} \qquad (2.31)$$

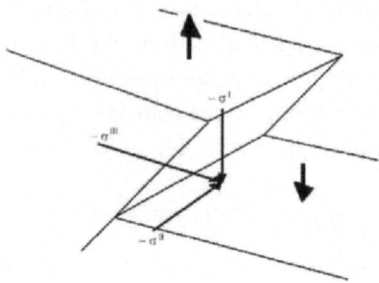

Figure 2.18: The principal stress values and axes for a reverse (thrust) fault.

Strike-slip faulting

In this case the absolute value of the vertical stress, due to the overburden, is intermediate and, using the usual factor of two latitude between maximum and minimum horizontal stresses (figure 2.19), the stress tensor is

$$\begin{vmatrix} -1.8 & 0 & 0 \\ 0 & -2.7 & 0 \\ 0 & 0 & -3.6 \end{vmatrix} \cdot 10^4 z \qquad \text{Pa} \qquad (2.32)$$

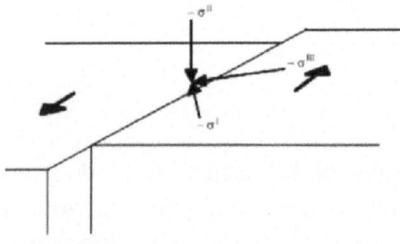

Figure 2.19: The principal stress values and axes for a strike-slip fault.

2.8 Earthquake energy balance

[By F. Mulargia, S. Castellaro, and M. Ciccotti]

The standard approach when considering the physics of a phenomenon is to explicitly describe the evolution of its energy. What, then, is the energy balance of an earthquake? It must be stated at the outset that the problem, due to both its inaccessibility to direct measurement and its complex nonlinear nature, is far from being well constrained. As a consequence, even order of magnitude estimates appear as a quite ambitious goal. This makes studying earthquakes different from studying most physical problems, in which writing the energy integrals is the first step when approaching a new phenomenon.

The picture we draw (cf. *Mulargia et al.*, 2003) will not be applicable to a macroscopic seismic source but rather to a portion of crust of a size sufficient to include all the microscopic and macroscopic effects.

2.8.1 Earthquake energy function

Earthquake phenomenology is complex. Its basic features have been reported in the opening of this chapter. Let us consider a portion of crust containing a fault 'patch' which is large enough to disregard 'grain' size. Since virtually all earthquakes occur on preexisting faults, the 'grain' size will be that of the fabric of *fault gouge*, which is the crushed and reworked incoherent material that surrounds the faults. This fabric has a power law distribution with no apparent upper cutoff for lengths up to the order of 10^{-3} m (*Sammis et al.*, 1986). Consistent with this, the thickness of the shear zone is taken to be of the order of 10^{-4} m. For reasons that will become clearer in the next chapter, we aim at a self-similar description, and the upper cutoff is represented by crustal thickness so that, since self-similarity on earthquake source linear dimension is found to occur experimentally over almost 4 decades (cf. section 3.6), we take our 'patch' as a geometrical entity of linear dimension of ~ 10 m. Note that this self-similar description includes all but the largest earthquakes.

Disregarding the gravitational term, the importance of which varies depending on individual fault mechanism and geometry, four terms can be considered in the energy function Ψ:

$$\Psi = E_E + E_S + E_R + E_T, \tag{2.33}$$

where E_E is the available elastic energy, E_S is the fracture energy, E_R is the radiated energy, and E_T is the thermal energy.

The elastic energy

Earthquakes are produced by deviatoric shear stresses (cf. section 2.7). The most simple case restricts the discourse to a single shear strain component ε, and, from the theory of elasticity, the elastic energy stored in the volume V is

$$E_E = \int_V \frac{1}{2}\mu\varepsilon^2 \, dV, \qquad (2.34)$$

where μ is the rigidity. The largest coseismic crustal strains are of the order of 10^{-3}, so that since in the earth's crust $\mu \sim 10^{10}$ Pa, the maximum available deviatoric elastic energy per unit volume is around 10^4 J/m^3. This can be taken as the energy of the strongest asperities. Lower values will be generally attained. Measuring them is in principle possible, albeit practically difficult (see previous section).

In order to proceed, we need to introduce a seismic source model. The standard seismic source model is a plane dislocation over a bidimensional plane surface (*Brune*, 1968) of linear dimension r, and area A proportional to r^2. Let us assume that the crust can be approximated as a three dimensional lattice of equal cubic cells of side r, and take self-similarity over the basic structure composed of two cubic cells with a face in common (see figure 2.20), which is the elementary sliding 'patch', subject to a shear stress σ on the sliding plane, assumed constant. Let us also assume that the frictional stress is a material constant equal to σ_f.

If the basic event is a slip s of the patch, where s is a small fraction $\eta \ll 1$ of $r, \eta \sim 10^{-3}$, then, consistent with the assumed value for r

$$s = \eta \, r \simeq 10^{-2} \text{ m.} \qquad (2.35)$$

According to Volterra's theorem, the work done by this slip is (*Landau and Lifshitz*, 1970; *Ben-Menahem and Singh*, 1981)

$$\Delta W = <\sigma> sA = E_E, \qquad (2.36)$$

where $<\sigma>$ is the stress during the slip, averaged between the initial and the final state. This slip involves contributions in the radiated, fracture and thermal energies. We will analyze how the work is partitioned among them by estimating the different contributions produced by a slip s.

The radiated energy

The radiated energy E_R can be directly estimated from the recorded seismic waves by considering that the ground displacement x at a given point reached at the time t by a seismic wave of amplitude a and period T is (*Gutenberg and Richter*, 1956)

$$x = a \, \cos\left(\frac{2\pi t}{T}\right). \tag{2.37}$$

The average kinetic energy imposed by the passing wave per unit volume is then

$$\frac{\rho}{2}\dot{x}^2 = \frac{\rho}{2T}\left(\frac{2\pi a}{T}\right)^2 \int_0^T \sin^2\left(\frac{2\pi t}{T}\right) dt = \rho\pi^2\left(\frac{a}{T}\right)^2 \tag{2.38}$$

and the total energy is twice as much since it includes an identical amount of potential energy. Integration in time and over the volume gives the measure of the total energy of the radiated waves. Alternatively, the energy density can be multiplied by the wave velocity, obtaining the surface energy flow, and then integrated twice over the wave front and in time. In practice, it is convenient to rely on more readily measured quantities. Considering the slip of the patch as whole, the basic quantity is the scalar seismic moment $M = \mu \Delta s A$ (see equation 2.1) yielding for the radiated energy

$$\Delta W_R = E_R = (<\sigma> - \sigma_{res})\frac{M}{\mu} = \Delta\sigma\frac{M}{2\mu}, \tag{2.39}$$

where σ_{res} is the residual stress after the slip s and $\Delta\sigma$ is the stress drop.

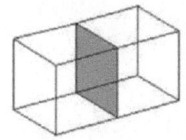

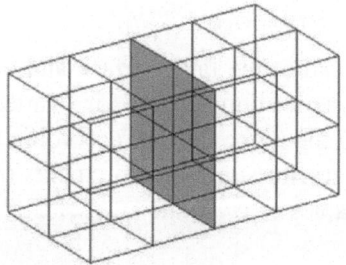

Figure 2.20: The self-similarity relation, with slip occurring at the contact face. Strain remains constant. The linear dimension of each cell is ~ 10 m.

The fracture energy

The basic method to estimate the fracture energy traditionally relies on energy conservation (*Griffith*, 1920). If a new crack is formed or an existing crack propagates, free surfaces are created by the breaking of bonds. It would appear tempting to generalize the Griffith approach, which works reasonably well on homogeneous and isotropic laboratory brittle specimens of simple geometry and with a single crack, to real systems, which regard a generic material with a population of cracks. Unfortunately, in the general case the simple Griffith method does not work, due to the presence of dissipative terms, which are often more important than the elastic terms themselves (*Herrmann and Roux*, 1990), and of important kinetic effects, which occur at propagation velocities comparable to those of the elastic waves. In addition, seismic faults occur in fault gouge, which has mechanical features very different than those of bulk rock (*Mora et al.*, 2000).

As a consequence, estimates of fracture energy E_S made using the approximation of a single propagating mode III crack in an elastic brittle continuum (*Kostrov*, 1966; *Eshelby*, 1969; *Freund*, 1998) must be regarded as speculative, and it appears more appropriate to rely directly on experimental fracture energy data relative to sliding experiments with realistic 'fault gouge'. The latter show that fracture energy is 3 to 4 orders of magnitude smaller than the total energy (*Yoshioka*, 1986), so that, since we are not considering the case of fresh faulting, fracture energy can be comfortably disregarded.

The thermal energy

For energy balance we have

$$\Delta W_T = \sigma_f s A = E_T, \tag{2.40}$$

where the average friction stress σ_f is equal to

$$\sigma_f = \sigma_n \phi, \tag{2.41}$$

with σ_n indicating the stress normal to the fault surface and ϕ the coefficient of friction, which under static and kinetic configuration for a variety of rocks at seismogenic depth conditions falls in the range 0.6–0.9 according to laboratory (*Byerlee*, 1978) as well as borehole in situ stress measurements (*McGarr and Gay*, 1978; *Brudy et al.*, 1997). The values of σ_n in the crust can be taken to be of the order of 10^8 Pa (*ibid.*).

The rate of heat q generated per unit area by fault slip can be roughly calculated by using the equation for a plane surface in an infinite medium

$$q = \sigma_n \phi v, \tag{2.42}$$

where v is the slip velocity and σ_n is the stress normal to the fault plane. If pore fluids are present, as is the general case for water in fault gouge (*Morrow et al.*, 1984), equation (2.42) reads as

$$q = (\sigma_n - P_p)\phi v, \tag{2.43}$$

where P_p is the fluid pore pressure, which at equilibrium can be assumed to be hydrostatic.

Assuming that heat generation starts at time $t = 0$ as a step function with a constant heat flux q, the temperature rise ΔT at a distance x from the emitting plane surface after a time t (*Carslaw and Jaeger*, 1986) is

$$\Delta T = \frac{q}{\rho C}\sqrt{\frac{t}{\pi K}}\exp\left(-\frac{x^2}{4Kt}\right) - \frac{q}{\rho C}\frac{|x|}{2K}\mathrm{erfc}\frac{|x|}{2\sqrt{Kt}}, \tag{2.44}$$

where C represents the specific heat. Close to the fault surface, i.e. when $|x| < \sqrt{Kt}$, equation (2.44) can be approximated as (*Sibson*, 1975)

$$\Delta T = \frac{q}{k}\sqrt{\frac{Ks}{\pi v}}, \tag{2.45}$$

where k is the thermal conductivity, K the thermal diffusivity, and $s = vt$ is the displacement at time t. Combining the latter with the above equations

$$\Delta T = \frac{\phi\sigma_n}{k}\sqrt{\frac{Ksv}{\pi}}. \tag{2.46}$$

Based on laboratory evidence, reasonable values of the parameters involved are $K = 10^{-6}$ m²/s, $k = 2$ J/(m s°C) (*Clark*, 1966). This shows that temperatures would raise easily above 10^3 K for velocities and displacements typical of seismic events since, consistent with our initial assumption for the size of the shear zone, sliding seems to be concentrated in a zone of 10^{-2} m or less (*Kanamori and Heaton*, 2000). However, the presence of melting products like pseudotachylytes appears to be rare (cf. section 2.1.3), and therefore some other mechanism capable of substantially reducing friction and frictional heating must exist. Note that melting itself might reduce friction (*Kanamori and Heaton*, 2000), if it were to occur much more extensively than it appears to do in exhumed faults (*Sibson*, 1992; *Magloughlin and Spray*, 1992).

The fault strength paradox

If seismic slip, which implies slip velocities of the order of $\sim 10^3$ m/s, occurs according to equation (2.43) with $\phi \simeq 0.6 \sim 0.9$, σ_n lithostatic and P_p approximately hydrostatic, very large amounts of frictional heat should be generated.

However, heat flow measurements along the San Andreas and other faults in California suggest heat values at least five times smaller (*Brune et al.*, 1969; *Lachenbruch and Sass*, 1992), and all indirect stress indicators suggest very low friction values (*Hickman*, 1991, and references therein). This set of conflicting evidence is termed the *fault strength paradox*.

Several possible mechanisms have been proposed to resolve this paradox by decreasing friction either statically or dynamically. Some of them, like 'ball bearing' effects of grain rolling or localization (cf. *Mora et al.*, 2000), appear weak against experimental evidence (cf. *Mair and Marone*, 2000). The others (see *Ben-Zion*, 2001, for a review), are based on detailed more or less *ad-hoc* formulations which suffer from a generalized incapability to estimate the relevant parameters. Here we will attempt to draw an approximate picture relying on simple dimensional analysis and taking into account the different time scales.

2.8.2   Earthquakes as a three stage process

Let us release the assumption shared by all the proposed mechanisms of friction reduction that slip occurs at a single time scale (*Ben-Zion*, 2001), and consider a first stage of high friction stick-slip similar to that encountered in the laboratory.

Stage I

The process starts when strain buildup induces on the fault patch a stress σ such that

$$\sigma > \sigma_f. \tag{2.47}$$

Sliding occurs, doing work

$$\Delta W = \Delta W_R + \Delta W_T \tag{2.48}$$

since, as we have seen, the fracture energy can be disregarded.

According to equations (2.40) - (2.41) the thermal work is

$$\Delta W_T = (\sigma_n - P_p)\phi s A. \tag{2.49}$$

In agreement with laboratory phenomenology slip occurs in stick-slip episodes, with $\phi \simeq 0.6 \div 0.9$, average velocity values of the order of 10^{-2} m/s, peak velocity values of the order of 10^{-1} m/s, and a radiated elastic energy which is a fraction of order 10^{-2} of the thermal energy (e.g. *Lockner and Okubo*, 1983), so that

$$\Delta W \simeq \Delta W_T. \tag{2.50}$$

In other words, virtually all the work done according to equation (2.42) is transformed into heat

$$\Delta W_T = \Delta Q = CAw\rho\Delta T \sim Cr^2 w\rho\Delta T, \tag{2.51}$$

where w is the width of the *thermal zone*, the latter being the volume surrounding the fault plane which is site of temperature increase. Rewriting equation (2.46), slip is accompanied by a temperature increase ΔT of the matter surrounding the slip plane equal to

$$\Delta T \simeq (\sigma_n - P_p)\frac{\phi}{k}\sqrt{\frac{Kv^2t}{\pi}}, \tag{2.52}$$

where t is time, which yields that temperatures of the order of 10^2 K are attained by a stick-slip episode with an average $v^2t \sim 10^{-4}$ m^2/s. Considering laboratory experiments, such an episode is likely to be constituted by a cluster of stick-slip events, with an average velocity v of a few cm/s over a time of 10^{-2} s, implying a total slip of the order of 10^{-3} m.

Stage II

Stage I has increased the temperature on the friction surfaces within the shear zone by a value of the order of 10^2 K with a slip episode of a duration of 10^{-2} s. The width w over which heat is carried by thermal diffusion is

$$w(t) = 2\sqrt{Kt}, \tag{2.53}$$

which means that the heat wave will propagate to the whole shear zone, which has a width w of the order of 10^{-2} m, in a time of the order of 10^2 s after the slip event. Experimental values for permeability in fault gouge after a sliding of the order of 10^{-3} m are around 10^{-9} Darcy (or $< 10^{-21}$ m^2) (*Morrow et al.,* 1981, 1984), and the resulting Darcy diffusivity K_D, of the order of 10^{-8} m^2/s, is smaller than thermal diffusivity K (of the order of 10^{-6} m^2/s). Therefore, the temperature increase due to frictional sliding diffuses faster than the thermally expanded pore fluid and, when it extends to the whole shear zone, it raises the fluid pressure from hydrostatic to lithostatic decreasing the effective normal stress, and thus friction, towards zero. Since in frictional sliding the largest asperities dominate the process (*Scholz,* 1990), only when the whole shear zone has been pressurized will a transition to friction values near zero occur. Further stick-slip episodes may take place during this stage, but they can reduce the time required to heat the whole shear zone only if their position is such that they generate thermal waves which 'fill' the shear zone before the original one.

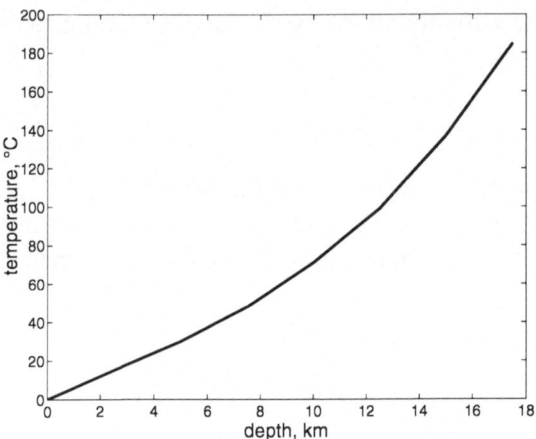

Figure 2.21: The temperature required to increase water pressure from hydrostatic to lithostatic (data are from *Burnham et al.*, 1969).

While propagating, the temperature of the heat wave decreases, but this is hardly a problem, since thermal energy is abundant and the temperature necessary to bring water, which is the main pore fluid constituent, from hydrostatic to the lithostatic pressure transition by differential thermal expansion is comparatively low (see figure 2.21; cf. also *Sibson*, 1977; *Lachenbruch*, 1980; *Mase and Smith*, 1987).

Stage III

The stick-slip episode of stage I has induced the generation of a frictional heat wave which in stage II has propagated to the whole shear zone. Once the shear zone has been pressurized, slipping can occur in a completely different way, and the process enters stage III.

Since at the beginning of stage I elastic forces are slightly larger than friction forces and since now friction has been reduced to ~ 0 (cf. *Kagan and Jackson*, 1998), the energy function is

$$\Psi = E_E - \Delta W_T > 0 \qquad (2.54)$$

and slip will occur with

$$E_E \simeq E_R \qquad (2.55)$$

with virtually all the residual deviatoric elastic energy radiated in seismic waves, and slip at velocities typical of seismic slip, i.e. on average of the order of 1 m/s.

2.8.3 The size of the earthquake

The above description regards a single patch, but a number of patches will be involved in the actual earthquake process. The size of the earthquake will depend on the number of patches which undergo a Stage III radiative slip. Once radiative slip occurs in a patch, it induces dynamic stress transfer in the optimally oriented neighboring patches, raising their elastic energy level. This may induce a stage I slip. However, no radiative stage III slip can occur on patches which are not already at the end of their stage II. The latter are the ones which entered stage I and II simultaneously with our patch, and over which the stage III can be assumed to be self-similar. The details of this process cannot be described by analytical equations due to the nonlinear multi-body character of the interactions, but following self-similarity, the number of patches in the cascade will be a function of the correlation length, i.e. of the length of largest connected patch (*Stauffer and Aharony*, 1994). This length will determine the upper limit of the self-similarity range, which has the single patch at its lower limit. In spite of their potential importance, here for simplicity border effects are disregarded, so that patches adjacent to unslipping patches experience no slip reduction.

Consistent with self-similarity, we assume that strain on the slip surface remains constant and equal to $\varepsilon \sim 10^{-3}$, so that calling l the linear length of the set of patches that slip, we have from equation (2.35) that global fault slip s is $\propto l$, while the global volume involved in the release of elastic energy is $\propto l^3$ (see figure 2.20). In light of this, the global strain release is constant at 10^{-3}, which means that slip s is

$$s = \eta l = 10^{-3} l. \tag{2.56}$$

In the limit of zero friction, the time scale of slip on each single patch is on the order of $s/v \sim 10^{-3}$ s for global 'fault' lengths of the order of 10^3 m. Although negligible with respect to the radiated energy, such a fast slip will also produce heat. The current knowledge of friction between two bodies which slip at velocities which are a significant fraction of the velocity of the elastic waves is very limited (cf. *Tsutumi and Shimamoto*, 1997; *Roder et al.*, 1998, 2000; *Hammerberg et al.*, 1998), but even considering a σ_f drop of two orders of magnitude, the increase in slip velocity should provide sufficient thermal energy (cf. equations 2.49 and 2.52) for thermal effects. However, in light of the short time scale involved, according to equation (2.53) the latter regards only a thin zone of $w < 10^{-3}$ m around the sliding plane. The persistence of impermeability possibly can induce further local friction abatement, as well as local melting (i.e. pseudotachylytes) in a highly fractured and crushed environment (see section 2.1).

2.9   References

Abercrombie, R., and Mori, J., 1994. Local observation of the onset of a large earthquake: 28 June 1992 Landers, California, *Bull. Seism. Soc. Am.*, **84**, 725–734.

Abercrombie, R. E., and Mori, J., 1996. Occurrence patterns of foreshocks to large earthquakes in the western United States, *Nature*, **381**, 303–307.

Allen, A. R., 1979. Mechanisms of frictional fusion in fault zones, *J. Struct. Geol.*, **1**, 231–343.

Anagnos, T., and Kiremidjian, A. S., 1984. Stochastic time-predictable model for earthquake occurrences, *Bull. Seism. Soc. Am.*, **74**, 2593–2611.

Anderson, J., and Chen, Q., 1995. Beginning of earthquake in the Mexican subduction zone on strong-motion accelerograms, *Bull. Seism. Soc. Am.*, **85**, 1107–1116.

Andrews, D. J., 1976a. Rupture propagation with finite stress in antiplane strain, *J. Geophys. Res.*, **81**, 3575–3582.

Andrews, D. J., 1976b. Rupture velocity of plane strain shear cracks, *J. Geophys. Res.*, **81**, 5679–5687.

Andrews, D. J., 1985. Dynamic plane-strain shear rupture with a slip weakening friction law calculated by a boundary integral methods, *Bull. Seism. Soc. Am.*, **75**, 1–21.

Andrews, D. J., and Hanks, T. C., 1985. Scarps degraded by linear diffusion: inverse solution for age, *J. Geophys. Res.*, **90**, 10193–10208.

Argus, D. F., and Lyzenga, G. A., 1994. Site velocities before and after the Loma Prieta and Gulf of Alaska earthquakes determined from VLBI, *Geophys. Res. Lett.*, **21**, 333–336.

Armigliato A., 2001. *A Two-dimensional Numeric Model to Calculate the Effect of Topography and the Mean Heterogeneity on the Coseismic Displacement and Strain in the Near Field* (in Italian), Ph. D. Thesis., Universit`a di Genova, Italy.

Arvidsson, R., and Ekström, G., 1998. Global CMT analysis of moderate earthquakes $M_w \geq 4.5$ using intermediate-period surface waves, *Bull. Seism. Soc. Am.*, **88**, 1003-1013.

Ashby, M. F., and Hallam, S. D., 1986. The failure of brittle solids containing small cracks under compressive stress states, *Acta Metall.*, **34**, 497–510.

Ashby, M. F., and Sammis, C. G., 1990. The damage mechanics of brittle solids in compression, *Pure Appl. Geophys.*, **133**, 489–521.

Atkinson, B. K., and Meredith, P. G., 1987. Theory of subcritical crack growth with application to minerals and rocks, in *Fracture Mechanics of Rock*, Atkinson, B. K. (ed.), Academic Press, London, 111–166.

Austrheim, H., and Boundy, T. M., 1994. Pseudotachylytes generated during seismic faulting and eclogitization of the deep crust, *Science*, **265**, 82–83.

Baumgartner, J., and Zoback, M. D., 1989. Interpretation of hydraulic fracturing pressure-time records using interactive analysis method, *Int. J. Rock Mech. Min. Sci.*, **26**, 461–469.

Belardinelli, M. E., Cocco, M., Coutant, O., and Cotton, F., 1999. Redistribution of dynamic stress during coseismic ruptures: evidence for fault interaction and earthquake triggering, *J. Geophys. Res.*, **104**, 14925–14945.

Bell, T. H., and Etheridge, M. A., 1973. Microstructure of mylonites and their terminology, *Lithos*, **6**, 337–348.

Ben-Menahem, A., and Singh, S. J. S., 1981. *Seismic Waves and Sources*, Springer-Verlag, New York.

Ben-Zion, Y., 2001. Dynamic ruptures in recent models of earthquake faults. *J. Mech. Phys. Solids*, **49**, 2209–2244.

Bird, P., Kagan, Y. Y., and Jackson, D. D., 2002. Plate tectonics and earthquake potential of spreading ridges and oceanic transform faults, in *Plate Boundary Zones, AGU Monograph*, Stein, S., and Freymueller, J. T. (eds.), 203–218.

Boullier, A. M., Ohtani, T., Fujimoto, K., Ito, H., and Dubois, M., 2001. Fluid inclusions in pseudotachylytes from the Nojima Fault, *J. Geoph. Res.*, **106**, 21965–21977.

Brodsky, E. E., and Kanamori, H., 2001. Elastohydrodynamic lubrication of faults, *J. Geophys. Res.*, **106**, 16357–16374.

Bruce, A., and Wallace, D., 1989. Critical point phenomena: universal physics at large length scales, in *The New Physics*, Davies, P. (ed.), Cambridge University Press, Cambridge, 236–267.

Brudy, M., Zoback, M. D., Fuchs, K., Rummel, F., and Baumgartner, J., 1997. Estimation of the complete stress tensor to 8 km depth in the KTB scientific drill holes: implications for crustal strength, *J. Geophys. Res.*, **102**, 18453–18475.

Brune, J. N., 1968. Seismic moment, seismicity, and rate of slip along major fault zones, *J. Geophys. Res.*, **73**, 777–784.

Brune, J. N., 1979. Implications of earthquake triggering and rupture propagation for earthquake prediction based on premonitory phenomena, *J. Geophys. Res.*, **84**, 2195–2198.

Brune, J. N., Henyey, T. L., and Roy, R. F., 1969. Heat fbw, stress and rate of slip along the San Andreas Fault, California, *J. Geophys. Res.*, **74**, 3821–3827.

Bufe, C. G., and Varnes, D. J., 1993. Predictive modeling of the seismic cycle of the greater San Francisco Bay region, *J. Geophys. Res.*, **98**, 9871–9883.

Bufe, C. G., Harsh, P. W., and Burford, A. O., 1977. Steady-state seismic slip–A precise recurrence model, *Geophys. Res. Lett.*, **4**, 91–94.

Burnham, C. W., Holloway, J. R., and Davis, N. F., 1969. Thermodynamic properties of water to $1000^{o}C$ and 10,000 bars, *Geol. Soc. Amer., Special Paper*, 132.

Byerlee, J. D., 1970. The mechanics of stick-slip, *Tectonophysics*, **9**, 475–486.

Byerlee, J.D., 1978. Friction of rocks, *Pure Appl. Geophys.*, **116**, 615–629.

Camelbeeck, T., and Meghraoui, M., 1996. Large earthquakes in northern Europe more likely than once thought, *Eos, Trans. Am. Geophys. Union*, **77**, 405–409.

Camelbeeck, T., and Meghraoui, M., 1998. Geological and geophysical evidence for large palaeo-earthquakes with surface faulting in the Roer Graben (northwest Europe), *Geophys. J. Int.*, **132**, 347–362.

Carlson, J. M., and Langer, J. S., 1989. Properties of earthquakes generated by fault dynamics, *Phys. Rev. Lett.*, **22**, 2632–2635.

Carslaw, H. S., and Jaeger, J. C., 1986. *Conduction of Heat in Solids*, 2nd ed., Oxford University Press, Oxford.

Castellaro, S., and Mulargia, F., 2001. A simple but effective cellular automaton for earthquakes, *Geophys. J. Int.*, **144**, 609–624.

Castellaro, S., and Mulargia, F., 2002. What criticality in cellular automata models for earthquakes?, *Geophys. J. Int.*, **150**, 483–493.

Cladouhos, T. T., 1999a. Shape preferred orientations of survivor grains in fault gouge, *J. Struct. Geol.*, **21**, 419–436.

Cladouhos, T. T., 1999b. A kinematic model for deformation within brittle shear zones, *J. Struct. Geol.*, **21**, 437–448.

Clark, S.P., (ed.), 1966. *Handbook of Physical Constants*, Geol. Soc. Am. Mem., New York.

Cochard, A., and Madariaga, R., 1994. Dynamic faulting under rate-dependent friction, *Pure Appl. Geophys..*, **142**, 419–445.

Costin, L. S., 1987. Time-dependent deformation and failure, in *Fracture Mechanics of Rock*, Atkinson, B. K. (ed.), Academic Press, London, 111–166.

Cowan, D. S., 1999. Do faults preserve a record of seismic faulting? A field geologist's opinion, *J. Struct. Geol.*, **21**, 995–1001.

Crespellani, T., Nardi, R., and Simoncini, C., 1988. *Earth's Liquefaction under Seismic Conditions* (in Italian), Ed. Zanichelli, Bologna.

Das, S., and Scholz, C. H., 1981a. Theory of time-dependent rupture in the Earth, *J. Geophys. Res.*, **86**, 6039–6051.

Das, S., and Scholz, C. H., 1981b. Off-fault aftershock clusters caused by shear stress increase?, *Bull. Seism. Soc. Am.*, **71**, 1669–1675.

DeMets, C., 1995. Plate motions and crustal deformation, US National Report to the International Union of Geodesy and Geophysics, 1991-1994, *Rev. Geophys.*, **33**, 365–369.

Deng, J. S., and Sykes, L. R., 1996. Triggering of 1812 Santa Barbara earthquake by a great San Andreas shock: implications for future seismic hazards in southern California, *Geophys. Res. Lett.*, **23**, 1155–1158.

Dieterich, J. H., 1972. Time-dependent friction as a possible mechanism for aftershocks, *J. Geophys. Res.*, **77**, 3771–3781.

Dieterich, J. H., 1979a. Modeling of rock friction 1, Experimental results and constitutive equations, *J. Geophys. Res.*, **84**, 2161–2168.

Dieterich, J. H., 1979b. Modeling of rock friction 2, Simulation of preseismic slip, *J. Geophys. Res.*, **84**, 2169–2175.

Dieterich, J. H., 1994. A constitutive law for rate of earthquake production and its application to earthquake clustering, *J. Geophys. Res.*, **99**, 2601–2618.

Dolan, J. F., Sieh, K., and Rockwell, T. K., 2000. Late quaternary activity and seismic potential of the Santa Monica fault system, Los Angeles, California, *Geol. Soc. Am. Bull.*, **112**, 1559–1581.

Du, Y., and Aydin, A., 1993. Stress transfer during three sequential moderate earthquakes along the central Calaveras fault, California, *J. Geophys. Res.*, **98**, 9947–9962.

Dziewonski, A. M., and Anderson D. L., 1981. Preliminary reference earth model (PREM), *Phys. Earth Planet. Int.*, **25**, 297–356.

Ellsworth, W. L., and Beroza, G. C., 1995. Seismic evidence for an earthquake nucleation phase, *Science*, **268**, 851–855.

Eshelby, J. B., 1957. The determination of the elastic field of an ellipsoidal inclusion, and related problems, *Proc. Roy. Soc. London*, **A241**, 376–396.

Eshelby, J. B., 1969. The elastic field of a crack extending nonuniformly under general anti-plane loading, *J. Mech. Phys. Solids*, **17**, 177–199.

Fabbri, O., Lin, A., and Tokushige, H., 2000. Coeval formation of cataclasite and pseudotachylyte in a Miocene forearc granodiorite, southern Kyushu, Japan, *J. Struct. Geol.*, **22**, 1015–1025.

Faure, G. C., 1986. *Principle of Isotope Geology*, 2nd ed., John Wiley & Sons.

Fernandez, M., Molina, E., Havskov, J., and Atakan K., 2000. Tsunamis and Tsunami Hazards in Central America, *Nat. Haz.*, **22**, 91–116.

Francis, P. W., 1972. The pseudotachylyte problem, *Comments Earth Sci. Geophys.*, **3**, 35–53.

Freund, L. B., 1998. *Dynamic Fracture Mechanics*, Cambridge University Pres, New York.

Fujii, Y., Kiyama, T., Ishijima, Y., and Kodama, J., 1998. Examination of a rock failure criterion based on circumferential tensile strain, *Pure Appl. Geophys.*, **152**, 551–577.

Gathercole, N., Reiter, H., Adam, T., and Harris, B., 1994. Life prediction for fatigue of T800/5245 carbon-fiber composites. 1. Constant-amplitude loading, *Int. J. Fatigue*, **16**, 523–532.

Geist, E. L., 1998. Local tsunamis and earthquake source parameters, *Adv. Geophys.*, **39**, 117–209.

Geller, R. J., 1997. Earthquake prediction: a critical review, *Geophys. J. Int.*, **131**, 425–450.

Geller, R. J., and Kanamori, H., 1977. Magnitudes of great shallow earthquakes from 1904 to 1952, *Bull. Seism. Soc. Am.*, **67**, 587–598.

Goldsby, D. L., and Tullis, T. E., 2002. Low frictional strength of quartz rocks at subseismic slip rates, *Geophys. Res. Lett.*, **29**(17), 1–4, doi: 10.1029/2002GL01240.

Grasso, J. R., and Sornette, D., 1998. Testing self-organized criticality by induced seismicity, *J. Geophys. Res.*, **103**, 29965–29987.

Griffith, A. A., 1920. The phenomena of rupture and flow in solids, *Phil. Trans. Roy. Soc. London*, **A221**, 163-198.

Griffith, A. A., 1922. The theory of rupture, in *Proc. 1 st Int. Congr. Appl. Mech., Delft: Tech. Boekhandel en Drukkerj*, Biezeno, C. B. and Burgers, J. M. (eds.), J. Waltman Jr., 54–63.

Gross, S. J., and Kisslinger, C., 1994. Stress and the spatial distribution of seismicity in the central Aleutians, *J. Geophys. Res.*, **99**, 15291–15303.

Gross, S., and Rundle, J., 1998. A systematic test of time-to-failure analysis, *Geophys. J. Int.*, **133**, 57–64.

Gutenberg, B., and Richter, C. F., 1941. Seismicity of the Earth, *Geol. Soc. Amer., Special Papers*, **34**, 1–131.

Gutenberg, B., and Richter, C. F., 1944. Frequency of earthquakes in California, *Bull. Seism. Soc. Am.*, **34**, 185–188.

Gutenberg, B., and Richter, C. F., 1956. Magnitude and energy of earthquakes, *Ann. Geofi s.*, **9**, 1–15.

Hammerberg, J. E., Holian, B. L., Roder, J., Bishop, A. R., and Zhou, S. J., 1998. Nonlinear dynamics and the problem of slip at material interfaces, *Physica D*, **123**, 330–340.

Hampton, M. A., Lee, H. J., and Locat, J., 1996. Submarine landslides, *Rev. Geophys.*, **34**, 33–59.

Hardebeck, J. L., Nazareth, J. J., and Hauksson, E., 1998. The static stress change triggering model: constraints from two southern California aftershock sequences, *J. Geophys. Res.*, **103**, 24427–24437.

Harris, R. A., 1998. Introduction to special section: stress triggers, stress shadows and implications for seismic hazard, *J. Geophys. Res.*, **103**, 24347–24358.

Harris, R. A., and Simpson, R. W., 1992. Changes in static stress on southern California faults after the 1992 Landers earthquake, *Nature*, **360**, 251–254.

Harris, R. A., and Simpson, R. W., 1996. In the shadow of 1857 - the effect of the great Ft. Tejon earthquake on subsequent earthquakes in southern California, *Geophys. Res. Lett.*, **23**, 229–232.

Harris, R. A., Simpson, R. W., and Reasenberg, P. A., 1995. Influence of static stress changes on earthquake locations in southern California, *Nature*, **375**, 221–224.

Hasegawa, H. S., and Kanamori, H., 1987. Source mechanism of the magnitude 7.2 Grand Banks earthquake of November 1929: double couple or submarine landslide?, *Bull. Seism. Soc. Am.*, **77**, 1984–2004.

Haskell, N. A., 1964. Total energy and energy spectral density of elastic wave radiation from propagating faults, *Bull. Seism. Soc. Am.*, **54**, 1811–1841.

Hast, N., 1969. The state of stress in the upper part of the Earth's crust, *Tectonophysics*, **8**, 169–211.

Herrmann, H. J., and Roux, S. (eds.), 1990. *Statistical Models for the Fracture of Disordered Media*, North-Holland, Amsterdam.

Hickman, S., Sibson, R., and Bruhn, R., 1995. Introduction to special section: mechanical involvement of fluids in faulting, *J. Geophys. Res.*, **100**, 12831–12840.

Hobbs, B. E., Ord, A., and Teyssier, C., 1986. Earthquakes in the ductile regime?, *Pure Appl. Geophys.*, **124**, 309–336.

Holland, T. H., 1900. The charnokite series, a group of Archean hypersthenic rocks in peninsular India, *India Geological Survey Memoirs*, **28**, 119–249.

Horii, H., and Nemat-Nasser, S., 1985. Compression induced microcrack growth in brittle solids: axial splitting and shear failure, *J. Geophys. Res.*, **90**, 3105–3125.

Huc, M., and Main, I. G., 2003. Anomalous stress diffusion in earthquake triggering: correlation length, time-dependence, and directionality, *J. Geophys. Res.*, **108**(B7), 1–12, doi: 10.1029/2001JB001645.

Hudnut, K. W., Seeber, L., and Pacheco, J., 1989. Cross-fault triggering in the November 1987 Superstition Hills earthquake sequence, southern California, *Geophys. Res. Lett.*, **16**, 199–202.

Hutton, J., 1788. Theory of the Earth, *Trans. Roy. Soc. Edinb.*, **1**, 209–304 (reprinted 1973, Hafner, New York).

Ide, S., and Takeo, M., 1997. Determination of constitutive relations of fault slip based on seismic wave analysis, *J. Geophys. Res.*, **102**, 27379–27391.

Iio, Y., 1995. Observations of the slow initial phase generated by microearthquakes: implications for earthquake nucleation and propagation, *J. Geophys. Res.*, **100**, 15333–15349.

Imamura, F., and Gica, E. C., 1996. Numerical model for tsunami generation due to subaqueous landslide along a coast, *Sci. Tsunami Hazards*, **14**, 13–28.

Ishimoto, M., and Iida, K., 1939. Observations of earthquakes registered with the microseismograph constructed recently, *Bull. Earthq. Res. Inst. Tokyo Univ.*, **17**, 443–478.

Iwasaki, S., 1987. On the estimation of a tsunami generated by a submarine landslide, *Proc. Int. Tsunami Symp.*, Vancouver, B.C., 134–138.

Iwasaki, S., 1997. The wave forms and directivity of a tsunami generated by an earthquake and a landslide, *Sci. Tsunami Hazards*, **15**, 23–40.

Jackson, J. A., 1987. *Glossary of Geology*, 4th ed., American Geological Institute, Alexandria, Virginia.

Jackson, J., and McKenzie, D., 1988. The relationship between plate motions and seismic moment tensors, and the active rates of deformation in the Mediterranean and Middle-East, *Geophys. J. Int.*, **93**, 45–73.

Janosi, I. M. and Kertész, J., 1993. Self-organised criticality with and without conservation, *Phys. Rev. A.*, **200**, 179–188.

Jaumé, S. C., and Sykes, L. R., 1992. Change in the state of stress on the southern San Andreas fault resulting form the California earthquake sequence of April to June 1992, *Science*, **258**, 1325–1328.

Jiang, L., and LeBlond, P. H., 1994. Three-dimensional modeling of tsunami generation due to a submarine mudslide, *J. Phys. Oceanogr.*, **24**, 559–573.

Kagan, Y. Y., 1993. Statistics of characteristic earthquakes, *Bull. Seism. Soc. Am.*, **83**, 7–24.

Kagan, Y. Y., 1997. Statistical aspects of Parkfield earthquake sequence and Parkfield prediction experiment, *Tectonophysics*, **270**, 207–219.

Kagan, Y. Y., 2002a. Seismic moment distribution revisited: I. Statistical results, *Geophys. J. Int.*, **148**, 520–541.

Kagan, Y. Y., 2002b. Seismic moment distribution revisited: II. Moment conservation principle, *Geophys. J. Int.*, **149**, 731–754.

Kagan, Y. Y., 2003. Accuracy of modern global earthquake catalogs, *Phys. Earth Planet. Inter.*, **135**, 173–209.

Kagan, Y. Y., and Jackson, D. D., 1991. Seismic gap hypothesis – ten years after, *J. Geophys. Res.*, **96**, 21419–21431.

Kagan, Y. Y., and Jackson, D. D., 1995. New seismic gap hypothesis – five years after, *J. Geophys. Res.*, **100**, 3943–3959.

Kagan, Y. Y., and Jackson, D. D., 1996. Statistical tests of VAN earthquake predictions: comments and reflections, *Geophys. Res. Lett.*, **23**, 1433–1436.

Kagan, Y. Y., and Jackson, D. D., 1998. Spatial aftershock distribution: effects of normal stress, *J. Geophys. Res.*, **103**, 24453–24467.

Kagan, Y. Y., and Jackson, D. D., 2000. Probabilistic forecasting of earthquakes, *Geophys. J. Int.*, **143**, 438–453.

Kanamori, H., 1977. The energy release in great earthquakes, *J. Geophys. Res.*, **82**, 2981–2987.

Kanamori, H., and Heaton, T. H., 2000. Microscopic and macroscopic physics of earthquakes, in *Geocomplexity and the Physics of Earthquakes*, Rundle, J. B., Turcotte, D.L., and Klein, W. (eds.), AGU Publ., Washington, 147–164.

Kelleher, J., Sykes, L., and Oliver, J., 1973. Possible criteria for predicting earthquake locations and their application to major plate boundaries of Pacific and Caribbean, *J. Geophys. Res.*, **78**, 2547–2585.

Kelleher, J., Savino, J., Rowlett, H., and McCann, W., 1974. Why and where great thrust earthquakes occur along island arcs, *J. Geophys. Res.*, **79**, 4889–4899.

Keilis-Borok, V. I., 1959. On estimation of a displacement in an earthquake source and of source dimensions, *Ann. Geofi s.*, **12**, 205–214.

Kikuchi, M., and Kanamori, H., 1991. Inversion of complex body waves - III, *Bull. Seism. Soc. Am.*, **81**, 2335–2350.

King, G. C. P., Stein, R. S., and Lin, J., 1994. Static stress changes and the triggering of earthquakes, *Bull. Seism. Soc. Am.*, **84**, 935–953.

Kiremidjian, A. S., and Anagnos, T., 1984. Stochastic slip-predictable model for earthquake occurrences, *Bull. Seism. Soc. Am.*, **74**, 739–755.

Knopoff, L. and Kagan, Y. Y., 1977. Analysis of the theory of extremes as applied to earthquake problems, *J. Geophys. Res.*, **82**, 5647–5657.

Kostrov, B. V., 1966. Unsteady propagation of longitudinal shear cracks, *J. Appl. Math. Mech. (transl. P. M. M.)*, **30**, 1241–1248.

Kostrov, V. V., 1974. Seismic moment and the energy of earthquakes and seismic flow of rock, *Izv. Acad. Sci. USSR Phys. Solid Earth*, **1**, 23–44.

Lachenbruch, A. H., 1980. Frictional heating, fluid pressure, and the resistance to fault motion, *J. Geophys. Res.*, **85**, 6097–6112.

Lachenbruch, A. H., and Sass, J. H., 1992. Heat-flow from Cajon Pass, fault strength, and tectonic implications, *J. Geophys. Res.*, **97**, 4995–5015.

Landau, L. D., and Lifshitz, E. M., 1970. *Theory of Elasticity*, 2nd ed., Pergamon Press, Oxford.

Lawn, B., 1993. *Fracture of brittle solids*, 2nd ed., Cambridge University Press, Cambridge.

Lay, T., and Wallace, T. C., 1995. *Modern Global Seismology*, Academic Press, New York.

Lee, J. C., Chen, Y. G., Sieh, K., Müller, K, Chen, W. S., Chu H. T., Chan, Y. C., Rubin,

C., and Yeats, R., 2001. A vertical exposure of the 1999 surface rupture of the Chelungpu fault at Wufeng, western Taiwan: structural and paleoseismic implications for an active trust fault, *Bull. Seism. Soc. Am.*, **91**, 914–929.

Lee, J., Spencer, J., and Owen, L., 2001. Holocene slip rates along the Owens Valley Fault, California: implications for the recent evolution of the Eastern California Shear Zone, *Geology*, **29**, 819–822.

Leonard, T., Papasouliotis, O., and Main, I. G., 2001. A Poisson model for identifying characteristic size effects in frequency data: application to frequency-size distributions for global earthquakes, 'starquakes' and fault lengths, *J. Geophys. Res.*, **106**, 13473–13484.

Levret, A., Combes, P., and Granier, T., 1996. Seismicity and Archaeology: a Multidisciplinary Approach (in French), Colloque National AFPS, Saint Remy les Chevreuses (France).

Lin, A., 1994. Glassy pseudotachylyte veins from the Fuyun fault zone, northwest China, *J. Struct. Geol.*, **16**, 71–83.

Lockner, D. A., and Okubo, P. G., 1983. Measurements of frictional heating in granite, *J. Geophys. Res.*, **88**, 4313–4320.

Lyakhovsky, V., Ben-Zion, Y., and Agnon, A., 2001. Earthquake cycle, fault zone and seismicity patterns in a rheologically layered lithosphere, *J. Geophys. Res.*, **106**, 4103–4120.

Maddock, R. H., 1983. Melt origin of fault-generated pseudotachylytes demonstrated by textures, *Geology*, **11**, 105–108.

Maddock, R. H., Grocott, J., and Van Nes, M., 1987. Vescicles, amygdales and similar structures in fault-generated pseudotachylytes, *Lithos*, **20**, 419–432.

Magloughlin, J. F., 1992. Microstructural and chemical changes associated with cataclasis and frictional melting at shallow crustal levels: the cataclasite-pseudotachylyte connection, *Tectonophysics*, **204**, 243–260.

Magloughlin, J. F., and Spray, J. G., 1992. Frictional melting processes and products in geological materials: introduction and discussion, *Tectonophysics*, **204**, 197–206.

Main, I. G., 1999. Applicability of time-to-failure analysis to accelerated strain before earthquakes and volcanic eruptions, *Geophys. J. Int.*, **139**, F1–F6.

Main, I. G. and Burton, P. W., 1984. Information theory and the earthquake frequency-magnitude distribution, *Bull. Seism. Soc. Am.*, **74**, 1409–1426.

Main, I. G., Sammonds, P. R., and Meredith, P. G., 1993. Application of a modified Griffith criterion to the evolution of fractal damage during compressional rock failure, *Geophys. J. Int.*, **115**, 367–380.

Main, I., Mair, K., Kwon, O., Elphick., S., and Ngwenya, B., 2001. Experimental constraints on the mechanical and hydraulic properties of deformation bands in porous sandstones: a review, in *The Nature and Significance of Fault Zone Weakening*, Holdsworth, R. E., Strachan, R. A., Magloughlin, J. F. and Knipe, R. J. (eds), Geol. Soc. Lond. Special Publications, **186**, 43–63.

Mair, K., and Marone, C., 2000. Shear heating in granular layers, *Pure Appl. Geophys.*, **157**, 1847–1866.

Maruyama, T., 1963. On the force equivalents of dynamical elastic dislocations with reference to the earthquake mechanism, *Bull. Earthq. Res. Inst. Tokyo Univ.*, **41**, 467–486.

Mase, C. W., and Smith, L., 1987. Effect of frictional heating on the thermal, hydrologic and mechanical response of a fault, *J. Geophys. Res.*, **92**, 6249–6272.

Massonnet, D., and Feigl, K. L., 1998. Radar interferometry and its application to changes in the earth's surface, *Rev. Geophys.*, **36**, 441–500.

Massonnet, D., Rossi, M., Carmona, C., Adragna, F., Peltzer, G., Feigl, K. and Rabaute, T., 1993. The displacement field of the Landers earthquake mapped by radar interferometry, *Nature*, **364**, 138–142.

Matsu'ura, M., and Hirata, N., 1982. Generalized least-squares solutions to quasi-linear problems with a priori information, *J. Phys. Earth*, **30**, 451–468.

Matsu'ura, M., Kataoka, H., and Shibazaki, B., 1992. Slip-dependent friction law and nucleation processes in earthquake rupture, *Tectonophysics*, **211**, 135–148.

McAdoo, B. G., Pratson, L. F., and Orange, D. L., 2000. Submarine landslide geomorphology, US continental slope, *Mar. Geol.*, **169**, 103–136.

McCalpin, J. P., 1996. *Paleoseismology*, International Geophysics Series, Academic Press, San Diego.

McCann, W. R., Nishenko, S., Sykes, L. R., and Krause, J., 1979. Seismic gaps and plate tectonics: seismic potential for major boundaries, *Pure Appl. Geophys.*, **117**, 1082–1147.

McGarr, A., and Gay, N. C., 1978. State of stress in the earth's crust, *Ann. Rev. Earth Planet. Sci.*, **6**, 405–436.

McKenzie, D., and Brune, J. N., 1972. Melting on Fault Planes During Large Earthquakes, *Geophys. J. Roy. Astr. Soc.*, **29**, 65–78.

Meghraoui, M., and Crone, A. J., 2001. Earthquakes and their preservation in the geological record, *J. Seism.*, **5**, 281–285.

Melosh J., 1996. Dynamic weakening of faults by acoustic fluidization, *Nature*, **397**, 601–606.

Michetti, A. M., and Hancock, P. L., 1997. Paleoseismology: understanding past earthquakes using quaternary geology, *J. Geodynamics*, **24**, 3–10.

Miller, S. A., Ben-Zion, Y., and Burg, J. P., 1999. A three-dimensional fluid-controlled fault model: behavior and implications, *J. Geophys. Res.*, **104**, 10621–10638.

Mora, P., Place, D., Abe, S., and Jaumé, S., 2000. Lattice solid simulation of the Physics of fault zones and earthquakes: the model, results, and directions, in *Geocomplexity and the Physics of Earthquakes*, Rundle, J. B., Turcotte, D. L., and Klein, W. (eds.), AGU Publ., Washington, 105–126.

Morgan, J. K., and Böttcher, M. S., 1999. Numerical simulations of granular shear zones using the distinct element method - 1. Shear zone kinematics and the micromechanics of localization, *J. Geophys. Res.*, **104**, 2703–2719.

Morgestern, N., 1967. Submarine slumping and the initiation of turbidity currents, *Proc. Int. Res. Conf. Marine Geotechnique*, Univ. Illinois Press, Urbana-Champaign.

Mori, J., and Kanamori, H., 1996. Initial rupture of earthquakes in the 1995 Ridgecrest, California sequence, *Geophys. Res. Lett.*, **23**, 2437–2440.

Morrow, C. A., Shi, L. Q., and Byerlee, J. D., 1981. Permeability and strength of San Andreas fault gouge under high pressure, *Geophys. Res. Lett.*, **8**, 325–328.

Morrow, C. A., Shi, L. Q., and Byerlee, J. D., 1984. Permeability and strength of San Andreas fault gouge under confining pressure and shear stress, *J. Geophys. Res.*, **89**, 3193–3200.

Mulargia, F., 2000. *An Introduction to the Mechanics of Faulting* (in Italian), CLUEB, Bologna.

Mulargia, F., and Gasperini, P., 1995. Evaluation of the applicability of the time- and slip-predictable earthquake recurrence models to Italian seismicity, *Geophys. J. Int.*, **120**, 453–473.

Mulargia, F., Castellaro, S., and Ciccotti, M., Earthquake energy balance, *Geophys. J. Int.*, in press, 2003.

Murty, T. S., 1979. Submarine slide-generated water waves in Kitimat Inlet, British Columbia., *J. Geophys. Res.*, **84**, 7777–7779.

Nalbant, S. S., Hubert, A., and King, G. C. P., 1998. Stress coupling between earthquakes in Northwest Turkey and the North Aegean Sea, *J. Geophys. Res.*, **103**, 24469–24486.

Nishenko, S. P., 1985. Seismic potential for large and great interplate earthquakes along the Chilean and southern Peruvian margins of South America: a quantitative reappraisal, *J. Geophys. Res.*, **90**, 3589–3615.

Nishenko, S. P., 1991. Circum–Pacific seismic potential – 1989-1999, *Pure Appl. Geophys.*, **135**, 169–259.

O'Hara, K. D., 2001. A pseudotachylyte geothermometer, *J. Struct. Geol.*, **23**, 1345–1357.

O'Hara, K. D., and Sharp, Z. D., 2001. Chemical and oxygen isotope composition of natural and artificial pseudotachylyte: role of water during frictional melting, *Earth Planet. Sci. Lett.*, **184**, 394–406.

Ohnaka, M., 1990. Nonuniformity of crack-growth resistance and breakdown zone near the propagating tip of a shear crack in brittle rock: a model for earthquake nucleation to dynamic rupture, *Can. J. Phys.*, **68**, 1071–1083.

Ohnaka, M., 1992. Earthquake source nucleation: a physical model for short-term precursors, *Tectonophysics*, **211**, 149–178.

Ohnaka, M., and Kuwahara, Y., 1990. Characteristic features of local breakdown near a crack-tip in the transition zone from nucleation to unstable rupture during stick-slip shear failure, *Tectonophysics*, **175**, 197–220.

Ohnaka, M., and Yamashita, T., 1989. A cohesive zone model for dynamic shear faulting based on experimentally-inferred constitutive relation and strong motion source parameters, *J. Geophys. Res.*, **94**, 4089-4104.

Ohtani, T., Fujimoto, K., Ito H., Tanaka, H., Tomida, N., and Higuchi, T., 2000. Fault rocks and past to recent fluid characteristics from the borehole survey of the Nojima fault ruptured in the 1995 Kobe earthquake, southwest Japan, *J. Geophys. Res.*, **105**,

16161–16171.

Okada, Y., 1985. Surface deformation due to shear and tensile faults in a half-space, *Bull. Seism. Soc. Am.*, **75**, 1135–1154.

Okada, Y., 1992. Internal deformation due to shear and tensile faults in a half-space, *Bull. Seism. Soc. Am.*, **82**, 1018–1040.

Okubo, P., 1989. Dynamic rupture modeling with laboratory-derived constitutive relations, *J. Geophys. Res.*, **98**, 12321–12335.

Okumura, K., 2001. Paleoseismology of the Itoigawa-Shizuoka tectonic line in central Japan, *J. Seismol.* **5**, 411–431.

O'Leary, D. W., 1993. Submarine mass movement, a formulative process of passive continental margins: the Munson-Nygren landslide complex and the southeast New England landslide complex. In *Submarine Landslides: Selected Studies in the U.S. Exclusive Economic Zone*, Schwab, W. C., Lee, H. J., and Twichell, D. C. (eds.), *U.S. Geol. Surv. Bull.*, **2002**, 23–39.

Olson, S. M., and Stark, T. D., 2002. Liquefied strength ratio from liquefaction fbw failure case histories, *Can. Geotech. J.*, **39**, 629–647.

Oskin, M., Sieh, K., Rockwell, T., Miller, G., Guptill, P., Curtis, M., McArdle, S., and Elliot, P., 2000. Active parasitic folds on the Elysian Park anticline: implications for seismic hazard in central Los Angeles, California, *Geol. Soc. Am. Bull.*, **112** 693–707.

Ozkan, G., and Ortoleva, P. J., 2000. Evolution of the gouge particle size distribution: a Markov model, *Pure Appl. Geophys.*, **157**, 449–468.

Papazachos, B. C., 1989. A time-predictable model for earthquake occurrence in Greece, *Bull. Seism. Soc. Amer.*, **79**, 77–84.

Parsons, T., Toda, S., Stein, R. S., Barka, A., and Dieterich, J. H., 2000. Heightened odds of large earthquakes near Istanbul: an interaction-based probability calculation, *Science*, **288**, 661–665.

Passchier, C. W., 1982. Pseudotachylyte and the development of ultramylonite bands in the Saint-Barth`ele`emy Massif, French Pyrenees, *J. Struct. Geol.*, **4**, 69–79.

Philpotts, A. R., 1964. Origin of pseudotachylytes, *Am. J. Science*, **262**, 1008–1035.

Place, D., and Mora, P., 2000. Numerical simulation of localisation phenomena in a fault zone, *Pure Appl. Geophys.*, **157**, 1821–1845.

Postpischl, D., Agostini, S., Forti, P., and Quinif, Y., 1991. Paleoseismicity from karst sediments: 'Grotta del Cervo" cave case study (central Italy), *Tectonophysics*, **193**, 33–44.

Poty, B., Menager, M., and Roth, E., 1990. *Nuclear Methods of Dating*, Roth, E., and Poty, B. (eds.), vol. 5, Solid Earth Sciences Library, Kluwer, Dordrecht.

Prior, D. B., and Coleman, J. M., 1979. Submarine landslides: geometry and nomenclature, *Z. Geomorph. N. F.*, **23**, 415–426.

Rabinowicz, E., 1965. *Friction and Wear of Materials*, John Wiley, New York.

Ramsey, J. G., and Huber, M. I., 1983. *The Techniques of Modern Structural Geology*, Academic Press, Harcourt Brace Jovanivich Publishers, New York.

Reasenberg, P. A., and Simpson, R. W., 1992. Response of regional seismicity to the static stress change produced by the Loma Prieta earthquake, *Science*, **255**, 1687–1690.

Reid, H. F., 1910. *The California Earthquake of April 18, 1906*, vol. 2: *The Mechanics of the Earthquake*, Carnegie Institution of Washington, Washington, D.C.

Reiter, L., 1990. *Earthquake Hazard Analysis*, Columbia University Press, New York.

Rice, J. R., 1993. Spatio-temporal complexity of slip on a fault, *J. Geophys. Res.*, **98**, 9885–9907.

Roder, J., Bishop A. R., Holian B. L., Hammerberg J. E., and Mikulla R. P., 2000. Dry friction: modeling and energy fbw, *Physica D*, **142**, 306–316.

Roder, J., Hammerberg J. E., Holian B. L., and Bishop A. R., 1998. Multichain Frenkel-Kontorova model for interfacial slip, *Phys. Rev. B*, **57**, 2759–2766.

Roeloffs, E., and Langbein, J., 1994. The earthquake prediction experiment at Parkfield, California, *Rev. Geophys.*, **32**, 315–336.

Rudnicki, J. W., 1988. Physical models of earthquake instability and the precursory process, *Pure Appl. Geophys.*, **126**, 531–554.

Rudnicki, J. W., and Chen, C.-H., 1988. Stabilization of rapid frictional slip on a weakened fault by dilatant hardening, *J. Geophys. Res.*, **93**, 4745–4757.

Ruina, A. L., 1983. Slip instability and state variable friction laws, *J. Geophys. Res.*, **88**, 10359–10370.

Rundle, J. B., 1989. Derivation of the complete Gutenberg-Richter magnitude frequency relation using the principle of scale invariance, *J. Geophys. Res.*, **94**, 12337–12342.

Rundle, J., Preston, E., McGinnis, S., and Klein, W., 1998. Why earthquakes stop: growth and arrest in stochastic fields, *Phys. Rev. Lett.*, **80**, 5698–5701.

Rydelek, P. A., Davis, P. M., and Koyanagi, R., 1988. Tidal triggering of earthquake swarms at Kilauea volcano, *J. Geophys. Res.*, **93**, 4401–4411.

Sammis, C. G., Osborne, R., Anderson, J., Banerdt, M. and White, P., 1986. Self-similar cataclasis in the formation of fault gouge, *Pure Appl. Geophys.*, **124**, 53–78.

Savage, J. C., 1991. Criticism of some forecasts of the national earthquake prediction evaluation council, *Bull. Seism. Soc. Am.*, **81**, 862–881.

Savage, J. C., 1993. The Parkfield prediction fallacy, *Bull. Seism. Soc. Am.*, **83**, 1–6.

Scherbaum, F., and Bouin, M. P., 1997. FIR filter effects and nucleation phases, *Geophys. J. Int.*, **130**, 661–668.

Scholz, C. H., 1968. Microfracturing and the inelastic deformation of rock in compression, *J. Geophys. Res.*, **73**, 1417–1432.

Scholz, C. H., 1990. *The Mechanics of Earthquakes and Faulting*, Cambridge University press, Cambridge.

Scholz, C. H., 1998. Earthquakes and friction laws, *Nature*, **391**, 37–42.

Scholz, C. H., Sykes, L. R., and Aggarwal, Y. P., 1973. Earthquake prediction – a physical basis, *Science*, **181**, 803–810.

Schwab, W. C., Lee, H. J., and Twichell, D. C., 1993. Submarine landslides: selected studies in the U.S. exclusive economic zone, *U.S. Geol. Surv. Bull. 2002*, U.S., Dept. of Interior, Washington, DC.

Schwartz, D. P., and Coppersmith, K. J., 1984. Fault behavior and characteristic earthquakes: examples from the Wasatch and San Andreas fault, *J. Geophys. Res.*, **89**, 5681–5698.

Segall, P., 1992. Induced stresses due to fluid extraction from axisymmetrical reservoirs, *Pure Appl. Geophys.*, **139**, 535–560.

Seismotectonic Atlas of India and its environs, 2000. Map & Cartography Division, Geological Survey of India, Calcutta, India.

Shand, S. J., 1916. The pseudotachylyte of Parjis (Orange Free State), *J. Geol. Soc. London*, **72**, 198–221.

Shen, P. Y., and Mansinha, L., 1983. On the principle of maximum entropy and the earthquake frequency-magnitude relation, *Geophys. J. Roy. Astr. Soc.*, **74**, 777–785.

Shimamoto, T., and Nagahama, H., 1992. An argument against the crush origin of pseudotachylytes based on the analysis of clast size distribution, *J. Struct. Geol.*, **14**, 999–1006.

Shimazaki, K., and Nakata, T., 1980. Time-predictable recurrence model for large earthquakes, *Geophys. Res. Lett.*, **7**, 279–282.

Sibson, R. H., 1973. Interactions between temperature and pore fluid pressure during earthquake faulting - a mechanism for partial or total stress relief, *Nature Phys. Sci.*, **243**, 66–68.

Sibson, R. H., 1975. Generation of pseudotachylyte by ancient seismic faulting, *Geophys. J. Roy. Astron. Soc.*, **43**, 775–794.

Sibson, R. H., 1977. Kinetic shear resistance, fluid pressures and radiation efficiency during seismic faulting, *Pure Appl. Geophys.*, **115**, 387–400.

Sibson, R. H., 1980. Transient discontinuities in ductile shear zones, *J. Struct. Geol.*, **2**, 165–168.

Sibson, R. H., 1992. Power dissipation and stress levels on faults in the upper crust, *J. Geophys. Res.*, **85**, 6239–6247.

Sieh, K., 1978. Prehistoric large earthquakes produced by slip on the San Andreas Fault at Pallet Creek, Southern California, *J. Geophys. Res.*, **83**, 3907–3939.

Sieh, K., 1984. Lateral offsets and revised dates of large prehistoric earthquakes at Pallet Creek, southern California, *J. Geophys. Res.*, **89**, 7641–7670.

Sieh, K., and Natawidjaja, D., 2000. Neotectonics of the Sumatran fault, Indonesia, *J. Geophys. Res.*. **105**, 28295–28326.

Sieh, K., Stuiver, M., and Brillinger, D., 1989. A more precise chronology of earthquakes produced by the San Andreas fault in southern California, *J. Geophys. Res.*, **94**, 603–623.

Sieh, K., Ward, S. N., Natawidjaja, D., and Suwargadi, B. W., 1999. Crustal deformation at the Sumatran subduction zone revealed by coral rings, *Geophys. Res. Lett.*, **26**, 3141–3144.

Simpson, R. W., and Reasenberg, P. A., 1994. Earthquake-induced static stress changes on central California faults, in The Loma Prieta, California earthquake of October 17, 1989 - Tectonic Processes and Models, Simpson, R. W. (ed.), *U.S. Geol. Surv. Prof. Pap.*, 1550-F, F55–F89.

Sleep, N. H., Richardson E., and Marone C., 2000. Physics of friction and strain rate localization in synthetic fault gouge, *J. Geophys. Res.*, **105**, 25875–25890.

Spray, J. G., 1987. Artificial generation of pseudotachylyte using friction welding apparatus: simulation of melting on a fault plane, *J. Struct. Geol.*, **9**, 49–60.

Spray, J. G., 1992. A physical basis for the the frictional melting of some rock forming minerals, *Tectonophysics*, **204**, 205–221.

Spray, J. G., 1995. Pseudotachylyte controversy: fact or friction?, *Geology*, **23**, 1119–1122.

Stauffer, D., and Aharony, A., 1994. *Introduction to Percolation Theory*, 2nd ed., Taylor & Francis, Philadelphia.

Steacy, S. J., and McCloskey, J. J., 1998. What controls an earthquake's size? Results from a heterogeneous cellular automaton, *Geophys. J. Int.*, **133**, F11–F14.

Stein, R. S., 1999. The role of stress transfer in earthquake occurrence, *Nature*, **402**, 605–609.

Stein, R. S., and Lisowski, M., 1983. The 1979 Homestead Valley earthquake sequence, California: control of aftershocks and postseismic deformation, *J. Geophys. Res.*, **88**, 6477–6490.

Stein, R. S., King, G. C. P., and Lin, J., 1992. Change in failure stress on the southern San Andreas fault system caused by the 1992 magnitude = 7.4 Landers earthquake, *Science*, **258**, 1328–1332.

Stein, R. S., King, G. C. P., and Lin, J., 1994. Stress triggering of the 1994 M = 6.7 Northridge, California, earthquake by its predecessors, *Science*, **265**, 1432–1435.

Stiros, S., and Jones, R. E., 1996. *Archaeoseismology*, Fitch Lab. Occas. Pap. 7, British School at Athens and Inst. Geol. Min. Explor, Oxbow Books, Oxford, UK.

Swanson, M. T., 1992. Fault structure, wear mechanism and rupture processes in pseudotachylyte generation, *Tectonophysics*, **204**, 223–242.

Sykes, L. R., and Quittmeyer, R. C., 1981. Repeat times of great earthquakes along simple plate boundaries, in *Earthquake Prediction: an International Review, Maurice Ewing Ser.*, **4**, Simpson, D. W., and Richards, P. G. (eds.), Am. Geophys. Un., Washington, D. C, 217–247.

Sykes, L. R., Shaw, B. E., and Scholz, C. H., 1999. Rethinking earthquake prediction, *Pure Appl. Geophys.*, **155**, 207–232.

Tappin, D. R., Watts, P., McMurtry, G. M., Lafoy, Y., and Matsumoto, T., 2001. The Sissano, Papua New Guinea Tsunami of July 1998. – Offshore Evidence on the Source Mechanism, *Mar. Geol.*, **175**, 1–23.

Thatcher, W., 1989. Earthquake recurrence and risk assessment in circum-Pacific seismic gaps, *Nature*, **341**, 432–434.

Thatcher, W., 1990. Order and diversity in the modes of circum-Pacific earthquake recurrence, *J. Geophys. Res.*, **95**, 2609–2623.

Thatcher, W., Marshall, G., and Lisowski, M., 1997. Resolution of fault slip along the 470-km-long rupture of the great 1906 San Francisco earthquake and its implications, *J. Geophys. Res.*, **102**, 5353–5367.

Toda, S., Stein, R. S., and Sagiya, T., 2002. Evidence from the AD 2000 Izu islands earthquake swarm that stressing rate governs seismicity, *Nature*, **419**, 58–61.

Toyoshima, T., 1990. Pseudotachylyte from the main zone of the Hidaka metamorphic belt, Hokkaido, northern Japan, *J. Metamorph. Geol.*, **8**, 507–523.

Trifunac, M. D., and Todorovska, M. I., 2002. A note on differences in tsunami source parameters for submarine slides and earthquakes, *Soil Dynamics and Earthquake Engineering*, **22**, 143–155.

Troise, C., De Natale, G., Pingue, F., and Petrazzuoli, S. M., 1998. Evidence for static stress interaction among earthquakes in the south-central Apennines (Italy), *Geophys. J. Int.*, **134**, 809.

Tsukahara, H., Ikeda, R., and Omura, K., 1996. In-situ stress measurement in an earthquake focal area, *Tectonophysics*, **262**, 281–290.

Tsutsumi, A., and Shimamoto, T., 1997. High-velocity frictional properties of gabbro, *Geophys. Res. Lett.*, **24**, 699–702.

Turcotte, D. L., 1991. Earthquake prediction, *Ann. Rev. Earth Planet. Sci.*, **19**, 263–281.

Twiss, R. J., and Moores, E. M. (eds.), 1992. *Structural Geology*, W. H. Freeman, New York.

Utsu, T., 1999. Representation and analysis of the earthquake size distribution: a historical review and some new approaches, *Pure Appl. Geophys.*, **155**, 509–535.

Valensise, G., and Pantosti, D., 2001. The investigation of potential earthquake sources in peninsular Italy: a review, *J. Seismol.*, **5**, 287–306.

Vere-Jones, D., Robinson, R. and Yang, W. Z., 2001. Remarks on the accelerated moment release model: problems of model formulation, simulation and estimation, *Geophys. J. Int.*, **144**, 517–531.

Vidale, J., Agnew, D., Johnston, M., and Oppenheimer, D., 1998. Absence of earthquake correlation with earth tides: an indication of of high preseismic fault stress rate, *J. Geophys. Res.*, **103**, 24567–24572.

Walker, D. A., and Bernard, E. N., 1993. Comparison of T-phase spectra and tsunami amplitudes for tsunamigenic and other earthquakes, *J. Geophys. Res.*, **98**, 12557–12565.

Watts, P., 1998. Wavemaker curves for tsunamis generated by underwater landslides, *J. Wtrwy, Port, Coast, and Oc. Engrg., ASCE*, **124**, 127–137.

Watts, P., 2000. Tsunami features of solid block underwater landslides, *J. Wtrwy, Port, Coast, and Oc. Engrg., ASCE*, **126**, 144–152.

Wenk, H. R., 1978. Are pseudotachylytes products of fracture or fusion?, *Geology*, **6**, 507–511.

Wenk, H. R., Johnson, L. R., and Ratschbacher, L., 2000. Pseudotachylytes in the Eastern Peninsular Ranges of California, *Tectonophysics*, **321**, 253–277.

Wesnousky, S. G., 1996. Reply to Yan Y. Kagan's comment on the Gutenberg-Richter or characteristic earthquake distribution, which is it?, *Bull. Seism. Soc. Am.*, **86**, 286–291.

White, J. C., 1996. Transient discontinuities revisited: pseudotachylyte, plastic instability and the influence of low pore fluid pressure on the deformation processes in the mid-crust, *J. Struct. Geol.*, **18**, 1471–1486.

Wiemer, S., and Wyss, M., 2002. Mapping spatial variability of the frequency-magnitude distribution of earthquakes, *Adv. Geophys.*, **45**, 259–302.

Wyss, M. (ed.), 1977. *Stress in the Earth*, Contrib. Cur. Res. Geophys., reprinted from *Pure Appl. Geophys.*, Birkh¨auser, Basel.

Yeats, R., Sieh, K., and Allen, C. (eds.), 1997. *The Geology of Earthquakes*, Oxford University Press, Oxford.

Yoshioka, N., 1986. Fracture energy and the variation of gouge and surface roughness during frictional sliding of rocks, *J. Phys. Earth*, **34**, 335–355.

Zachariasen, J., Sieh, K., Taylor, F. W., and Hantoro, W. S., 2000. Modern vertical deformation above the Sumatran subduction zone: paleogeodetic insights from coral microatolls, *Bull. Seism. Soc. Am.*, **90**, 897–913.

Zoback, M. D., and Harjes, H. P., 1997. Injection-induced earthquakes and crustal stress at 9 km depth at the KTB deep drilling site, Germany, *J. Geophys. Res.*, **102**, 18477–18491.

Chapter 3

The physics of complex systems: applications to earthquake

[by F. Mulargia, I. Main, M. Ciccotti, S. Castellaro, and J. Kertész]

We have seen in the previous two chapters that the classical approach to earthquake physics provides an intuitively reasonable model of earthquake occurrence. This theory can be successfully applied to explain the propagation of seismic waves radiated by earthquake sources. However, the classical theory is not able to account globally for the following basic features of earthquake occurrence (cf. section 1.1),

1. Earthquakes are rare events.

2. Earthquakes are clustered in both space and time.

3. Earthquakes are rupture events which occur mostly on preexisting faults.

4. Earthquakes have a quasi-constant stress drop which is, on average, much smaller than ambient stress (*Abercrombie and Leary*, 1993).

5. The external forcing function, i.e. tectonic strain, is small and constant, inducing extremely low strain rates.

6. Fault traces are power law distributed in length.

7. Faults are rough surfaces, with power law distributed roughness.

8. The spatial distribution of hypocentral locations of earthquakes and laboratory acoustic emissions are power law distributed in both space and time (*Kagan and Knopoff*, 1980; *Hirata et al.*, 1987).

9. Earthquakes are power law distributed in size (Gutenberg-Richter law).

10. Earthquakes have aftershock sequences that decay with a power law in time (Omori law).

11. Seismicity can be induced by stress perturbations smaller than the stress drop of individual events. These may be due to previous earthquakes occurring at relatively great distances (see section 2.6), or to changes in local pore fluid pressure through man-made activity.

The inability of the classical theory to explain the above phenomenology suggests that it should be scrapped and replaced by an entirely new approach. This might seem to be a radical statement, but the history of physics contains many instances (e.g. the advent of quantum mechanics or relativity) where apparently reasonable theories had to be discarded because they failed to agree with observations outside the parameter range for which they were originally defined.

3.1 Phase transitions, criticality, and self-similarity

What is the physical meaning of the ubiquity of power law distributions, which appears to be an important clue to earthquake phenomenology? Why is the classical approach broadly incapable of reproducing this picture? Finally, is there a way to use the above phenomenological picture to develop a new class of earthquake models? As discussed below, the answer is positive.

A new class of models exists which appear broadly capable of explaining all the above evidence, power law behavior in particular. This new class of models takes a completely different perspective than the classical mechanistic view. Acknowledging the complexity of earthquakes, it abandons the deterministic view of the classical approach and turns to the tool that physics uses to deal with systems with a very large number of degrees of freedom: *statistical mechanics*. By taking such a view, though, one can only achieve average descriptions of ensembles of similar events, rather than deterministic models of individual earthquakes.

This nevertheless opens a very important question: the classical statistical-mechanical approach aims at representing thermodynamic properties of a system in a state of *equilibrium*, which is reached after a sufficiently long *relaxation time*. How can earthquakes, which are by no means equilibrium phenomena, be reconciled with this view?

Most undergraduate courses are confined to *equilibrium* thermodynamics and statistical physics and only during specialization do students learn about *non-equilibrium* phenomena. This is historically motivated and didactically natural, but it does not reflect the importance of *Non-Equilibrium Statistical Physics*

(NESP). Boltzmann's kinetic theory, and Einstein's relation expressing what we call today *Fluctuation Dissipation Theorem* (FDT) are the first landmarks of NESP. In fact, FDT beautifully shows the intimate relationship between linear nonequilibrium phenomena and equilibrium. However, very often, due to some external drive, we are out of the linear regime, in a situation termed 'far from equilibrium' which is the topic of intensive ongoing research with much relevance to earthquake physics.

Before discussing some aspects of the physics of systems that are far from equilibrium, we review results from equilibrium statistical physics on phase transitions which revolutionized our way of thinking concerning many-component systems and had much impact on NESP as well. Starting a little over three decades ago, several ideas began to converge to a thoroughly 'new' perspectives on some 'old' problems. These new insights were applicable to a number of diverse phenomena.

We have to recall that physics concentrates on the laws of interactions (gravity, electromagnetism, etc.) and on the behavior resulting from them. The latter can be unexpectedly rich because of the nonlinear couplings and/or because of the large number of interacting units where the properties of the ensemble is by far more complicated than a simple combination of the properties of the units. The study of such systems has been very successful in the last two decades leading to what is called the *Physics of Complex Systems* (PCS) (cf. *Mallamace and Stanley*, 2000). The PCS with its obvious interdisciplinary applications ranging from biology to social sciences, and, definitely, to geosciences (cf. *Gadomski et al.*, 2000) is an important part of what is called the 'New Physics' by *Davies* (1989). Earthquakes were not among the phenomena considered by Davies, but were added to this list a short time later (e.g. *Bak and Tang*, 1989). There have been continuing attempts to apply the PCS to earthquakes and we shall review them in this chapter. We should, however, emphasize from the very beginning that what we feel most important in the PCS in this context is its approach, tools, and concepts which are very promising for our subject.

At this point we should also emphasize another difference between the classical approach, which was discussed in the last chapter, and the PCS approach. In the former, one attempts to understand earthquakes by creating models of earthquakes based on the classical theory of elasticity. Conversely, in the latter, we consider earthquakes as just one of many seemingly different physical systems that in fact exhibit common properties.

We begin here by considering in general the phenomenon of 'phase transitions', using relatively simple physical systems as examples. After this discussion we then will turn to the topic of the underlying similarities between these simple examples of phase transitions and earthquakes.

The most common example of a phase transition is the 'change' of water from the liquid to the vapor phase. This transition, which at room pressure occurs at 100°C, is accompanied by an abrupt change in density, with a decrease by a factor of approximately 1600 going from the liquid to the vapor phase. As the ambient pressure is increased the phase transition occurs at progressively higher temperatures (see figure 3.1). At the same time, the jump in density between the two phases also decreases progressively, until at a pressure of 21.8 MPa and at a temperature of 374°C this difference vanishes altogether. This point is called the *critical point*, and the corresponding temperature the *critical temperature, T_c*. At the same time, some peculiar phenomena can be observed: critical opalescence indicates that the *correlation length* ξ, the distance over which fluctuations 'know' about each other, becomes macroscopically large. In fact, for an infinite system it diverges at T_c,

$$\xi \propto |T - T_c|^{-\nu}, \tag{3.1}$$

where ν is a *critical exponent*. This diverging characteristic length has severe consequences, e.g., the compressibility κ also diverges, and there is a singularity in the specific heat C:

$$\kappa \propto |T - T_c|^{-\gamma}, \tag{3.2}$$

$$C \propto |T - T_c|^{-\alpha}, \tag{3.3}$$

where γ and α are also critical exponents. The same argument applies to the time scale.

A similar divergence is observed in many phenomena, even far from equilibrium. One of these is rupture, which is at the basis of earthquake occurrence. The link between criticality and divergence of the characteristic length is illustrated by a simple model of criticality, the *Ising spin model*, which was originally introduced to study the existence of a spontaneous (zero field) magnetization in ferromagnetic materials, giving them the capacity to attract or repel other similar materials. The average magnetic moment is the sum of the atomic spins and decreases smoothly up to a certain temperature at which it vanishes altogether, defining a critical point in the transition between ferromagnetic and non-ferromagnetic phases. Central to this phenomenon is the *cooperative* character of the system, which is disregarded altogether by the classical approach. To deal with this it is necessary to account for a very large number of degrees of freedom, which is hardly amenable to analytical formulations. A *mean field approximation*, i.e. replacing the values of the variables with their thermal equilibrium mean values, can be attempted in some cases, but this approximation is often problematic (cf. *Ashcroft and Mermin*, 1981; *Yeomans*, 1992). In physics sometimes the mean field or Landau theory is called 'classical', but it does not disregard cooperativity. What it ignores are the fluctuations, and this is what sets the limits of this approch.

A more effective tool is computer simulation, illustrated using the example of the square lattice Ising model, in which each cell can take either a value of -1 or 1, depicted respectively in black or white color with an energy function preferring the same colors in neighboring positions. The model can be taken as a two dimensional picture of a solid in which each cell represents an atom with its spin 'up' or 'down' value. However, it can also be taken as the picture of a rupture surface on which -1 and +1 represent respectively intact and fractured patches. Note that this Ising model of rupture is of course very crude. Rupture is a nonequilibrium phenomenon, while the phase transition of the Ising model is treated as equilibrium. Two parameters are useful to characterize the system: the *order parameter Q*, which stems from the fractional difference between the populations of the -1 and +1 cells, and the correlation length ξ, which here is roughly the linear dimension of the largest correlated spatial structure (i.e. the size of the largest 'white' or 'black' islands).

Figure 3.2 shows three configurations of such a lattice respectively below, at, and above the critical point. The white areas represent the spin up state. In terms of the correlation length it is clear that below the critical point one configuration prevails and that the correlation length (approximately the size of the largest cluster) is finite. We are interested in the correlation of the fluctuations, therefore below T_c the average magnetization (the huge black sea) is disregarded. Above T_c the islands have finite characteristic sizes leading to finite correlation length. At the critical point none prevails, the order parameter is zero, and the correlation length

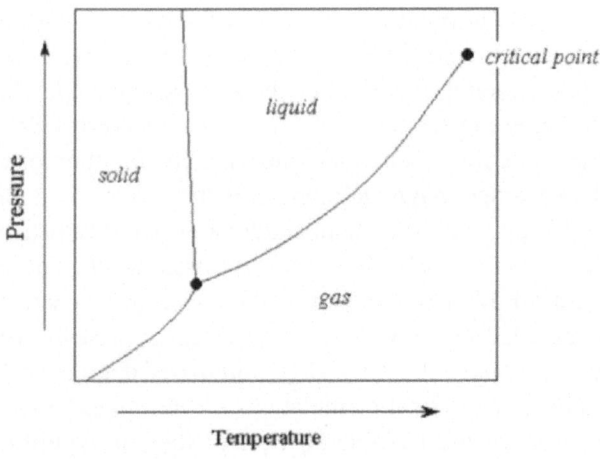

Figure 3.1: Phase diagram of H_2O.

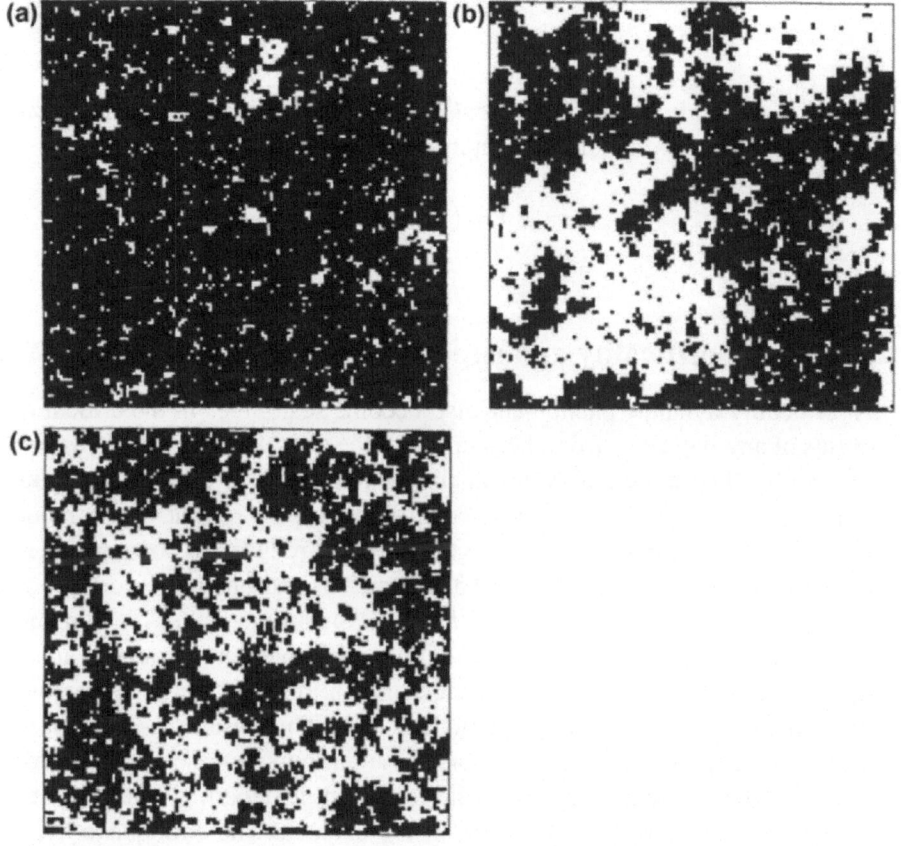

Figure 3.2: Three configurations of an Ising lattice respectively (a) below, (b) at, and (c) above the critical point.

is equal to the maximum linear dimension of the system, ideally infinite. That is, there is no characteristic length associated with the system. In other words, the system becomes *scale invariant*, or *self-similar* at *all scales*. This means that changing scale and disregarding the fine detail (an operation termed *coarsening*) leaves the picture statistically the same. This is the basic feature of any critical system.

A second keypoint is that 'sufficiently close' to the critical point, both the order parameter and the correlation length show power law behavior

$$Q \propto |t|^{\beta}, \qquad (3.4)$$

and

$$\xi \propto |t|^{-\nu}, \tag{3.5}$$

where β and ν are called the *critical exponents* and t is the *reduced temperature* with respect to the critical temperature T_c:

$$t = (T - T_c)/T_c. \tag{3.6}$$

3.1.1   Subcriticality and supercriticality

At criticality dynamic phenomena also become scale free. In the critical regime, events of any size exist, from the smallest to the ones which span the whole model area with a distribution of events in size which is strictly power law (figure 3.3). If the model events involve areas small with respect to that of the model, the probability of occurrence of the largest events is reduced in comparison with an extrapolation of the power law trend (figure 3.3) and one refers to these models as *subcritical* (cf. *Stauffer and Aharony*, 1994; *Main*, 1996). A third possible case is that the area involved is not only comparable with the entire area of the model, but that the probability of occurrence of the largest events is greater than the extrapolated power law trend (figure 3.3). This case is called *supercritical*. Transferring these types of behavior to the earthquake case is straightforward. The subcritical case appears to be the rule, with its tapering at large size, while the supercritical case is reminiscent of the characteristic earthquake hypothesis (section 2.4).

3.1.2   Universality

Many systems — not all — that show critical behavior have the same values of the critical exponents, i.e. $\beta = 0.33$ and $\nu = 0.63$ in three dimensions, independent of whether the transitions are magnetic, fluid-vapor, or chemical (other groups of transitions are characterized by other sets of critical exponents). This remarkable result, which occurs totally independent of the nature of the processes, has been termed *universality* (and the groups are called *universality classes*). If this concept were applicable to earthquakes, it would allow an exceedingly simple description. A question is then in order: are universal critical models applicable to earthquakes, which are non-equilibrium phenomena? And, more generally, what is the role of critical phenomena in earthquake physics? After some attempts to model earthquakes as thermodynamic equilibrium systems (e.g. *Caputo*, 1977, 1982), the problem is still open. A clear picture is starting to emerge, although a definite answer is lacking for a number of reasons. First, experimentally the

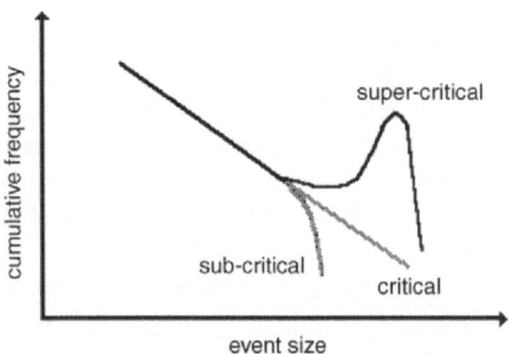

Figure 3.3: A frequency-size distribution of events showing sub-critical, critical, and supercritical behavior.

source of earthquakes is widely inaccessible to direct measurement. Second, realistic 'computer experiments' are also beyond present computing capabilities, since to analyze the detail of a general critical system one has to deal with three typical lengths: (1) the correlation length ξ; (2) the largest microscopic length of interactions (i.e. lattice spacing for nearest neighbors interacting systems) L_{min}; (3) the maximum linear dimension of the system L_{max}.

As a consequence, to realize in simulation $L_{max}/\xi \simeq \xi/L_{min} \simeq 10^2$ would require handling a number of variables which in two dimensions is equal to 10^8 and in 3 dimensions is equal to 10^{12}, beyond the capability of the most powerful present day computers.

Is all hope lost and should this road be abandoned? Fortunately not: First of all one can force the system to be at $L_{max} \sim \xi$ and then use the theory of finite size scaling (cf. *Privman*, 1990; see also below) to evaluate the simulation results. In this way one can omit the inequality $L_{max} \gg \xi$ to gain 4 (6) orders of magnitude in 2 (3) dimensions. Second, there are some issues that allow one to deal with simplified 'reduced' problems. All of these issues hinge on the assumption that *scale invariance* is the source of the observed power law behavior (cf. section 3.6).

3.2 Scale invariance: the analytical approach

In general, a scale invariant phenomenon is one for which the distribution of the properties at different length scales of a given variable can be obtained from one another by a similarity transformation. It is worth noting that all classical macroscopic physics is inherently scale invariant. Continuum mechanics, which totally ignores the existence of atoms and, in general, of any physical cutoff, is the typical example. The typical tools of a continuum mechanics approach are the use of asymptotic expansions to ∞ and 0 to achieve workable analytic expressions. The postulate of this classical approach is of a reductionist type: irrespective of the fact that physical cutoffs obviously do exist, one can isolate various ranges of scale length within which the fundamental laws do not depend on the scale considered. This reductionist approach is more or less tacitly and intuitively assumed by all practical applications. If one has to model the motion of a ship on the scale from 1 to 1000 kilometers, the internal deformation of the boat is disregarded, and only the position of its centroid in a geographic frame of reference is used. But if one has to describe the motion of the ship in the length scale range 1 millimeter to 1 meter it would be necessary to use a frame of reference attached to the undeformed ship, neglecting its geographic position altogether.

The practical consequences of scale invariance within a given range between physical cutoffs remained unexplored till the 1940's. Then, it was of particular interest that establishing self-similarity with respect to time had the great advantage of transforming time-dependent problems into time-independent ones, thus turning the partial differential equations of the system into ordinary differential equations which could be readily solved. A typical example of this (cf. *Barenblatt*, 1979) is the heat conduction equation, describing a finite amount of heat E which is provided instantaneously in a medium with specific heat c and diffusivity k. The temperature increment T at a time instant t at a point at a distance r is

$$T = \frac{E}{c(2\sqrt{\pi kt})^3} \exp\left(-\frac{r^2}{4kt}\right) \tag{3.7}$$

Rewriting this equation in the adimensional form $T/T_0 = f(r/r_0)$ and, in particular, if

$$T_0 = \frac{E}{c(kt)^{3/2}} \tag{3.8}$$

and

$$r_0 = \sqrt{kt} \tag{3.9}$$

the equation takes the form

$$\frac{T}{T_0} = \frac{1}{8\pi^{3/2}} \exp\left[-\frac{1}{4}\left(\frac{r}{r_0}\right)^2\right] \tag{3.10}$$

which does not depend on time.

Note how in this case, rewriting the equation in self-similar form requires that we first rewrite it in adimensional form. This is a general requirement, and the basic tool in writing the equations in adimensional form is therefore the simple classical *dimensional analysis*. The latter expresses each physical variable in terms of monomials of the basic quantities length, mass, time, electric charge. Its use to rewrite the equations in adimensional form is also the classical tool to reduce the number of independent variables, and thus the number of data needed to make an experiment meaningful, to the minimum possible (cf. *Bridgman*, 1931).

Unfortunately, dimensional analysis does not allow us to write self-similar equations in all cases, and further detailed analysis is in general required, usually leading to nonlinear eigenvalue problems (*Barenblatt and Sivashinskii*, 1969). The general existence of self-similarity only under particular values clarifies its real meaning: self-similarity is equivalent to *intermediate asymptotics*, which means (under these conditions) the independence of the initial and/or boundary conditions, with the system far from its limiting state.

This tool proved exceptionally effective in studying phenomena like nuclear detonations (which are also an example of a supercritical process), and the shock waves that they induce in the atmosphere. Formulated in terms of the *renormalization group* theory (*Wilson*, 1983; *Binney et al.*, 1992) it also allowed successful unified statistical mechanical descriptions. However, this tool leads in general to difficult mathematical problems, and its (possibly great) potential in the study of earthquakes has not yet been exploited. Meanwhile, another simpler aspect of self-similarity has progressively gained popularity. It relies on a very common empirical property, which is *geometrical scale invariance*.

3.3   Scale invariance: the geometrical approach

In geology and geophysics it is very common to encounter morphologies that are scale invariant. For instance, in a photograph of geological outcrops one cannot detect the scale of length unless a reference object, like a hammer or a lens cover, is included in the image.

It is worth noting that in this case self-similarity is identical to the analytic case, but concerns a geometric variable V which depends on a parameter x which, under an arbitrarily change in scale

$$x \to ax, \tag{3.11}$$

transforms as:

$$\dot{V} \to bV \tag{3.12}$$

where both a and b are numbers. Its solution, as can be immediately verified by substitution, is

$$V = cx^{-[\log b/\log a]}, \tag{3.13}$$

where c is a constant. As a consequence, the relation

$$\frac{V(ax)}{V(x)} = a^{-[\log b/\log a]} \tag{3.14}$$

is also valid. This illustrates the fundamental link between power laws and self-similarity. In fact, in logarithmic scale the transformation (3.11) is equivalent to

$$\log x \to \log x + \log a \tag{3.15}$$

and

$$\log V(\log x) \to \log V(\log x + \log a) + \log b \tag{3.16}$$

which means that a *scale transformation* of $\log x$ is simply equivalent to a *translation* of $\log V$ (cf. *Feder*, 1988; *Sornette*, 1997).

A new geometry based on this morphological property was developed a few decades ago. Its basic element is the increase in length as a function of the decrease of the scale of observation (*Mandelbrot*, 1977, 1982).

$$L(\varepsilon) \propto \varepsilon^{1-D} \tag{3.17}$$

where $L(\varepsilon)$ is the length of a line measured using a segment of length ε and D is termed the *fractal dimension* . Unlike the Euclidean dimension, the value of D can be a real number rather than just an integer (cf. also the *Caputo fractional derivative* in *Caputo*, 1969). The latter equation is valid but not helpful generally (see e.g. the Cantor dust example). Instead one has to cover the object with E dimensional spheres of size ε (where E is the *embedding dimension*, i.e. the dimension of the Euclidean space in which the fractal is contained) and count the number N of spheres needed. Then $N \sim \varepsilon^{-D}$ (e.g. for a square we would have $N \sim \varepsilon^{-2}$). This general relation holds true for any self-similar object, and these obviously include the special case of Euclidean objects, for which case the fractal dimension is an integer and equal to the Euclidean one. The prototype of connected fractal forms is the *Koch curve* (figure 3.4), while the prototype of the unconnected (point-like) fractal forms is the *Cantor set* (figure 3.5), which can be constructed through self-explanatory iterative procedures.

An immediate extension of self-similarity is self-affinity, i.e. the appearance of self-similarity only after anisotropic magnification. Also in this case it is possible to use a fractal description.

3.3.1   Measuring an object's fractal dimension

While the fractal dimension D for synthetic fractal curves can be calculated exactly from its recursion relations, for real objects D must be estimated using special algorithms. At their root there is a *log-log plot* together with the well-known but often uncritically applied mathematical technique: *linear regression*.

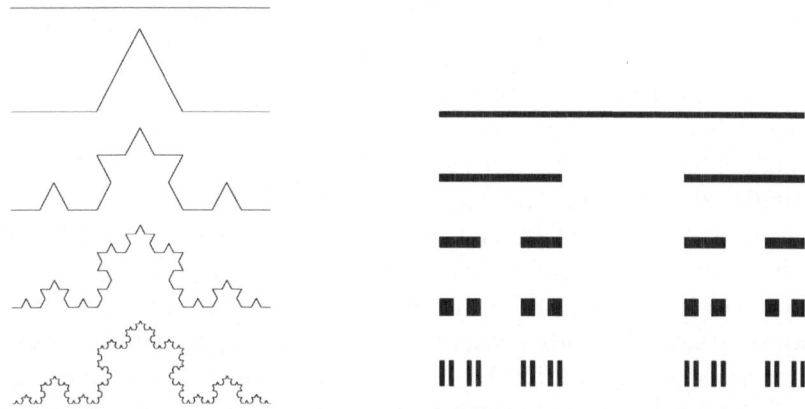

Figure 3.4: A Koch curve.   **Figure 3.5:** A Cantor dust.

The most common algorithm for measuring the fractal dimension of a given picture is *box counting* (BC) (e.g., *Mandelbrot*, 1982). This procedure is based on the notion of 'covering', i.e. if we cover the object with $N(\eta)$ boxes of linear dimension η, then in the limit $\eta \to 0$ it should scale as:

$$N \propto \eta^{-D}, \tag{3.18}$$

where D is the fractal dimension. The latter equation is easily rewritten in the linear form

$$\log N(\eta) = a - D \log \eta. \tag{3.19}$$

The covering is obtained by considering a uniform partition (generally dyadic) made of nonoverlapping boxes of radius η, and then counting the number of boxes $N(\eta)$ that have a nonempty intersection with the set. Automation of BC requires the digitization of the sets (cf. *Gonzato et al.*, 1998, 2000) and the fractal dimension D is generally estimated through a least squares linear fit.

Starting in the early 80s (*Von Seggern*, 1980; *Aki*, 1981; *Caputo*, 1981), the applicability of fractals to geology and geophysics has been the subject of a burgeoning literature, including textbooks (e.g., *Turcotte*, 1992; *Xie*, 1993). Fractals have been inferred for a wide variety of natural morphologies.

One of the main problems in fractal analysis is the lack of bandwidth in the primary observations. Rarely do reported values of D have the three orders of magnitude needed to soundly support a power law behavior (*Feder*, 1988; *Malcai et al.*, 1997; *Bonnet et al.*, 2001; *Ciccotti and Mulargia*, 2002).

For earthquakes in particular, the hypocenters (as well as the loci of fracture emission in the laboratory) have been found to be distributed fractally (*Kagan and Knopoff*, 1980; *Hirata et al.*, 1987), with some mild tendency towards alignment (*ibid.*). Earthquake data typically cover several orders of magnitude in terms of source length and hence can be regarded as comparatively well defined fractal sets, at least for small and intermediate sized events. Real faults have also a fractal geometry, with a typical fractal dimension of 2.6–2.7 for systems and 2.2 for single faults (*Sahimi et al.*, 1992). However, the latter appear to be only approximately self-similar. Furthermore, the fractal dimension of faults seems to approach 2 (i.e. the Euclidean geometry of a plane) in the range around 1 km, although even in this case the faults have a fractal and not planar geometry (*Aviles et al.*, 1987; *Okubo and Aki*, 1987). A complex fractal behavior is apparent on restricted ranges interrupted by several cutoffs given by characteristic lengths, which correspond to well identified geologic units: 1.4–1.8 m thickness of the outcropping sandstone beds, 700 m thickness of the sandstone formation, 6.5 km thickness of the sedimentary basin, 21 km bottom of the semi-brittle crust and 43 km bottom of the plastic region (*Ouillon et al.*, 1996). Some fractal models of faults have been proposed, although in the physical rather than geophysical literature (cf. *DeRubeis et al.*, 1996; *Hallgass et al.*, 1997).

3.3.2 Multifractals

A generalization of the fractal concept is the *multifractal*. The definition of the latter can be tied to the *box counting* method of measuring fractal dimension, and attaches a measure $p_i(l)$ to each nonempty box, i.e. to each box i of side l containing some part of the object (see above). Then the moments of order q can be calculated

$$M_q(l) = \sum_{i=1}^{n(l)} p_i^q(l),$$ (3.20)

where $n(l)$ is the number of nonempty boxes and $\sum_{i=1}^{n(l)} p_i(l) = 1$. For self-similar measures, $M_q(l)$ scales as

$$M_q(l) \propto l^{(q-1)D_q},$$ (3.21)

which define the generalized fractal dimension D_q, with D_0, D_1, and D_2 representing respectively the *capacity* (or simply *fractal*), *information* and *correlation* dimensions. As q increases the moments m_q are controlled by the most densely filled boxes since their relative 'weight' increases provided the measure used is simply the density. Therefore, the D_q provide information about the clustering properties of the fractal set. As a consequence, the prime use of multifractals is in quantifying the degree of concentration, clustering properties, and intermittency in the fractal set of interest. Multifractal measures have been profitably used to describe the geometrical properties of faults, fractures, and earthquakes (*Geilik-man et al.*, 1990; *Legrand et al.*, 1996; *Bonnet et al.*, 2001; *Herrmann and Roux*, 1990) and also in the laboratory (*Davy et al.*, 1992).

3.3.3 The empirical origin of fractality

In phenomenological terms, the basic issue is that a fractal morphology seems to spontaneously originate from hierarchical fracture processes, the prototype of which is repeated creep on a homogeneous brittle material (*King and Sammis*, 1983). The basic experiment in this case was performed on an ice sample which originally did not contain any internal flaws (*Weiss and Gay*, 1998). It was observed that stress concentrations develop from elastic and even more so, plastic anisotropy in grain size. At the beginning the distribution of crack length showed just a peak corresponding to the grain size, then it became a power law on a limited scale range. As fracturing proceeded during secondary creep, a hierarchical organization emerged progressively, leading to a full scale self-similar behavior.

Another interesting attempt was made by Vicsek and collaborators who studied the morphogenesis of mountains due to erosion (*Czirok et al.*, 1993) through the following model experiment: a granular pile with the shape of a ridge was placed on a table which was then watered by sprayers. The water loosened the pile and avalanches slid down leaving a new morphology for the pile. A statistical analysis of the video recordings showed that all phenomena were governed by power laws: the distribution of sizes, velocities and correlations in the resulting landscape. This finding is just another realization of fractality. Surprisingly enough the scaling exponent of the correlations agreed well with that of the Dolomites (*ibid.*).

3.3.4 Deterministic low-dimensional chaos: hope for predictability?

A concept closely tied to fractal geometry is that of *deterministic chaos*, which asserts that some simple nonlinear deterministic equations, under particular conditions, may yield a highly erratic behavior which might appear similar to the 'usual' stochastic chaos, but which is actually predictable.

The temporal evolution of a dynamical system is best studied in phase-space. Most dynamical systems evolve asymptotically with time into simple geometrical loci in phase-space, called *attractors*, which are either points, circular orbits, or toroidal orbits, and can also be projected in lower dimensions to give *Poincaré sections*.

Lorenz (1963) discovered a new type of attractor that develops in a simplified model of thermal convection in the atmosphere. His set of nonlinear differential equations had no analytical solutions, but numerical integration shows that some particular values of the parameters of the system produce a diverging amplification of tiny differences in the initial state, leading to the famous conclusion that a butterfly wing flip in Africa might drastically affect the weather in South America a month later. As in all systems exhibiting deterministic chaos, the attractor of his equations had a fractal geometry and was called *strange*, in contrast to the Euclidean 'point' or 'limit cycle' attractors for more linear systems.

Deterministic chaos originates due to high sensitivity to the initial conditions which are affected by unavoidable uncertainties. A typical example is the *logistic equation* for a generic variable X

$$X_{n+1} = aX_n(1 - X_n) \qquad 0 \le a \le 4, \tag{3.22}$$

which yields chaotic behavior when the parameter a exceeds 3.5699.

Most of the processes in Earth sciences are characterized by an 'erratic' time evolution. Up to the last decade, this irregular behavior could only be ascribed to the stochastic nature of the system, i.e. interpreted as a system with a large number of degrees of freedom. The discovery of deterministic chaos suggested a possible alternative interpretation: the irregular behavior might be due to some particular nonlinear deterministic system with a few degrees of freedom. If this were the case, such a system would have the appealing property of being predictable, at least in the near future, if it were possible to decipher its rules. For this reason, the discovery of deterministic chaos aroused considerable enthusiasm in the scientific community (c.f. *Tsonis*, 1992). Unfortunately, however, hopes for linking such low-order mathematical systems to reality have progressively faded. The reasons appear to be that (1) detecting deterministic chaos in a real system is difficult, even if it exists at all, and (2) even if it were positively identified it would be extremely difficult to guess the constitutive equations for the simplified low-dimensional system.

Attempts to detect deterministic chaos in a time series are generally based on a phase-space reconstruction aimed at calculating the properties of a supposed underlying attractor, such as the fractal dimension and the *Lyapunov exponents* (similar to the critical exponents; see e.g. *Grassberger and Procaccia*, 1983) or are directly tied to the predictability of deterministic-chaotic systems. The latter has the peculiar property of depending on time. At short times it is similar to that of deterministic systems and allows reliable deterministic predictions. At longer times it progressively degrades to match that of stochastic systems (e.g. *Farmer and Sidorovich*, 1987; *Sugihara and May*, 1990). In this sense the predictability of the systems showing a deterministic-chaotic behavior are similar to autocorrelated processes, the best example of which in everyday life is probably meteorological weather. It is therefore worth emphasizing that, contrary to what has sometimes appeared in the geophysical literature, the presence of deterministic chaos would imply *some* degree of predictability which could be achieved either empirically, by carefully studying the series of past events to find the sequence closest to the present one, or, in a more optimistic way, by correctly guessing the constitutive equations.

In geophysics the presence of deterministic chaos has been suggested for earthquakes (*Huang and Turcotte*, 1990), volcanic eruptions (*Sornette et al.*, 1991), and geomagnetic inversions (*Dubois and Pambrun*, 1990). In general, detecting deterministically chaotic behavior in a geophysical time series is difficult because real sets of data are contaminated by unavoidable experimental errors and contain therefore an unavoidable, possibly strong, stochastic component. Failing to explicitly account for this can lead to the fallacious identification of deterministic chaos as an artifact of statistical fluctuations. The use of appropriate methods for detecting deterministic chaos has so far always produced negative results for real data (*Marzocchi et al.*, 1997). In short, there is no case in which deterministic chaos has been found to play a role in a seismic or volcanic data series. As a consequence, deterministic chaos remains an interesting and stimulating concept, but so far has not been applied successfully to the earthquake problem. One reason for this may be that earthquakes result from the co-operative behavior of a system with many degrees of freedom, rather than a few.

3.4 Characterizing scale-invariant systems

Several tools are available to study scale invariance. We have already seen one of them, the analytical approach, which consists of studying the specific form of the adimensionalized constitutive equations of the process, and cannot therefore be discussed in general. The only application which has apparently been so far attempted to the earthquake problem (*Anifrani et al.*, 1995; *Saleur et al.*, 1996) made use of the *renormalization group* approach (see e.g. *Binney et al.*, 1992), and obtained a solution predicting that the close approach of a major event was characterized by a cluster of smaller events with a rate increasing as a power law with a superimposed lognormal oscillation. This behavior was apparently observed in the acoustic emission events in the close proximity of the failure of pressurized spherical tanks made of different materials ranging from composite fibers to various metals (*ibid.*). If this were applicable to large earthquakes it would in principle open some hope for predictability. However, the above behavior shows up only in the close proximity of failure (within a few percent of the failure load) and in the Earth's crust the domain that will fail is neither determined in advance nor accessible to direct observations. As a consequence, detecting such a behavior appears to be an unrealistic goal; in practice, when confronted with real earthquake data, this hypothesis did not pass formal validation testing (*Gross and Rundle*, 1998).

3.4.1   Log-log plots

In the empirical approach, we have seen above that the most general tool is inter-scale analysis of appropriate functions (like box occupancy in the Box Counting, cf. section 3.3) by log-log plots and linear regression analysis. The latter, simple as they may be, nevertheless hide a few traps.

In any case, even if appropriately performed, log-log plots are inadequate when a detailed characterization of scaling properties, and especially of multi-fractals, is required. A more effective tool, in this case, is the *wavelet transform*.

3.4.2   Wavelets

The wavelet transform (WT) was firstly developed to provide a representation of seismic traces (*Goupillaud et al.*, 1984) with the properties of keeping the different frequency bands reasonably separated without excessive loss of resolution in the time domain and to allow a reconstruction of the original function with arbitrarily high precision.

The representation provided by the WT is a compromise between the pure time content of the signal itself and the pure frequency content provided by the Fourier transform. Instead of representing the signal on a basis made of per-fectly localized delta functions, or of infinite plane waves of given frequencies, the wavelet transform provides a basis constituted of scaled fast decaying oscillat-ing functions named *wavelets*. The time and frequency resolution of each wavelet are mutuated by an equivalent Heisenberg uncertainty principle for the space-time domain $\Delta t \cdot \Delta \omega \geq 1/2$.

The WT appears very suitable for analyzing complex signals with fractal or multifractal properties, characterized by localized features at different scales. This is often the case for geophysical signals (*Kumar and Foufoula-Georgiou*, 1997). In the very simple example of a Cantor signal (see figure 3.6) the WT allows one to identify the characteristic scales and features of the branching process that generates this fractal signal.

This principle has been developed by *Muzy et al.* (1993) to provide an alterna-tive method for determining the multifractal spectrum of singularities of a signal. The method consists of calculating a suitable partition function on the wavelet coefficients taken along the modulus maxima lines and using the thermodynamic link between the singularity spectrum $f(\alpha)$ and the multifractal moment function $\tau(q)$ (e.g. *Vicsek*, 1992).

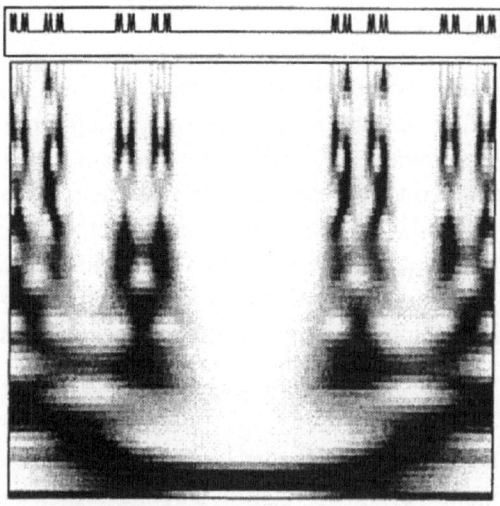

Figure 3.6: Modulus of the wavelet transform of a Cantor set. The horizontal axis represents location, while the vertical one represents scale.

Due to the high computational cost of a continuous wavelet transform (CWT), a discrete version (DWT) has been developed with the great advantage of using orthonormal bases of wavelets and very fast computational algorithms (*Daubechies*, 1992). This representation is optimal for multiscale filtering, image processing, feature identification, data compression (it is presently the standard for image coding on the World Wide Web) and also provides a suitable basis for solving non-linear partial differential equations on irregular domains (*Benedetto and Frazier*, 1994). Geophysical applications are widespread, including downhole logging measurements, petrophysical core measurements, ocean wind waves, land surface topography, seafloor bathymetry, marine seismic data, transport in heterogeneous porous media (*Foufoula-Georgiou and Kumar*, 1994).

Other multiscale analysis methods are under development, like for example the Detrended Fluctuation Analysis (DFA, *Kantelhardt et al.*, 2002), and these are often more efficient than wavelets.

3.5 Modeling scale invariant systems

There are several possible options for modeling scale invariance. We will start with the simplest, which is *percolation*.

3.5.1 Percolation

Rather than being an ordinary model, percolation is a consistent description of the clustering properties of a random binary system (cf. *Stauffer and Aharony*, 1994). Let us take the Ising spin model (cf. section 3), where in the earthquake analogy the white areas are assumed to represent a rupture surface. For each cell one state corresponds to the fractured condition (white) and the other to an intact condition (black). In a percolation problem, elements in a d dimensional lattice are filled at random with a fixed probability p so that a variety of different geometrical forms may result. In two dimensions the number n of connected clusters, each one of size s, is:

$$n(s) \propto s^{-\tau} f(cs), \tag{3.23}$$

where f is a function with a sharp characteristic cutoff similar to an exponential function and c is a function of the correlation length ξ, which can be written as:

$$\xi^2 = \frac{\sum r^2 g(r)}{\sum g(r)}, \tag{3.24}$$

where $g(r)$ is the *correlation function*, that is the probability that a site at distance r from an occupied site is also occupied and belongs to the same finite cluster. As p increases there is eventually a finite probability that a cluster crosses the whole model space. At the *critical point*, p_c, the correlation length diverges as (cf. equation 3.1):

$$\xi \propto |p - p_c|^{-\nu}, \tag{3.25}$$

while

$$c \propto |p - p_c|^{1/\sigma}, \tag{3.26}$$

where ν and σ are the critical exponents of the system. At this point rupture events of any size can occur. It can be shown that the fractal dimension of the largest (incipient infinite) cluster D is

$$D^{-1} = \sigma \nu \tag{3.27}$$

which yields $D = 1.896$ for a two-dimensional system and $D = 2.5$ for a three-dimensional one. Based on deep analogies, one can identify the percolating phase with the ordered one and the non-percolating with the disordered one. This is also expressed by the percolation order parameter which is the probability that a site belongs to the infinite cluster. From (3.23) the rupture area in percolation theory and hence the local energy take the form of a gamma distribution, i.e. a power law for low s or E and a decreasing exponential at higher values (cf. *Main et al.*, 2000).

As a modeling tool percolation is very simple, and describes essentially the statistical properties of a system showing a phase transition while completely ignoring the local physical interactions.

Another simple model which has its roots in percolation is the *epidemic model* or *invasion percolation* (*Herrmann*, 1986). The basic concept here is a cluster of sites which is grown probabilistically on a regular lattice, each site being either occupied (fractured) or intact (barrier). Growth takes place along the boundary of the cluster and stops when the entire cluster is surrounded by barriers. If the probability of fracture depends on the state of its neighbors, then we have a correlated epidemic model. Epidemic models have properties similar to those of thermodynamical systems at criticality. The cluster size distribution at criticality has the same exponents as percolation. Along this line are the Epidemic Type Aftershock Sequence (ETAS) models (*Ogata*, 1999). The main difference is that in the ETAS models power-law statistics is assumed rather than emerging as a property of the model.

3.5.2 Cellular automata

While percolation is able to describe the basic features of a system in a critical state, its application to the dynamics of the Earth's crust with respect to earthquake generation appears too schematic. Analytic models, which consist of a compact set of differential equations, appear unsuitable to represent the many-body interactions which govern rupture dynamics. Therefore, the key to modeling earthquake criticality relies on computer simulation using *Cellular Automata* models. The latter, hereafter denoted CA, are conceptual (or physical) devices composed of a discrete number of components which can individually assume certain states and respond to given stimuli (inputs) according to predefined rules. Typically, a CA is a computer algorithm which simulates a lattice of cells the behavior of which is ruled by an appropriate set of laws describing their interaction and the external forcing.

Let us assume that the grids represent either a fault surface, or a portion of the Earth's crust. The set of cells should then in principle reproduce the geometric space, that is they should be three-dimensional, but this is not realizable due to present limitations in computing power. As a consequence, CA models typically consist of two-dimensional square grids of some $10^3 \times 10^3$ elements.

Let us assume that the value of each cell represents the proximity to rupture (originated by strain or any other physical or chemical effect), and let us assume that the cells are loaded according to some function at each time step, up to a given threshold. When this threshold is reached, they rupture, losing a given amount of energy which is redistributed on the neighboring cells. The redistribution might involve only the nearest neighbors, following the rationale that fracture is a local phenomenon, or might involve long range interactions, following the rationale that elastic interactions decrease slowly with distance r as $1/r$. In addition, the behavior of the cells can be based on empirical rules or follow the constitutive equations of some given phenomenon, typically those of stick-slip (see below), while in the following time steps there can be additional redistribution of strain according to time-independent (elastic) or time-dependent (anelastic) rules. In this process, a certain amount of energy can also be assumed to be lost by thermal dissipation. In general, some parameters of the system must be tuned to achieve critical behavior. If this is the case and the parameters are set precisely to the critical values, the system is said to be externally *organized* or tuned *critically* (OC).

Slider-block cellular automata models

Cellular automata for earthquakes have been used primarily in attempts to reproduce the Gutenberg-Richter law. The first model to be proposed, known as the *slider-block* model (*Burridge and Knopoff*, 1967), consists of a set of spring-mass harmonic oscillators undergoing stick-slip (see below), for which the constitutive rules are given by the theory of elasticity and the laws of friction. An appropriate set of differential equations can be derived and solved.

This model was originally developed to provide an analog for the lithosphere as an arrangement of discrete spring-block slider elements, sandwiched between two rigid plates that are driven at a constant velocity V. The two plates are connected by leaf springs to the discrete fault elements with constitutive laws which represent the elastic-brittle frictional properties of a pre-existing two dimensional fault. The blocks are also connected to one another by coil springs (see figure 3.7), so that the stress drop caused by the failure of one element is immediately redistributed to its four nearest neighbors. The size of an event can be counted as simply the number of the connected slipped blocks or, to mimic seismic moment, as the total area of connected elements multiplied by the slip and the rigidity of the leaf springs. This model was the first, with its strongly nonlinear nature, capable of reproducing a Gutenberg-Richter type law.

Many variants of this model have been proposed and their performance compared to various aspects of observed seismicity (*Carlson and Langer*, 1989; *Carlson*, 1991; *Bak and Tang*, 1989; *Nakanishi*, 1990, 1991; *Brown at al.*, 1991; *Rundle and Jackson*, 1977; *Steacy and McCloskey*, 1999; *Main et al.*, 2000).

In no case was statistical validation performed; the above models produce only a semi-qualitative agreement with reality. Various attempts to improve the agreement by adding 'realistic features' like rate- and state-dependent friction, proxies for crustal fluid migration, lattice heterogeneities to model barriers or asperities (*Lomnitz-Adler and Lemus-Diaz*, 1989; *Knopoff et al.*, 1992; *Rundle and Klein*,

1993; *Ben-Zion and Rice*, 1995) have been made. Other efforts along this line attempt to mimic the spatiotemporal clustering of the number of earthquakes and stress release, the scaling of average slip with seismic moment, and the dependence of seismicity on depth, length, and structural heterogeneity of the faults, by adding realistic features like viscous coupling to asthenosphere (*Hainzl et al.*, 1999; *Pelletier*, 2000).

The bridge models

As an asymptote to the latter line of approach, a class of CA models has been also proposed which directly attempt an application to macroscopic scale, that is they do not attempt to study the basic microscopic processes of rupture, but rather they assume that these are known and follow the classical continuum mechanics models, and use the CA only to mimic the spatial complexity of real crustal regions. As such, they represent a 'bridge' between the classic and the new physics approaches. The prototype of such models are those of *Ward* (1991, 1996, 1997).

Rather than starting from a model region of didactic geometry (typically bidimensional and square), they hypothesize that the basic physics is known and proceed to model a crustal region by assuming that it is completely known in detail, regarding structure, faults, stresses and strains, and externally applied loads. In this case the cells are therefore only a discretization of the real Earth. Computer simulation is then used to mimic its evolution, accounting for fault interaction over many earthquake cycles. Quite obviously, such models suffer from all of the drawbacks of the physical model they adopt and result in an amazingly large number of parameters (with 10–20 parameters per fault segment, and a total of 600 segments, a total number of parameters of the order of 10^4 results), which occasionally allow outstanding fits. However, the latter are achieved solely in ret-

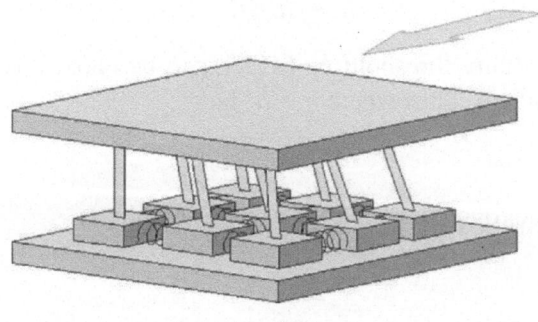

Figure 3.7: The Burridge and Knopoff model.

rospect and therefore are not statistically significant. Note also that such a high number of parameters would *de facto* inhibit any validation project unless the system were to self-organize into one with simple scaling laws (see also below). This would, in turn, make the reliable prediction of individual events an inherently untenable goal.

Self-organization

CA models other than slider-block have been developed, focused on the basic physical ideas allowing the exploration of a wide variety of options. An important observation which can be made on self-similar systems is that as the length scale diverges when self- similarity appears, there is always a related time scale which should also go to infinity (see section 3.1). That means we have temporal scale-independence as well as spatial scale-independence. Self-similarity might therefore appear to be a quite peculiar behavior, which can be achieved in general only by fine tuning the system. It is, however, a quite general phenomenon in nature even if possibly not as common as stated in the literature (see section 3.6). Also temporal scale invariance is quite common ($f^{-\alpha}$ noise, avalanche frequencies, etc.). A possible resolution of this apparent contradiction is that the critical point is a thermal equilibrium phenomenon while our world is mostly far from equilibrium. It is a driven system which operates as a class of simple CA models tending to self-organize on a critical state called *Self-Organized Criticality* or SOC.

A typical self-organized CA model (*Bak and Tang*, 1989), assumes that: i) the system is slowly driven, ii) spatial and temporal scaling go together, iii) activity occurs in bursts. The model laws in the simplest formulation are the following. Take a lattice with an integer variable h_i over each site i. 'Grains'[1] are added to the system at a randomly chosen site i:

$$h_i \rightarrow h_i + 1.$$

There is a local stability threshold h_c (which can be chosen as the number of neighbors on the lattice). Sites where $h \geq h_c$ fail:

$$h_i \rightarrow h_i - h_c$$

and the grains are distributed among the nearest neighbors:

$$h_{i+nn} \rightarrow h_{i+nn} + 1,$$

where $i + nn$ means the neighbors to i. Of course, it may happen due to the failure at i that a failure in turn occurs at a neighbor, and so on. This is how avalanches

---

[1] Grain is just an arbitrary word; the model has nothing to do with granular systems.

occur. A metastable steady state can be achieved if grains can leave the system at the boundaries (free boundary conditions). The observed behavior is indeed characterized by power laws (*Bak et al.*, 1988):

$$P(s) \sim s^{-\tau}, \tag{3.28}$$

$$P(t) \sim t^{-\alpha}. \tag{3.29}$$

Here s (t) means the number of failures (the duration) of an avalanche and $P(\cdot)$ is the probability. One can also define an average duration t_s of avalanches of size s which shows a power law dependence:

$$t_s \sim s^{\gamma}$$

where the three exponents are not independent:

$$\gamma(1 - \alpha) = 1 - \tau.$$

The numerical values of the exponents on the square lattice are: $\tau = 1.1$ and $\gamma = 0.68$. Equations (3.28) and (3.29) express the criticality of the system, i.e., its scale invariance. Note that spatial and temporal scaling do not necessarily mean there will be a nontrivial frequency dependence in the power spectrum (*Jensen et al.*, 1989; *Kertész and Kiss*, 1990). In fact, the sandpile model does have a trivial (f^{-2}) power spectrum in spite of the power laws (3.28, 3.29). What is seen, however, is that we do have criticality without fine tuning.

In other words, the model system organizes itself spontaneously to the critical point (the critical 'angle of repose') and then remains there apart from dynamic fluctuations (the avalanches). Although driven far from equilibrium by the constant flux of 'sand grains', the 'sandpile' remains in a stationary state represented by a relatively constant angle of repose and power law scaling of the size distribution of avalanches. An interesting feature is that small events can start a chain reaction that can affect any number of elements in the system. Individual avalanches are not predictable, but the average properties (angle of repose, size distribution of avalanches, etc.) remain relatively constant in time. Over long time-scales the angle of repose averages out at a constant value to maintain the critical slope, but short-term dynamic fluctuations are fundamental to maintain this 'long-term' metastability. One of the problems is the definition of the time scale, and this will be discussed in section 3.7.

It is not only sandpiles (*Carreras*, 2002; *Ceva and Luzuriaga*, 1998; *Medvedev and Diamond*, 1998) that have been taken as prototypes and modeled through CA, but also landslides (*Di Gregorio et al.*, 1999; *Turcotte et al.*, 2002) and lava flows (*Matos and Duarte*, 1999; *Spezzano et al.*, 1996). An analogy between the sandpile models and earthquakes was soon recognized by *Bak and Tang* (1989) who applied massless cellular automata to earthquakes, finding that they could reproduce the power-law statistics with a much simpler model than the slider-block models.

3.5.3   Earthquakes as SOC

Based on the above discussion, it is appealing to approach earthquakes from the point of view of SOC. Burridge–Knopoff models were transformed into the slowly driven framework of SOC by *Olami et al.* (1992), which can be formulated in terms of a sandpile model (*h* here represent forces):

$$h_i \rightarrow h_i + vt \quad \text{for all } i \tag{3.30}$$

is the growth step and

$$h_i \rightarrow 0 \tag{3.31}$$

$$h_{i+nn} \rightarrow h_{i+nn} + h_i(1-\Delta)/z \tag{3.32}$$

is the relaxation step. Here z is the coordination number (i.e. the number of neighbors) and Δ the loss or dissipation parameter; for $\Delta = 0$ we have conservation. This is a *deterministic* model where randomness comes solely from the initial conditions.

The model indeed leads to criticality, so the corresponding activity distribution is a power law like (3.28) with an exponent τ which is between 1.8 and 2.7. There is an ongoing debate on whether the exponent depends on the actual value of Δ; in any case, it is certain that τ is different from that observed in the model with conservation. If the activity is identified with the energy of an earthquake, a Gutenberg-Richter type law is obtained with $\tau = B + 1$ which leads to a quite reasonable value of $B \approx 0.8$. Similarly, temporal scaling leads to the Omori law with an exponent $p = 1$. The moments of the distributions are determined by the upper (finite size) cutoffs. An important consequence of this model is that it suggests *unpredictability* of single occurrences.

It should be emphasized at this point that there is ongoing research related to this model (*Christensen et al.*, 2001; *Bak et al.*, 2002). We have already mentioned the dependence of the exponents on Δ. Another relevant and nontrivial issue is that of the boundary conditions. It was shown some time ago that disorder in the transition thresholds drives the system off criticality, though one would expect some robustness in this respect from an adequate model (*Jánosi and Kertész*, 1993).

Other cellular automata models which have been remarkably successful in describing fracture growth are the random fuse network and the Diffusion Limited

Aggregation (DLA) models (see e.g. *Feder*, 1988; *Herrmann and Roux*, 1990; *Zapperi et al.*, 2000).

3.6 The origin of power laws and fractality

We have seen that power laws and fractal geometry appear to be ubiquitous. What produces this behavior? Is criticality and its related scale invariance the only candidate? The answer is no. We discuss this problem below.

3.6.1 Scale invariance: artifacts and reality

One possibility is that the observed behavior is an artifact. Scale invariance in natural datasets often occurs in narrow ranges of at most two decades (*Malcai et al*, 1977). Earthquakes seem to be one of the notable exceptions (see below).

Let us remember that since the most common algorithms for fractal analysis (walker's ruler, box counting, Minkowsky sausage, etc.) ultimately resort to a linear fit to experimental data points, it is always possible to obtain an apparent fractal dimension over some small range. However, this result might be wrong or physically meaningless if one or more of the following conditions apply (*Gonzato et al.*, 1998, 2000; *Hamburger et al.*, 1996; *Ciccotti and Mulargia*, 2002): (1) inadequate linear fitting to the data with either too few data points, or nonrandom trends in the residuals; (2) bias inherent to the methods of fractal analysis (remainders, pixelization, saturation, omission of tracing and zooming); (3) bias induced by the presence of physical cutoffs.

The first problem is obvious, since any curve can be approximately fitted by a linear segment over a restricted range. However, an incorrect fit will be revealed by an appropriate rigorous statistical test. The second problem arises from the process of image analysis due to framing and digitization. The third problem is that power law scaling over a finite range is meaningful only if the range is large enough. The latter depends on the application. Consider for example the following case, in which an integer dimension is of interest. The scaling of the surface of a table is two dimensional over a finite range from about 0.1 mm (the size of machining asperities) to 1 m (the size of the table). Although the scaling range in which a table is described as an Euclidean plane is limited to four decades, the utility of such a model for most practical applications is unquestionable. Consider now a table covered with small pebbles of 10 mm radius cemented by a loose paste. Then, the applicability of an Euclidean plane model would be reduced to a couple of decades.

In general, any real object has different scaling properties over different ranges of scales, some of which may be modeled by a fractal dimension, which occasionally has an integer value related to an Euclidean geometry (see the above example). The scales that separate different behaviors are generally related to some characteristic lengths of the object (such as the size of the atoms, the size of some characteristic structure or the extension of the object) or of the measuring process (such as the spatial resolution of the instruments, the minimum size of the features that were chosen in the drawing process or the size of the measured region). The transition points, which correspond to physical cutoffs, are not sharp, and it takes about one or two decades before the scaling properties can be represented by a different dimension.

As a consequence, several problems can arise from applying a linear fit to a short range of data in the transition zones, which can result in a fractional dimension that is useless for modeling. For example, *Hamburger et al.* (1996) have shown that an apparent fractal dimension extending over a range up to two decades emerges as an artifact when a set of randomly distributed balls is analyzed by the box counting and Minkowsky sausage methods in the limit of low coverage. The effect is more general and typically spans a range of two decades in the linear dimension L (*Ciccotti and Mulargia*, 2002). The conclusion is of a rule of thumb type: caution should be exercised in attaching significance to morphologies for which a power law scaling is observed over less than three decades on the metric L; this is hardly the case in many published studies (obviously, using L^2 as yardstick would double the scale invariant range). One should therefore apparently conclude that power law scaling, and therefore fractality is a very interesting concept which should, however, be used with extreme care.

Global CMT data are complete only for magnitudes of about 5.5 and greater and, as discussed in chapter 2 and by *Kagan* (2002), do not follow a strict linear power law, especially for events with $m > 7.5$. In order to study the Gutenberg-Richter law over a wide magnitude range we therefore use data from the Southern California Earthquake Center (SCEC) catalogue `http://www.scecdc.scec.org/ftp/catalogs/SCEC_DC/` which is well suited to this purpose (cf. *Bak et al.*, 2002).

The local magnitude, m_L, is typically used to quantify earthquake size in the SCEC catalogue. We write the empirical relation between m_L and the seismic moment M (see equation 2.1) in the form

$$m_L = C_1 \log_{10} M + C_2. \tag{3.33}$$

Ben Zion and Zhu (2002), using data for events in Southern California with $m_L \geq 3.5$ (seismic moment is not routinely measured for smaller events), found $C_1 = 0.74 \pm 0.02$. We will use the approximation $C_1 = 3/4$ in the following discussion.

The Gutenberg-Richter law for earthquake size (2.3) is not a Pareto law but since magnitude is a logarithmic estimator of the wave amplitude, it should be interpreted as a Pareto law in terms of the variable 10^m which is found to span more than five decades (see figure 3.8).

Magnitude is not a 'length' variable, but its relation to the metric s is tied to the fact that the linear source dimension s can be derived by considering that the scalar seismic moment M scales as s^3 for all but the largest earthquakes (cf. section 2.3.2). Combining this with equation (3.33), we deduce that the magnitude yardstick 10^m scales as $s^{9/4}$ and that scale invariance for the seismic source dimension s is therefore apparent over a little more than 2 decades.

Let us now consider equation (2.12):

$$\log N = -3\frac{b}{c}\log s + \text{const} \tag{3.34}$$

This equation, together with the definition of fractal dimension D (cf. 3.19)

$$\log N(s) = -D\log s + \text{const} \tag{3.35}$$

yields the result that, for the 'classical' values $b \approx 1$ and $c = 2/3$ (cf. equations 2.16–2.18 and figure 2.10), the seismic sources are distributed with a fractal dimension equal to 2, which is equal to the Euclidean dimension, i.e. they are distributed on a plane. While this is obviously consistent with the moment tensor formulation which is based on Euclidean geometry, it is inconsistent with the evidence that seismic sources are generally distributed according to more complex geometries (cf. *Von Seggern*, 1980; *Caputo*, 1981; *Fisher et al.*, 1997). While the comparatively small set of data analayzed demands further confirmations, the experimental values measured for the SCEC catalog appear to provide a consistent solution to this apparent paradox. The Gutenberg-Richter b value is again equal to about 1, which, together with the above value of C_1 gives

$$\frac{b}{c} = bC_1 \approx \frac{3}{4} \tag{3.36}$$

with a genuinely non-integer dimension for the distribution of seismic sources, $D \approx 9/4$.

3.6.2 Do power laws always mean geometrical scale invariance?

Scale invariance is not the only physical mechanism that can cause power law distributions in size. They have been found for so many and such a diverse group of phenomena, such as the areas burned in the largest fires, web file transfers,

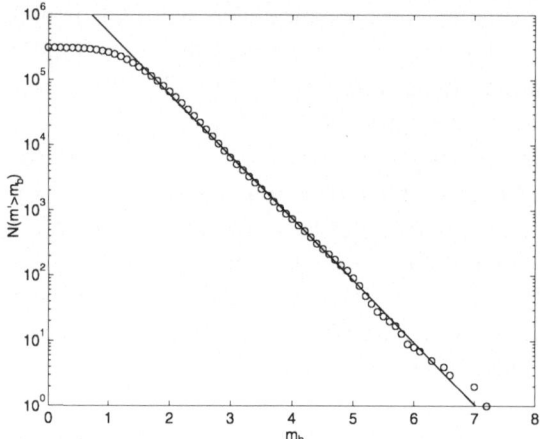

Figure 3.8: Cumulative magnitude distribution for South California earthquakes between years 1984 and 2000 (SCEC catalogue). The solid line is the Gutenberg-Richter law fitted for $m_b > 2$: $\log_{10} N(m' > m_b) = a - bm_b$ with $b = 1.0$.

power outages, bibliometric statistics, species extinction, social conflict, automotive traffic jams, air traffic delays, financial market volatility, etc., that a single origin seems very unlikely. In addition to this, geometrical scale invariance appears not to be applicable to several of the above cases.

A recently proposed and well rooted alternative explanation for the ubiquity of power laws is Highly Optimized Tolerance (HOT) (*Carlson and Doyle*, 1999, 2000), which is based on the fact that a power law behavior is apparent in the catastrophic failures of highly optimized designs. Such systems achieve a great flexibility and robustness against the various disturbances which they are designed to withstand, but are fragile under unexpected disturbances.

The main argument of HOT is readily apparent in the Forest Fire models, which are two dimensional cellular automata models with a coupling to external disturbances represented by 'sparks' that impact individual sites on the lattice. Sparks initiate 'fires' when they hit a cell occupied by a tree, burning through the associated connected cluster. Fires are the rapid cascading failure events, all starting at one cell, but capable of extending to a wide range of sizes, depending on the configuration and the site that is hit. If the tree planting and spark throw occur at random, the spark-fire sequence leads to a SOC scale invariant system (see *Turcotte*, 1992, for references).

However, if one defines the yield Y as the remaining density after one spark, one can optimize the system to maximize it by planting trees not at random but under precisely designed configurations, which is what is done in practice for real forests through firebreaks. With an appropriate design a much higher density is achieved in this way with respect to the critical state and the system is much more robust against sparks. Yet, fires may occur and these may be very large if the system fails under unpredicted disturbances, like a defect in a firebreak. What is found is that a power law distribution in the size of fires applies in this case just as in the random case, but with a much higher density.

HOT differs substantially from SOC because it describes systems of inherently complex self-dissimilar geometry not amenable to fractals. This difference is very substantial, since SOC applies to renormalizable geometric structures function of a geometric variable V. SOC is thus able to describe simple systems for which no design optimization whatsoever is used, and large events are the result of random internal fluctuations characteristic of the self-similar onset of systemwide connectivity at the critical state. In HOT systems the power law statistics are one symptom of 'robust, yet fragile' behavior, reflecting tradeoffs in systems characterized by high densities and throughputs, where many internal variables have been tuned to favor smaller losses in common events, at the expense of larger losses when subject to rare or unexpected perturbations.

HOT applies therefore to systems which have been carefully designed, as occurs in industrial engineering practice. It remarkably occurs also in natural systems by biological species selection. However, HOT appears, in general, not to be applicable to natural systems such as earthquakes and other crustal processes, in which the random components dominate, and the experimentally observed geometrical scale invariance suggests SOC as a strong candidate.

3.6.3 General features of self-organizing cellular automata earthquake models

Let us attempt to summarize the common features of the variety of cellular automata models for earthquakes which have been so far presented according to the comparative studies which explored the effects of a large, if not exhaustive, variety of parameters, including initial conditions (homogeneous or heterogeneous), loading mode (homogeneous, random or wave-like), failure mode (total or partial stress drop), load redistribution (only nearest neighbors or also the other neighboring shells), and the presence and the amount of local dissipation (*Lomnitz-Adler* 1993, *Pelletier*, 2000; *Castellaro and Mulargia* 2001, 2002).

The following were found to be the common properties:

1. Systems starting from a homogeneous state as well as systems starting from a random heterogenous state evolve spontaneously into a metastable state.

2. The geometrical pattern of ruptured cells is fractal.

3. A power law scaling in size is exhibited by most CA models in a restricted range, with the largest events smaller than system size. Hence, sub-critical behavior appears in most cases.

4. The mean system 'energy' (i.e. the cell level averaged over the whole grid) shows a first transient (of up to $\sim 10^6$ realizations for $10^2 \times 10^2$ grids) during which the system is not under stationary conditions and then tends to a stationary or quasi-stationary metastable state.

5. Rupture occurs in clusters of events culminating with a mainshock, and around it there is a left tail of foreshocks and a right tail of aftershocks. Each event cluster is separated from the other clusters by a period of inactivity during which rupture does not occur.

3.7　Problems in applying CA models to earthquakes

There are several problems in applying CA models to earthquakes, beyond the fact, as mentioned above, they have not so far passed appropriate statistical validations.

The first problem of CA models is the difficulty of correctly reproducing the time scale of the phenomena involved. As we have seen, the CA models aim at reproducing essentially two realities: (1) the dynamics of a single fault and (2) the dynamics of a crustal region with several faults. The time scales involved in the two cases are different. Let us look at case 1, which involves times of $\tau_1 =$ one to several seconds for the rupture duration. There is then a second time scale, related to post-seismic stress redistribution, and related effects, which involves different and partly unknown relaxation processes and a time scale which can be empirically inferred from aftershock occurrence as of the order of hours to months, i.e. $\tau_2 = 10^4 - 10^7$ s. There is then a third time scale, τ_3, related to the reloading of the same fault patch by tectonics, and this one involves a time scale equal to the recurrence time, which palaeoseismology indicates as (cf. *Pantosti et al.*, 1993) thousands years, i.e. $\tau_3 = 10^{11}$ seconds. This implies the need to simulate the evolution of a system over 11 decades, an intractable task for any computer.

Case 2, i.e. the use of a CA to model the dynamics of a crustal region with several faults, disregards the rupture processes within each single fault and deals with the interaction among different faults under the forcing of a steadily increasing tectonic load. The period analyzed should include several recurrences on the same fault sufficient to infer some reasonable statistical estimates, and should therefore feasibly regard an interval of 10,000 to 100,000 yr. Note that in this case only τ_2 and τ_3 are involved, implying simulations over 8–9 decades. Provided that the model is not too complex and the number of operations required at each time step is not extravagant, this appears to be a realistic, albeit very intensive, task. In turn, what appears unrealistic in this case is the detail with which the geometry and the loads must be known (cf. the bridge models in section 3.5.2).

The existence of the above problems has been hardly acknowledged by published studies, in which the CA models generally use only two neighboring time scales. The first time scale, within a single time step of the algorithm, regards the occurrence of a rupture event; the other one, at the following iteration in the computer code, involves both the stress relaxation and the 'tectonic' reloading. This extreme 'compression' of the time scales over several orders of magnitude crucially distorts the dynamic picture, over-weighting the elastic with respect to the anelastic effects (cf. *Rice*, 1993; *Ward*, 1996). As a consequence, CA models are expected to be able to reproduce correctly only the size distribution of the events, and not its dynamics.

Finally, it is worth noting that laboratory work shares a similar problem and using laboratory data to match or calibrate CA models does not provide the correct time scale for *in situ* applications. In fact (cf. section 2.6), while typical strain rates for tectonic processes are on the order of 10^{-15} s^{-1}, laboratory strain rates are about 10^{-5} s^{-1}. This means that time-dependent processes may be severely underestimated, just as happens in CA modeling. The agreement between the latter two, and their collective distance from *in situ* results, is therefore not a surprise.

The second problem is the limited size of the models due to the limitation of present computing power. Grids up to about 10^6 cells can be analyzed, which in two dimensional sets avoids the most evident instabilities (much less so in three dimensional ones), but does not eliminate the questions about the dependence of the results on grid size (*Kinouchi and Prado*, 1999). In this respect, it is important to note that the implementation of CA models on parallel machines provides a comparatively limited improvement. The reason is that the process of energy redistribution following rupture is inherently sequential and cannot be fully parallelized. Nonetheless, sharing the memory resources among different processors makes it possible to simulate grids one order of magnitude larger than the sequential algorithms, but there is still no known way to significantly enhance the computational speed (*Castellaro and Mulargia*, 2002).

In addition to this, when simulating models, the computer capacities (memory and available CPU time) have to be taken into account. This is especially crucial for models where large characteristic sizes come into play as it is the case for critical phenomena. Since the correlation length ξ diverges at the critical point, no computer would be big enough to cope with the problem. Here physicists make a virtue of necessity: by looking at the dependence of the behavior *on the system size* one can extract important information. The related theory is called finite size scaling (*Privman*, 1990) and it constitutes the basis of any simulation of critical or scale free systems.

It is advisable to study the size dependence even if the system is not critical, since this is an efficient way of exploring the different characteristic lengths. As long as the simulated system size L is smaller than the characteristic size ξ we expect a strong dependence of the behavior on L; only as L considerably exceeds ξ will the results become independent of L.

A third problem is the difficulty of reconciling the continuum limit required by elasticity with the coarsening required by the SOC CA (*Rice*, 1993), although acceptable solutions can be found by either working on the model details (cf. *Cochard and Madariaga*, 1994) or adding viscous terms (*Shaw*, 1994).

Finally, another minor problem regards the movement of the driver plates relative to the movement of slipped blocks. Since the relative motion of the plates mimics the strain buildup between two tectonic plates, its time scale is much longer than that of the elastic rebound of the slipped blocks. In all computer models motion takes place in discrete jumps, and the most simple dynamics moves the driver plate forward by equal discrete amounts at successive time steps and relaxes all blocks whose elastic stress exceeds the frictional restoring force after the driver plate has been moved. It is possible for two or more unconnected patches of the fault surface to become unstable during the same time step and be counted as one earthquake (*Brown et al.*, 1991; *Rundle and Klein*, 1993). One way to correct this is to advance the driver plate only until the next block in the model becomes unstable.

3.8   Dynamical implications

What are the implications of the PCS approach? Does it yield new insights into the earthquake problem? Does it bring about a new practical capability to predict earthquakes? Let us now discuss these problems.

3.8.1   Intermittent criticality

The general criticality issue disregards the finite size effects which are present in any real system. As a consequence, (1) scale invariance can be expected on a restricted range of scales and (2) events which are large with respect to the system can be expected to change its dynamics. The first issue has been already discussed above. The second one has been formalized as the *intermittent criticality* issue (*Sornette and Sammis*, 1995), which gained some popularity in the geophysical literature because it is reminiscent of the characteristic earthquake hypothesis and of the seismic cycle. It postulates that a system is driven away from critical-ity by large events and then goes back to it after a considerably long reloading time. However, criticality indicates the asymptotic long term stable metastabil-ity of a system, and fluctuations around the critical point are physiological with a spectral amplitude distributed as $f^{-\alpha}$, where f is frequency. Establishing the exis-tence of deviations from criticality would imply the existence of abnormally large and systematic fluctuations in the correlation length. For the reasons discussed above, to have practically relevant results, this would require more detailed and extended CA simulations than those presented so far. When transferred to the Earth, intermittent criticality would imply systematic fluctuations in seismicity, with a marked decrease for recharging following a large event. However, estab-lishing the spatial domain on which this fluctuation should occur is extremely dif-ficult even retrospectively (cf. section 2.3 and *Kagan and Jackson*, 1999) and the use of formal statistical mechanical concepts fails to provide definite answers with currently available data (*Main and Al-Kindy*, 2002). Approaching the problem in prospective mode, as would be required for evaluating a hypothetical prediction capability, appears unrealistic.

3.8.2   Power law evolution before failure - Voight's law

The PCS approach provides a theoretical basis for the observed empirical ten-dency of earthquakes to cluster and postulates a power law increase in seismicity before a major event. Again, a major difficulty is to frame the spatial domain in which it should occur. If this effect could be identified in prospective time, it would allow the prediction of the time, but not necessarily the magnitude, of a future earthquake. Possibly, some applications of this effect are to be expected for other geophysical phenomena, like volcanic eruptions and landslides. An empir-ical equation for such power law precursors or behavior has been formulated by *Voight* (1988, 1989).

Voight's empirical equation, for which the PCS provides theoretical support, is

$$\dot{\Omega}^{-\alpha}\ddot{\Omega} - a = 0,\tag{3.37}$$

where Ω is any experimental variable and a is a constant. By taking logarithms, this gives

$$\log(\dot{\Omega}^{-\alpha}\ddot{\Omega}) = \log a\tag{3.38}$$

or

$$-\alpha\log\dot{\Omega} + \log\ddot{\Omega} = \log a.\tag{3.39}$$

Reordering, we obtain

$$\log\ddot{\Omega} = \log a + \alpha\log\dot{\Omega}\tag{3.40}$$

i.e. a linear fit.

i.e. a linear fit.

An effect of this type is indeed well known to engineers but it occurs in the failure of a) heterogeneous materials and b) at high strain laboratory rates (cf. section 1.1). Whether it is possible to observe this for earthquakes, where the strain rates are many orders of magnitude lower, is still unproven. A few identification claims have been published (*Bufe and Varnes*, 1993; *Bowman et al.*, 1998), but these are all retrospective identifications based on selecting from all the available data, and very rarely apply statistical tests to validate the case studies. As such, there is hardly a reason to conclude that they are unaffected by the retrospective selection bias and false discovery problem that we have discussed in section 1.2. This, in addition to the fact that this issue failed to pass validation testing even in retrospective mode (*Gross and Rundle*, 1998), casts heavy doubts about its practical applicability to tectonic earthquakes.

Albeit in retrospective mode and without any significance testing, this law has been found applicable to a number of cases of geodetic measurements and seismicity preceding volcanic eruptions (*Voight*, 1988, 1989; *Kilburn and Voight*, 1998; *Main*, 1999). Conversely, statistical tests have shown (*Gross and Rundle*, 1998) that tectonic earthquakes do not show any consistent power law increase. This result may appear at odds with the PCS approach, but it is not. The very long time scale of tectonic earthquakes calls for a very large number of superimposing effects which determine a generalized tendency towards randomness, which shows up in the ubiquitous small deviations from a Poisson process (cf. *Kagan and Jackson*, 1991). In other words, the physics of earthquakes is likely to be that of a critical system, but with so much superimposed noise that the signature of criticality is apparent only in the long-term behavior and in large-scale averaging. This possibly gives hope for future development, since detecting signals embedded in noise has been performed successfully in many fields of modern science and technology, and a variety of techniques exist, although their success largely depends on the *a priori* knowledge of the features of the signal to be extracted. However, it must also be considered that demonstrating the existence of a given process buried in the noise requires a very careful statistical analysis, since fluctuations could be incorrectly interpreted as real effects. The problem is clearly illustrated by an analysis (*Main*, 2000; see also *Sornette et al.*, 1996) which showed that a modest addition of Gaussian noise to a pure power law Gutenberg-Richter could lead to a variety of artifacts, ranging from a second power law branch with a larger *b* coefficient to a hump, reminiscent of the characteristic earthquake hypothesis (cf. section 2.4.3). A very careful statistical analysis of the data is therefore necessary before drawing any conclusions.

3.9 Statistical implications

What are the statistical implications of the PCS approach?

The principal one is related to the power-law behavior itself, which implies that the populations are distributed with a comparatively large number of samples in the (right-hand) tail of the distribution in respect to the most common distributions, like the normal, which have exponential tails. Populations showing this property are termed *heavy tailed* (or *fat tailed*).

There are several heavy tailed distributions but the typical one representing a power law behavior is the *Pareto* distribution, which has probability density function form

$$f(x) = Kx^{-\alpha-1} \qquad x > c, \alpha > 0, \tag{3.41}$$

where α is called the tail index. In theory, a Pareto distribution extends to infinity and has an undefined average and an infinite variance. However, our world is finite and we face the presence of physical cutoffs (cf. section 3.6) which make the latter parameters both defined and finite. Namely, what applies is a truncated (cf. *Kagan*, 2002) or tapered (cf. *Johnson*, 1994; *Vere-Jones et al.*, 2001) Pareto distribution which has both defined and finite parameters. An alternative is also the gamma distribution (cf. *Kagan*, 2002; *Main and Burton*, 1984 and section 2.3) which is able to approximate a power law behavior over a wide range, and which has also defined and finite parameters.

The most important conclusions are therefore methodological. If the PCS is a good approach, the distributions of observables are not normal, and the departures from normality tend to have a higher proportion of extreme observations. If techniques designed for normal distributions (for example linear regression analysis by least squares fitting, see section 3.3.1) are used on such data, the results will tend to be unreliable. In particular, uncertainties will be underestimated, and results can appear to be more significant than they really are.

The use of each one of the above mentioned distributions appears appropriate. In addition, there are also special statistical techniques called *robust* (or *resistant*) and *nonparametric*, which allow to effectively treat non-normal data (e.g. *Lehmann*, 1975; *Huber*, 1981; *Hampel et al.*, 1986; *Leonard et al.*, 2001).

3.10 Implications for predictability

The PCS approach has first of all had considerable impact from a cultural point of view. A stationary state maintained near criticality corroborates the basic *a priori* assumption of long-term asymptotic stationarity in the earthquake process, shared by all current practices in seismic hazard estimation (e.g., *Reiter*, 1990), which assumes that past occurrences should be used to predict the long-term future hazard. The PCS approach gives a rationale for the empirical finding that earthquake clustering occurs at all scales in time and space. Therefore, since fluctuations in both time and space appear as a main feature of earthquake occurrence, the application of stationarity is only asymptotically valid, and requires a very long statistical sample compared with the recurrence rates of the largest events. In numerical terms, recurrence time over the same fault patch of large earthquakes is of the order of 10^3–10^5 yr (*Wallace*, 1987; *Galli and Bosi*, 2002), which, for making any credible statistical estimate, would require palaeoseismic catalogues complete over inordinate amounts of time. What the PCS suggests is that, due to fractal clustering, fluctuations occur at all time scales, and hazard estimates should always be time-dependent and based on catalogues of the time length appropriate to capture the fluctuations (*Kagan and Jackson*, 2000). Note in this respect that the current standard assumption in hazard analysis is that large earthquakes occur as a random Poisson process (e.g., *Reiter*, 1990), which is at odds with fractal clustering. However, seismic catalogues are so poor in sampling the recurrence of large events that data from neighboring regions are usually combined to provide a set sizable enough to be analyzed. The consequence is that this smooths out the clustering and makes the Poisson model appear to still be the best solution (*Gross and Rundle*, 1998).

Long-term asymptotic estimates can be better constrained by the long-term seismic moment release rate $\dot{M}$ determined from geodetic or plate tectonic studies (cf. section 6.1) synthesized with the paleoseismic data into a single distribution with the shorter-term seismicity catalogue via the mean magnitude and the seismic event rate (*Main and Burton*, 1984). Implicit in this procedure is the assumption of stationarity in the flux of energy or moment, as required by a gamma distribution (cf. section 2.3), which exhibits power law scaling of energy at small magnitudes.

While triggering by other earthquakes is obvious for aftershocks, its quantitative characterization is trivial in retrospective mode, but so far it has not proved effective in the prospective mode. In regional studies, a triggering signal has been inferred several decades after a large event (e.g. *Stein et al.*, 1997). In fact, the notion of triggering has been applied by the State of California to issue low-level alerts following the occurrence of larger events. If triggering is a general feature of earthquakes, then the null hypothesis for predictability at a greater level, say from observed precursors, should include allowance for the additional clustering due to triggering inherent in an avalanche process of a near-critical system. In practice, once an appropriate null hypothesis for testing a time-dependent model (*Kagan and Jackson*, 2000) is selected, the hypothesis can easily be tested formally in prospective mode.

In conclusion, the PCS approach brings about renewed hopes for understanding earthquakes and assessing any degree of predictability. However, in light of the nature of the earthquake process, such predictability will be only of statistical type. It is the clustering feature at all scales, typical of self-similar processes, that dominates the picture. A forecasting algorithm based on this has been developed by Kagan and Jackson (see section 5.2) and this appears as the best possible option at present. This represents a significant step towards incorporating a degree of predictability in earthquake populations that can be validated in an unambiguous way, but is far from the 'unrealistic' idea of prediction of individual events discussed in the Introduction and in chapter 1.

Further progress is likely to come from ultra-low strain rate laboratory studies, which thanks to self-similarity will possibly allow to study the onset of earthquake rupture under controlled conditions and at a scaled dimension.

However, another field looks the most mature and promising for further progress. This field is based on satellite geodesy. Crustal stress cannot be extensively measured, but crustal strain can, thanks to the spatially distributed satellite geodesy techniques. This is a matter that deserves detailed discussion and will be treated in detail in section 6.1. The strains measured by such techniques will provide much more accurate estimates of the long term asymptotes for models of earthquake occurrence.

3.11   References

Abercrombie, R., and Leary, P., 1993. Source parameters of small earthquakes recorded at 2.5 km depth, Cajon Pass, Southern California - implications for earthquake scaling, *Geophys. Res. Lett.*, **20**, 1511–1514.

Aki, K., 1981. A probabilistic synthesis of precursory phenomena, in *Earthquake Prediction, an International Review*, D. W. Simpson and P. G. Richards (eds.), Maurice Ewing Series: 4, American Geophysical Union, Washington.

Anifrani, J. C., Le Floch, C., Sornette, D., and Souillard, B., 1995. Universal log-periodic correction to renormalization group scaling for rupture stress prediction from acoustic emission, *J. Phys. I France*, **5**, 631-638.

Ashcroft, N. W., and Mermin, N. D., 1981. *Solid State Physics*, Holt-Saunders Int. Edition, Philadelphia.

Aviles, C. A., Scholz, C. H., and Boatwright, J., 1987. Fractal analysis applied to characteristic segments of the San Andreas fault, *J. Geophys. Res.*, **92**, 331–344.

Bak, P., and Tang, C., 1989. Earthquakes as a self-organized critical phenomenon, *J. Geophys. Res.*, **94**, 635–637.

Bak, P., Tang, C., and Wiesenfeld, K., 1988. Self-organized criticality, *Phys. Rev. A*, **38**, 364–374.

Bak, P., Christensen, K., Danon, L., and Scanlon, T., 2002. Unified scaling law for earthquakes, *Phys. Rev. Lett.*, **88**, art. no. 178501.

Barenblatt, G. I., 1979. *Similarity, Self-Similarity and Intermediate Asymptotics*, Translation from Russian, Consultants Bureau, New York.

Barenblatt, G. I., and Sivashinskii, G. I., 1969. Self-similar solutions of the second kind in nonlinear filtration, *Appl. Math. Mech. PMM*, **33**, 836–845.

Benedetto, J. J., and Frazier, M. W., 1994. *Wavelets. Mathematics and Applications*, CRC Press, Boca Raton.

Ben-Zion, Y., and Rice, J. R., 1995. Slip patterns and earthquake populations along different classes of faults in elastic solids, *J. Geophys. Res.*, **100**, 12959–12983.

Ben-Zion, Y., and Zhu, L., 2002. Potency-magnitude scaling relations for southern California earthquakes with $1.0 < M_L < 7.0$, *Geophys. J. Int.*, **148**, F1–F5.

Binney, J. J., Dowrick, N. J., Fisher, A. J., and Newman, M. E. J., 1992. *The Theory of Critical Phenomena. An Introduction to the Renormalization Group*, Clarendon Press, Oxford.

Bonnet, E., Bour, O., Odling, N. E., Davy, P., Main, I., Cowie, P., and Bekowitz, B., 2001. Scaling of fracture systems in geological media, *Rev. Geophys.*, **39**, 347–383.

Bowman, D. D., Ouillon, G., Sammis, C. G., Sornette, A., and Sornette, D., 1998. An observational test of the critical earthquake concept, *J. Geophys. Res.*, **103**, 24359–24372.

Bridgman, P. W., 1931. *Dimensional Analysis*, Yale University Press, New Haven.

Brown, S. R., Scholz, C. H., and Rundle, J. B., 1991. A simplified spring-block model for earthquakes, *Geophys. Res. Lett.*, **18**, 215–218.

Bufe, C., and Varnes, D. J., 1993. Predictive modeling of the seismic cycle of the greater San Francisco Bay region, *J. Geophys. Res.*, **98**, 9871–9883.

Burridge, R., and Knopoff, L., 1967. Model and theoretical seismicity, *Bull. Seism. Soc. Am.*, **57**, 341–371.

Caputo, M., 1969. *Elasticity and Dissipation* (in Italian), Zanichelli, Bologna.

Caputo, M., 1977. A mechanical model for the statistics of earthquakes, magnitude, moment, and fault distribution, *Bull. Seism. Soc. Am.*, **67**, 849–861.

Caputo, M., 1981. A note on a random stress model for seismicity statistics and earthquake prediction, *Geophys. Res. Lett.*, **8**, 485–488.

Caputo, M., 1982. On the reddening of the spectra of earthquake parameters, *Earthq. Pred. Res.*, **1**, 173–178.

Carlson, J. M., 1991. Two-dimensional model of a fault, *Phys. Rev. A*, **44**, 6226–6232.

Carlson, J. M., and Doyle, J., 1999. Highly optimized tolerance: a mechanism for power laws in designed systems, *Phys. Rev. E*, **60**, 1412–1427.

Carlson, J. M., and Doyle, J., 2000. Highly optimized tolerance: robustness in complex systems, *Phys. Rev. Lett.*, **84**, 2529–2532.

Carlson, J. M., and Langer, J. S., 1989. Properties of earthquakes generated by fault dynamics, *Phys. Rev. Lett.*, **22**, 2632–2635.

Carreras, B. A., Lynch, V. E., Newman, D. E., and Sanchez R., 2002. Avalanche structure in a running sandpile model, *Phys. Rev. E*, **66**, art. no. 011302.

Castellaro, S., and Mulargia, F., 2001. A simple but effective cellular automaton for earthquakes, *Geophys. J. Int.*, **144**, 609–624.

Castellaro, S., and Mulargia, F., 2002. What criticality in cellular automata for earthquakes?, *Geophys. J. Int.*, **150**, 483–493.

Ceva, H., and Luzuriaga, J., 1998. Correlations in the sand pile model: from the lognormal distribution to self-organized criticality, *Phys. Lett. A*, **250**, 275–280.

Christensen, K., Hamon, D., Jensen, H. J., and Lise, S., 2001. Self-organized criticality in the Olami-Feder-Christensen model and reply, *Phys. Rev. Lett.*, **87**, 39801–39802.

Ciccotti, M., and Mulargia, F., 2002. Pernicious effect of physical cutoffs in fractal analysis, *Phys. Rev. E*, **65**, 37201–37204.

Cochard, A., and Madariaga, R., 1994. Dynamic faulting under rate-dependent friction, *Pure Appl. Geophys..*, **142**, 419–445.

Czirok, A., Somfai, E., and Vicsek, T., 1993. Experimental-evidence for self-affine roughening in a micromodel of geomorphological evolution, *Phys. Rev. Lett.*, **71**, 2154–2157.

Daubechies, I., 1992. *Ten Lectures on Wavelets*. SIAM, Philadelphia.

Davies, P. (ed.), 1989. *The New Physics*, Cambridge University Press, Cambridge.

Davy, P., Sornette, A., and Sornette, D., 1992. Experimental discovery of scaling laws relating fractal dimensions and the length distribution exponent of fault systems, *Geophys. Res. Lett.*, **19**, 361–363.

DeRubeis, V., Hallgass, R., Loreto, V., Paladin, G., Pietronero, L., and Tosi, P., 1996. Self-affine asperity model for earthquakes, *Phys. Rev. Lett.*, **76**, 2599–2602.

Di Gregorio, S., Rongo, R., Siciliano, C., Sorriso-Valvo, M., and Spataro, W., 1999. Mount ontake landslide simulation by the cellular automata model SCIDDICA-3, *Phys. Chem. Earth A*, **24**, 131–137.

Dubois, J., and Pambrun, C., 1990. Reversals of the Earth's magnetic field in the last 165 Myr. Attractor in the dynamic system, (in French), *C. R. Acad. Sci. Paris*, **311**, 643–650.

Farmer, J. D., and Sidorovich, J. J., 1987. Predicting chaotic time series, *Phys. Rev. Lett.*, **59**, 845–848.

Feder, J., 1988. *Fractals*, Plenum Press, New York.

Fisher, D. S., Dahmen, K., Ramanathan, S., and Ben-Zion, Y., 1997. Statistics of earthquakes in simple models of heterogeneous faults, *Phys. Rev. Lett.*, **78**, 4885–4888.

Foufoula-Georgiou, E., and Kumar, P. (eds.), 1994. *Wavelets in Geophysics*, Academic Press, New York.

Gadomski, A., Kertész, J., Stanley, H. E., and Vandewalle, N. (eds.), 2000. *Applications of Statistical Physics*, North Holland, Amsterdam.

Galli, P., and Bosi, V., 2002. Paleoseismology along the Cittanova fault: implications for seismotectonics and earthquake recurrence in Calabria (southern Italy), *J. Geophys. Res.*, **107** (B3), ETG 1–19, doi: 10.1029/2001JB000234

Geilikman, M. B., Golubeva, T. V., and Pisarenko, V. F., 1990. Multifractal patterns of seismicity, *Earth Planet. Sci. Lett.*, **99**, 127–132.

Gonzato, G., Mulargia F., and Marzocchi, W., 1998. Practical application of fractal analysis: problems and solutions, *Geophys. J. Int.* **132**, 275–282.

Gonzato, G., Mulargia, F., and Ciccotti, M., 2000. Measuring the fractal dimension of ideal and actual objects: implications for application in geology and geophysics, *Geophys. J. Int.*, **142**, 108–116.

Goupillaud, P., Grossmann, A, and Morlet, J., 1984. Cycle-octave and related transforms in seismic signal analysis, *Geoexploration*, **23**, 85–102.

Grassberger, P., and Procaccia, I., 1983. Measuring the strangeness of strange attractors, *Physica D*, **9**, 189–208.

Gross, S., and Rundle, J. B., 1998. A systematic test of time-to-failure analysis, *Geophys. J. Int.*, **133**, 57-64.

Hainzl, S., Zoller, G., and Kurths, J., 1999. Similar power laws for foreshock and aftershock sequences in a spring-block model for earthquakes, *J. Geophys. Res.*, **104**, 7243–7253.

Hallgass, R., Loreto, V., Mazzella, O., Paladin, G., and Pietronero, L., 1997. Earthquake statistics and fractal faults, *Phys. Rev. E*, **56**, 1346–1356.

Hamburger, D., Biham, O., and Avnir, D., 1996. Apparent fractality emerging from models of random distributions, *Phys. Rev. E*, **53**, 3342–3358.

Hampel, F. R., Ronchetti, E. M., Rousseeuw, P. J., and Stahel, W. A., 1986. *Robust Statistics: the Approach Based on Influence Functions*, Wiley, New York.

Herrmann, H. J., 1986. Geometrical cluster growth models and kinetic gelation, *Phys. Rep.*, **136**, 153–227.

Herrmann, H. J., Hovi, J. P., and Luding, S. (eds.), 1998. *Physics of Dry Granular Materials*, NATO ASI Proceedings, Kluwer, Dordrecht.

Herrmann, H. J., and Roux, S., 1990. *Statistical Models for the Fracture of Disordered Media*, North Holland, Amsterdam.

Hirata, T., Satoh, T., and Ito, K., 1987. Fractal structure of spatial distribution of microfracturing in rock, *Geophys. J. Roy. Astr. Soc.*, **90**, 369–374.

Huang, J., and Turcotte, D. L., 1990. Are earthquakes an example of deterministic chaos?, *Geophys. Res. Lett.*, **17**, 223–226.

Huber, P. J., 1981. *Robust Statistics*, Wiley, New York.

Jánosi, I. M., and Kertész, J., 1993. Self-organized criticality with and without conservation, *Physica A*, **200**, 179.

Jensen, H. J., Christensen, K., and Fogedby, H. C., 1989. $1/f$ noise, distribution of lifetimes, and a pile of sand, *Phys. Rev. B*, **40**, 7425–7427.

Johnson, N. L., Kotz, S., and Balakrishnan, N., 1994-1995. *Continuous Univariate Distributions*, 2 vols., 2nd ed., Wiley, New York.

Kagan, Y. Y., 2002. Seismic moment distribution revisited: I. Statistical results, *Geophys. J. Int.*, **148**, 520–541.

Kagan, Y. Y., and Jackson, D. D., 1991. Long term earthquake clustering, *Geophys. J. Int.*, **104**, 117–133.

Kagan, Y. Y., and Jackson, D. D., 1999. Worldwide doublets of large shallow earthquakes, *Bull. Seism. Soc. Am.*, **89**, 1147–1155.

Kagan, Y. Y., and Jackson, D. D., 2000. Probabilistic forecasting of earthquakes, *Geophys. J. Int.*, **143**, 438–453.

Kagan, Y. Y., and Knopoff, L., 1980. Spatial distribution of earthquakes: the two-point correlation function, *Geophys. J. Roy. Astr. Soc.*, **62**, 303–320.

Kantelhardt, J. W., Zschiegner, S. A., Koscielny-Bunde, E., Havlin, S., Bunde, A., and Stanley, H. E., 2002. Multifractal detrended fluctuation analysis of nonstationary time series, *Physica A*, **316**, 87–114.

Kertész, J., and Kiss, L. B., 1990. The noise spectrum in the model of self-organized criticality, *J. Phys. A-Math. Gen.*, **23**, 1433–1440.

Kilburn, C. R., and Voight, B., 1998. Slow rock fracture as eruption precursor at Soufriere Hills volcano, Montserrat, *Geophys. Res. Lett.*, **25**, 3665–3668.

King, G. P., and Sammis, C. G., 1983. The mechanism of finite brittle strain, *Pure Appl. Geophys.*, **138**, 611-640.

Kinouchi, O., and Prado, C. P. C., 1999. Robustness of scale invariance in models with self-organized criticality, *Phys. Rev. E.*, **59**, 4964–4969.

Knopoff, L., Landoni, J. A. and Abinante, M. S., 1992. Causality constraint for fractures with linear slip weakening, *J. Geophys. Res.*, **105**, 28035–28043.

Kumar, P., and Foufoula-Georgiou, E., 1997. Wavelet analysis for geophysical applications. *Rev. Geophys.*, **35**, 385–412.

Legrand, D., Cisternas, A., and Dornbath, L., 1996. Multifractal analysis of the 1992 Erzincan aftershock sequence, *Geophys. Res. Lett.*, **23**, 933–936.

Lehmann, E. L., 1975. *Nonparametrics: Statistical Methods Based on Ranks*, Holden Day, San Francisco.

Leonard, T., Papasouliotis, O., and Main, I. G., 2001. A Poisson model for identifying characteristic size effects in frequency data: application to frequency-size distributions for global earthquakes, starquakes and fault lengths, *J. Geophys. Res.*, **106**, 13473–13484.

Lomnitz-Adler, J., 1993. Automaton models of seismic fracture: constraints imposed by the magnitude-frequency relation, *J. Geophys. Res.*, **98**, 17745–17756.

Lomnitz-Adler, J., and Lemus-Diaz, P., 1989. A stochastic model for fracture growth on a heterogeneous seismic fault, *Geophys. J. Int.*, **99**, 183–194.

Lorenz, E. N., 1963. Deterministic nonperiodic fbw, *J. Atmos. Sci.*, **20**, 130–141.

Main, I. G., 1996. Statistical physics, seismogenesis, and seismic hazard, *Rev. Geophys.*, **34**, 433-462.

Main, I. G., 1999. Applicability of time-to-failure analysis to accelerated strain before earthquakes and volcanic eruptions, *Geophys. J. Int.*, **139**, F1–F6.

Main, I. G., 2000. Apparent breaks in scaling in the earthquake cumulative frequency-magnitude distribution: fact or artifact?, *Bull. Seism. Soc. Am.*, **90**, 86–97.

Main, I. G., and Burton, P. W., 1984. Information theory and the earthquake frequency-magnitude distribution, *Bull. Seism. Soc. Am.*, **74**, 1409–1426.

Main, I. G., Leonard, T., Papasouliotis, O., Hatton, C.G., and Meredith, P. G., 1999. One slope or two? Detecting statistically-significant breaks of slope in geophysical data, with application to fracture scaling relationships, *Geophys. Res. Lett.*, **26**, 2801-2804.

Main, I. G., O'Brien, G., and Henderson, J. R., 2000. Statistical physics of earthquakes: comparison of distribution exponents for source area and potential energy and the dynamic emergence of log-periodic energy quanta, *J. Geophys. Res.*, **105**, 6105–6126.

Malcai, O., Lidar, D. A., Biham, O., and Avnir, D., 1997. Scaling range and cutoffs in empirical fractals, *Phys. Rev. E*, **56**, 2817–2828.

Mallamace, F., and Stanley, H. E. (eds.), 1996. *The Physics of Complex Systems*, Proc. of the International School of Physics 'Enrico Fermi', Course CXXXIV, SIDF 2000.

Mandelbrot, B. B., 1977. *Fractals: Form, Chance and Dimension*, W. H. Freeman, San Francisco.

Mandelbrot, B. B., 1982. *The Fractal Geometry of Nature*, W. H. Freeman, New York.

Marzocchi, W., Mulargia, F., and Gonzato, G., 1997, Detecting low-dimensional chaos in geophysical time series, *J. Geophys. Res.*, **102**, 3195–3209.

Matos, J. A. O., and Duarte, J. A. M. S., 1999. On a conservative lava fbw automaton, *Int. J. Modern Phys. C*, **10**, 321–335.

Medvedev, M. V., and Diamond, P. H., 1998. Self-organized states in cellular automata: exact solution., *Phys. Rev. E*, **58**, 6824–6827.

Muzy, J. F., Bacry, E., and Arneodo A., 1993. Multifractal formalism for fractal signals: the structure-function approach versus the wavelet-transform modulus-maxima method, *Phys. Rev. E*, **47**, 875–884.

Nakanishi, H., 1990. Cellular automaton model of earthquakes with deterministic dynamics, *Phys. Rev. A*, **41**, 7086–7089.

Nakanishi, H., 1991. Statistical properties of the cellular automata model for earthquakes, *Phys. Rev. A*, **43**, 6613–6621.

Ogata, Y., 1999. Seismicity analysis through point process modeling, *Pure Appl. Geophys.*, **155**, 471–507.

Okubo, P. G., and Aki, K., 1987. Fractal geometry in the San Andreas fault system, *J. Geophys. Res.*, **92**, 345–355.

Olami, Z., Feder, H. J. S., and Christensen, K., 1992. Self-organized criticality in a continuous, nonconservative cellular automaton modeling earthquakes, *Phys. Rev. Lett.*, **62**, 1244.

Ouillon, G., Castaing, C., and Sornette, D., 1996. Hierarchical geometry of faulting, *J. Geophys. Res.*, **101**, 5477–5487.

Pantosti, D., Schwartz, D. P., and Valensise, G., 1993. Palaeoseismology along the 1980 surface rupture of the Irpinia fault: implications for earthquake recurrence in Southern Apennines, Italy, *J. Geophys. Res.*, **98**, 6561–6577.

Pelletier, J. D., 2000. Spring-block models of seismicity: review and analysis of structrually heterogeneous model coupled to viscous Asthenosphere, in *Geocomplexity and the Physics of Earthquakes*, Rundle, J. B., Turcotte, D. L., and Klein, W. (eds.), AGU Publ., Washington, 17–42.

Privman, V., (ed.), 1990. *Finite Size Scaling and Numerical Simulation of Statistical Systems*, World Scientific, Singapore.

Reiter, L., 1990. *Earthquake Hazard Analysis*, Columbia University Press, New York.

Rice, J. R., 1993. Spatiotemporal complexity of slip on a fault, *J. Geophys. Res.*, **98**, 9885–9907.

Rundle, J. B., and Jackson, D. D., 1997. Numerical simulation of earthquake sequences, *Bull. Seism. Soc. Am.*, **67**, 1363–1377.

Rundle, J. B., and Klein, W., 1993. Scaling and critical phenomena in a cellular automaton slider-block modle for earthquakes, *J. Stat. Phys.*, **72**, 405–413.

Sahimi, M., Robertson, M. C., and Sammis, C. G., 1992. Relation between the earthquake statistics and fault patterns and fractals and percolation, *Physica A*, **191**, 57–68.

Saleur, H., Sammis, C. G., and Sornette, D., 1996. Discrete scale invariance, complex fractal dimensions, and log-periodic fluctuations in seismicity, *J. Geophys. Res.*, **101**, 17661–17677.

Shaw, B. E., 1994. Complexity in a spatially uniform continuum fault model, *Geophys. Res. Lett.*, **21**, 1983–1986.

Sornette, A., Dubois, J., Cheminee, J. L., and Sornette, D., 1991. Are sequences of volcanic eruptions deterministically chaotic?, *J. Geophys. Res.*, **96**, 11931–11945.

Sornette, D., 1997. Discrete scale invariance, in *Scale Invariance and Beyond*, Dubrulle, R., Graner, F., and Sornette, D. (eds.), Le Houches Workshop, Springer-EDP Science, Berlin, 235–247.

Sornette, D., and Sammis, C. G., 1995. Complex critical exponents form renormalization group theory of earthquakes: implications for earthquake prediction, *J. Phys. I*

France, **5**, 607–619.

Sornette, D., Knopoff, L., Kagan, Y. Y., and Vanneste, C., 1996. Rank-ordering statistics of extreme events: application to the distribution of large earthquakes, *J. Geophys. Res.*, **101**, 13883–13893.

Spezzano, G., Talia, D., Di Gregorio, S., Rongo, R., and Spataro, W., 1996. A Parallel Cellular Tool for Interactive Modeling and Simulation, *IEEE Comp. Sci. Engin.*, **3**, 33–43.

Stauffer, D., and Aharony, A., 1994. *Introduction to Percolation Theory*, 2nd ed., Taylor and Francis, Philadelphia.

Steacy, S. J., and McCloskey, J., 1999. What controls an earthquake's size? Results from a heterogeneous cellular automaton, *Geophys. J. Int.*, **133**, F11-F14.

Stein, R. S., Barka, A. A., and Dieterich, J. M, 1997. Progressive failure on the North Anatolian fault since 1939 by earthquake stress triggering, *Geophys. J. Int.*, **128**, 594–604.

Sugihara, G., and May, R. M., 1990. Nonlinear forecasting as a way of distinguishing chaos from measurement error in time series, *Nature*, **344**, 734–741.

Tsonis, A. A., 1992. *Chaos: from Theory to Applications*, Plenum Press, New York.

Turcotte, D. L., 1992. *Fractal and Chaos in Geology and Geophysics*, Cambridge University Press, Cambridge.

Turcotte, D. L., Malamud, B.D., Guzzetti, F., Reichenbach, P., 2002. Self-organization, the cascade model, and natural hazards, *Proc. Nat. Acad. Sci. USA*, **99**, 2530–2537.

Vere-Jones, D., Robinson, R., and W. Z. Yang, 2001. Remarks on the accelerated moment release model: problems of model formulation, simulation and estimation, *Geophys. J. Int.*, **144**, 517-531.

Vicsek, T., 1992. *Fractal Growth Phenomena*, 2nd ed., World Scientific, Singapore.

Voight, B., 1988. A method for prediction of volcanic eruptions, *Nature*, **232**, 125–130.

Voight, B., 1989. A relation to describe rate-dependent material failure, *Science*, **43**, 200–289.

Von Seggern, D., 1980. A random stress model for seismicity and earthquake prediction, *Geophys. Res. Lett.*, **7**, 637–640.

Wallace, R. E., 1987. Grouping and migration of surface faulting and variations in slip rates on faults in the great basin province, *Bull. Seism. Soc. Am.*, **77**, 868–876.

Ward, S. N., 1991. A synthetic seismicity model for the Middle America Trench, *J. Geophys. Res.*, **96**, 21433–21442.

Ward, S. N., 1996. A synthetic seismicity model for southern California: cycles, probabilities, and hazard, *J. Geophys. Res.*, **101**, 22393–22418.

Ward, S. N., 1997. Dogtails versus rainbows: synthetic earthquake rupture models as an aid in interpreting geological data, *Bull. Seism. Soc. Am.*, **87**, 1422–1441.

Weiss, J., and Gay, M., 1998. Fracturing of ice under compression creep as revealed by a multifractal analysis, *J. Geophys. Res.*, **103**, 24005–24016.

Wilson, K. G., 1983. The renormalization group and critical phenomena (Nobel Address), *Rev. Mod. Phys.*, **55**, 583–608.

Xie, H., 1993. *Fractals in Rock Mechanics*, A. A. Balkema, Rotterdam.

Yeomans, J. M., 1992. *Statistical Mechanics of Phase Transitions*, Clarendon Press, Oxford.

Zapperi, S., Herrmann, H. J., and Roux S., 2000. Planar cracks in the fuse model. *Eur. Phys. J. B*, **17**, 131–136.

Chapter 4

Time-independent hazard

4.1 Seismic hazard assessment and site effect evaluations at regional scale

[By D. Albarello and M. Mucciarelli]

Seismic hazard assessment is the estimation of the expected level of ground motion at a site due to the occurrence of possible earthquakes during a fixed future time interval (exposure time). Such estimates require a good knowledge of the relevant seismogenic processes (both in terms of fracture geometries of future earthquakes and of time evolution of tectonic loading), of seismic energy propagation features in the area under study, and of the Earth structure near the site of interest. In fact, the complexity of seismic phenomena requires a modeling capability which is still beyond our reach and would imply in any case the quantification of a large number of relevant parameters dependent on local geological and dynamic conditions. On the other hand, since at least rough seismic hazard estimates are necessary for engineering and regional scale planning in seismic areas, the use of simplified and 'robust' approaches is thus necessary.

Effective procedures for Seismic Hazard Assessment (SHA) should therefore be calibrated on the data set actually available rather attempting to take into account the full extent of mechanical complexity. The implications of this point of view are illustrated below by the discussion of two specific problems: the development of reliable SHA procedures and the assessment of experimental tools for fast site effect evaluations.

4.1.1 Seismic hazard estimates

As stated above, an *ab initio* physical modeling approach to SHA is not appropriate. This, together with the fact that the data available to constrain the relevant parameters are affected by significant uncertainty, reduces seismic hazard assessment to the comparative evaluation of several possible hypotheses about the maximum expected ground motion level.

In general, having fixed the exposure time Δt, the degree of belief associated with a specific hypothesis about the expected earth shaking level S can be expressed in terms of the probability density function $r_{\Delta t}(S)$. On the basis of these probabilities, the expected value S_e of S is chosen such that

$$R_{\Delta t}(S_e) = \int_{S_e}^{S_{max}} r_{\Delta t}(s)\, ds \leq R_0 \tag{4.1}$$

where $R_{\Delta t}$ is the survivor function associated with $r_{\Delta t}$, S_{max} is the maximum possible level of earth shaking and the probability R_0 is generally fixed by the specific code considered and depends on the adopted level of 'conservatism' (*Reiter*, 1990). As an example, in the case of the Italian seismic code R_0 is 10 per cent for Δt equal to 50 years (*Slejko et al.*, 1998).

In general, discussions focus on S_e estimates only and this obscures the essential probabilistic character of hazard estimates. Actually, hazard estimates can be considered as a quantification of our lack of knowledge about future seismicity. This implies that, being essentially a convolution of different uncertainties, the results of the SHA procedure reflect the uncertainty level of the underlying knowledge. In particular, at least four sources of uncertainty can be identified (*Egozcue and Ruttener*, 1997): (a) uncertainty due to the discrepancy between natural seismic process and the mathematical models (i.e., geometry of the seismic sources or earthquake recurrence models); (b) uncertainty from the statistical estimation of parameters using finite samples; (c) uncertainty corresponding to the probabilistic models themselves (i.e., earthquakes recurrence models, etc.); (d) uncertainty arising from data (inaccuracy, heterogeneity, incompleteness, etc.). It should be clear that a complete evaluation of the effects of these sources of uncertainty will be reflected in the broadening of the distribution $r_{\Delta t}$ which implies, via equation (4.1), higher S_e values (see, e.g., *McGuire*, 1993a). The reverse is also true. Introducing 'deterministic' aspects into seismic hazard computations would in principle reduce uncertainty and result in less conservative hazard estimates. However, the drawback is that underestimates or incorrect evaluations of the uncertainty associated with such information, could produce substantial underestimates of the hazard level. The problem in general remains open. In this regard, Monte Carlo (e.g., *Musson*, 1999, 2000) procedures and logic-tree approaches based on the probabilistic combination of multiple expert opinions (see, e.g., *Electr. Pow. Res.*

Inst., 1986; *Bernreuter et al.*, 1986) have been considered. Both approaches have advantages but also have important limitations (see e.g., *McGuire*, 1993a). For example, in a Monte Carlo procedure, it is relatively easy to take into account the effect of uncertainty on the parameters of the relevant probability distributions (e.g., those relative to earthquake inter-event times) but it becomes difficult to consider the effects of the possible different choices for the form of such distributions or to consider the effects of uncertain seismotectonic regionalizations. A critical discussion on major drawbacks of the logic-tree approach can be found in *Krinitzky* (1993, 1995).

A number of different procedures for SHA have been proposed so far. These approaches range from purely phenomenological statistical analyses of past seismicity (see, e.g., *Veneziano et al.*, 1984) to para-deterministic estimates (see, e.g., *Ward*, 1994) passing through more or less advanced hybridizations between these two approaches (e.g., *WGCEP*, 1995; *Wu et al.*, 1995). However, despite the availability of several alternative models, the procedures actually used in common practice throughout the world are quite few (see, e.g., *McGuire*, 1993b). This is mainly due to a basic lack of knowledge about many relevant aspects of seismic process for most regions of the world. As an example, time-independent statistical characterizations of seismicity and the mutual independence of seismic sources activity are widespread assumptions. This seems to contradict most recent knowledge on the intrinsic complexity of earthquake process (e.g. *Main*, 1996) and increasing empirical evidence on earthquake space/time clustering well beyond the classical foreshock/main-shock/aftershock sequence (see, e.g., *Kagan and Jackson*, 1991; *Lomnitz*, 1996). However, since more sophisticated statistical modeling would require too complex a parameterization in relation to the data actually available in most cases, time-independent modeling remains a basic tool for SHA (see *McGuire*, 1993b; *Giardini and Basham*, 1993).

It is common that the amount of available data increases quite slowly as compared to the growing sophistication of computational procedures. Thus, effective future developments should be devoted much less to developing new sophisticated models for SHA and much more to making the best use of available data and to setting up procedures for checking the actual reliability of hazard estimates provided by the different approaches in relation with observed seismicity.

As concerns the first point, it is important to evaluate the consistency of available procedures with the basic features of available data and their capability to capture the actual uncertainty present in the basic data set in the final hazard estimate. In particular, it could be of interest to examine in this respect, the SHA procedure introduced by *Cornell* (1968) which, in a more sophisticated implementation (*Bender and Perkins*, 1987), has been recently adopted as an international 'standard' (see, e.g., *Giardini and Basham*, 1993). This procedure can be summarized as follows: (1) seismicity data, extracted from epicentral catalogues, are spatially disaggregated into n separate seismic sources A_i represented by lines (faults) or areas (fault areas) identified on the basis of regional seismotectonic features; (2) for each i_{th} seismic source, the average number v_i of earthquakes above the threshold m_0 is computed per unit area starting from the complete part of the seismic catalogue relative to the zone considered; (3) using the same data set, the probability density function $f_i(m)$ is defined; this represents the probability per unit area that an earthquake with magnitude m will be generated in the i_{th} seismogenic zone; depending on the specific numerical code considered, $f_i(m)$ is defined in terms of a continuous function (the Gutenberg-Richter power law truncated at m_{max} is often used) or discretized for a finite set of magnitude values; (4) the probability density function $g_i(r|m)$ is defined which represents the probability that an earthquake of magnitude m occurs in the i_{th} zone at a distance r from the site under study; thus $g_i(r|m)$ represents the seismic activity rate within the i_{th} zone. In common practice, g_i is assumed to be uniform within the seismogenic zone; (5) the attenuation (i.e. the rate at which seismic ground motion S or seismic effects decay with respect to epicentral distance r) as a function of magnitude m is defined in the form of a probability distribution $P(S > s|r,m)$; in general, P is assumed to be lognormal; (6) the average annual rate $\lambda(S)$ of earthquakes able to shake the site with a ground motion at least equal to S is computed as

$$\lambda(S) = \sum_{i=1}^{N} \int_{A_i} \int_{m_0}^{m_{max}} P_i(S|r,m)g_i(r|m)f_i(m)dmdr; \qquad (4.2)$$

(7) the survivor function $R_{\Delta t}$ is computed on the assumption that seismicity is a stationary poissonian process and thus

$$R_{\Delta t} = \exp(-\lambda(s)\Delta t). \qquad (4.3)$$

The procedure described above is a mix of statistical modeling (e.g., the Poisson earthquake recurrence model, power-law magnitude distribution) and deterministic constraints (e.g., seismogenic zoning).

Although the above method was developed for instrumental data (epicentral location, magnitude, Peak Ground Acceleration, etc.), it has also been widely adopted for the analysis of macroseismic information deduced from documentary sources. This implies that intensity data (discrete and ordinal in nature) have to be 'forced' via empirical conversion rules to supply para-instrumental information, thus increasing the uncertainty affecting input parameters. This problem becomes of great importance when the basic data set is mostly constituted, as in the case of many countries such as Eastern Europe or Eastern Asia, by epicentral information deduced from macroseismic information. As an example of the forcing exerted on intensity data it can be recalled (*Mucciarelli*, 1998a) that most of the different empirical relationships between magnitude and epicentral intensity are the artifact of a statistical bias induced by insufficient sampling and the saturation affecting both of the variables considered. Furthermore, since local information provided by macroseismic data are completely discarded, site effects are not considered as part of the input information. A further drawback of this standard approach is hidden in the procedure adopted for the computation of earthquake occurrence rates, which uses only the part of the catalogue assumed as 'complete', discarding in this way precious information about older seismicity.

In summary, the 'standard' approach is suitable for situations where good geological information is available and the basic data are primarily instrumental information available for a time span which is long relative to the average duration of the relevant seismic cycle. On the other hand, where tectonic deformation rates are relatively low, geological information and instrumental data alone do not allow us to univocally constrain the geometry of source zones or to characterize the recurrence of major earthquakes. In these cases, the bulk of the available data is mostly represented by ill-defined macroseismic information. The use of the 'standard' approach in these cases may lead to misleading results. In fact, while the estimate of seismic hazard is essentially a site-oriented problem, the 'standard' approach basically relies on source information (seismogenic zones, rates of earthquake generation, etc.) reduced at the site via empirical attenuation rules. This induces a tendency to discard available local data (e.g., macroseismic information on the local effects of past earthquakes) in favor of more or less constrained epicentral information only indirectly related to this basic information. The paradox is that, to produce hazard evaluations, such 'epicentral' information has to again be reduced at the sites from which original data were obtained. In these situations (quite common in Europe, for example), it is far preferable to adopt a different computational scheme. An example of the latter is the statistical methodologies recently proposed by *Mucciarelli et al.* (1992), *Magri et al.* (1994), and *Albarello and Mucciarelli* (2002), which extensively use intensity data derived from documentary sources now available for most Italian localities (*Monachesi and Stucchi,* 1997; *Boschi et al.,* 2000). This technique, outlined below in the appendix (section 4.1.3), has been developed to allow the inclusion of macroseismic information and thereby allow a formally correct treatment of such data. The seismic history of each site, reconstructed from documentary data taking into account the different levels of completeness and uncertainty, is the basic information considered for seismic hazard evaluation. This permits the full exploitation of information on past seismic effects (a particularly large dataset in Italy) and, consequently, it could provide more reliable hazard estimates. Moreover, unlike the standard technique, no *a priori* hypothesis is requested about the statistical properties of seismicity (e.g., Poissonian distribution of inter-event time) and the geometry of seismogenic zones. A comparison between the results of the 'site' approach and those provided by standard procedures for the Italian territory (*Mucciarelli et al.,*

2000), has shown that the latter provides less conservative hazard estimates, possibly as an effect of a less complete uncertainty treatment.

As concerns the problem of validation of seismic hazard estimates, it is worth noting that available data are generally not sufficient to constrain, at least from the statistical point of view, many basic assumptions of the adopted SHA procedure. In most cases, personal attitudes play a major role. On the other hand, different choices (e.g., those concerning models for time recurrence or energy partition of earthquakes) may dramatically affect final estimates (e.g., for the Italian situation, see *Romeo and Pugliese* (2000). To control this problem, distribution-free approaches could be adopted (*Kagan and Jackson*, 2000; *Kijko et al.*, 2001 and references therein). However, at least at our knowledge, the only application of such kind of approaches to hazard estimates at regional scale is that proposed for Southern California by *Frankel et al.* (1996).

A sensitivity analysis would in principle allow to sort the key features of the adopted procedure. For example, *Romeo and Pugliese* (2000) have shown that if one does not consider the effect of seismotectonic zoning, hazard estimates are mainly sensitive to the choice of empirical rules used to compute decay with distance of earth shaking level (attenuation laws) and, at progressively decreasing level, to the estimate of seismicity rates relative to each seismic source zone, magnitude upper bounds and magnitude thresholds. However, it can be shown (*Rabinowitz and Steinberg*, 1991) that the importance of several input parameters depends critically on the settings of other parameters. In this sense, the adopted SHA procedure should be considered (and checked) as a whole. A basic tool for the determination of relative reliability of alternative procedures, which could avoid endless discussions between the involved scientists, could be the comparison of seismic hazard estimates provided by the different methodologies with seismicity actually observed during a control period (see, e.g., pioneering works by *McGuire*, 1979 and *McGuire and Barnhard*, 1981). Results of an analysis of this type, applied to seismicity rate evaluations, are at present the object of debate in the North American community of seismologists (see, e.g., *Petersen et al.*, 2000; *Kagan and Jackson*, 2000). With respect to such approaches, the procedure proposed by *Grandori et al.* (1998) based on numerical simulations appears less effective due to the intrinsic difficulty to capture in the numerical modeling the features of actual seismicity.

4.1.2 Site effects estimates: how precise should they be?

Studies on site effects encompass several disciplines and are located in a vague borderland between seismology, soil dynamics and civil engineering. There is a widening gap between cutting edge research and the practitioner's needs. On the one hand, more and more precise physical models are being developed taking into

account the elasto-visco-plastic properties of a 3D multi-phase medium subjected
to complex wavefields, while on the other hand the seismic codes aim at uniform
spectra for the widest possible set of soil characteristics. At the same time, great
effort is made towards complete a posteriori understanding of one or more strong-
motion recordings for a single quake, while on the other hand there is a demand
for average, reliable and repeatable ground motion characteristics.

The predictive capability of soil amplification models have been tested *a priori*
just in two cases: the blind prediction experiment at Turkey Flat, California and
at Ashigara Valley, Japan. For soil-structure interaction the only known blind test
was carried out at Lotung, Taiwan.

The ever increasing complexity of models raises questions about the possibility of obtaining significant data to feed the model itself. Just as an example: consider a 3D cubic model of a structure with sides measuring 1 km, just for S-waves
and assuming homogeneous behavior; suppose that modeling should be carried
out up to 10 Hz. If the S-wave velocity ranges down to 200 m/s, the number
of elements of a mesh for finite element modeling could reach 125,000. Having
to consider just the shear modulus, the density and the damping, this means that
375,000 parameters are required. The latter are obviously not independent, but
what is a reasonable lower limit? Are few, sparse drillings sufficient to prepare a
sound model or are we just playing a computer game? How much does our model
depend on the unknown characteristics of the next real earthquake? We can perform a lot of parametric analyses on directivity, moment release, stress drop and
so on, but very few end users could afford such an effort.

On the other side of the scale of complexity, simple techniques like the Horizontal to Vertical Spectral Ratio (HVSR) are gaining enormous popularity among
users due to the very low cost. There is a seemingly wide consensus about the fact
that the HVSR technique is able to predict the resonance frequency of soils but
fails to correctly estimate the amplification of ground motion. However, a closer
scrutiny reveals a quite different reality. The foundations of the method itself are
not so clear: the author of the original work (*Nakamura*, 1989) returned ten years
later (*Nakamura*, 2000) with a paper entitled "Clear identification of fundamental
idea of Nakamura's technique and its applications". Most of the papers referring
to the first work of Nakamura are simply related to practical application for micro-
zoning purposes, and the theoretical background as well as the limitations of the
method are not of interest to many authors and are tackled only by a few papers
dealing with these topics. The main problems one may encounter studying the
applicability of the HVSR technique can be summarized as follows. (1) There is
no general agreement about a standard for data collection and processing. Most
papers do not even mention which equipment was used. (2) There are multiple
views about which seismic phases are we dealing with and interested in, but nevertheless we can obtain a satisfactory estimate of elastic soil behavior in presence

of direct S waves. (3) There is a very poor statistical treatment of data. The first test proposed to discriminate spurious peaks from statistical significant ones in microtremor HVSR was published very recently (*Albarello*, 2001). (4) The HVSR method gives very stable results in time. Even if not yet completely modeled, HVSR is related to a permanent feature of the investigated sites. (5) The method seems to work better (especially in amplitude) when applied to buildings, where the original hypotheses on the theory behind it do not apply, and gives interesting results when applied as a prospecting technique.

The main criticisms of this technique are that (1) it is not able to correctly predict amplification; (2) it does not take into account soil non linearity; (3) it fails to reproduce the exact amplification observed during a specific earthquakes.

The answers to these questions opens up interesting methodological issues:

1. Which is the method that has to be used as a standard reference? The Reference Site Spectral Technique attracts criticisms as well, since it is not clear what a reference site should be, and engineers may ask about site response in a town which is in the middle of an alluvial plain, with the nearest hard rock outcropping some hundred kilometers away or buried several kilometers at depth.

2. Soil non-linearity is favored by engineers since it translates to a reduction of input soil motion. But the vast majority of existing poorly built structures may suffer significant damage well before the soil motion reaches a level needed to start significant non-linear soil behavior. Moreover, non-elastic response means non-instantaneous response, and before a significant attenuation due to anelastic, non-linear behavior is reached by soils, the buildings may suffer significant damage.

3. The aim of a simple technique like HVSR is to estimate an average response (*Mucciarelli*, 1998b) before a generic earthquake occurs. Of course, the *a posteriori* examination of strong motion may lead to a more accurate picture of local soil dynamics, but was the information needed for a more complex model available before the earthquake? Who guarantees that the next earthquake will have the same characteristics? Will the soil characteristics be unaltered and will the next amplification be the same? The answer to all of these questions is, unfortunately, no. Incidentally, the presence of buildings is not taken into account in soil dynamic models, but recent results (*Wirgin and Bard*, 1996; *Gueguen et al.*, 2000; *Gallipoli et al.*, 2003) show that the vibration induced by a building may transfer back to the ground a quantity of energy able to adversely affect nearby structures.

The conclusion of this discussion is that we need complex models to fully understand soil dynamics under seismic loading, but at the same time we need simple models that are valid on average, whose results can be easily transferred to end users without prohibitive expenditure.

4.1.3   Conclusions

In general, a cost/benefit evaluation should be performed to address future research on the development of reliable SHA procedures and to define effective tools for assessment of site effects applicable to large-scale studies. In general, the common assumption "the more complete and complex is the reference model the more effective is the resulting procedure" may be misleading.

As concerns SHA, more complex models require a greater number of relevant parameters to be estimated from available data and these are correspondingly more numerous (and heterogeneous) sources of uncertainties to be considered and controlled. Since hazard estimates are essentially a convolution of such uncertainties, results obtained from increasingly complex models become more and more sensitive to the correct parametrization of a basic lack of knowledge, which, in many cases, leads to highly unstable results. It is worth noting that an incorrect parameterization of the uncertainty will result in incorrect hazard estimates and, in particular, due to the peculiar features of seismicity, underestimated uncertainty directly reflects in hazard underestimates. Thus, the paradoxical effect is that the addition of new constraints not associated with reliable uncertainty estimates could make resulting hazard evaluation less reliable. As a consequence, future improvements in SHA will come more by the development of new procedures for the treatment of the relevant uncertainty and though validating the results by considering observed seismicity and less from the increasing complexity of models aimed at capturing the actual features of the seismic phenomenon. The same also holds true for estimates of site effects; very complex models require a large amount of data that can be obtained at a very large cost or may be impossible to get at all. It is not feasible to start extensive drilling projects under old urban centers of Europe. Another paradox is that while seismologists try to reproduce the subtlest features of acceleration time histories, engineers are shifting toward models relating the seismic resistance of a building to its ability to withstand displacements. This may also have some beneficial effects. Starting from source models and going through propagation and site effects dealing with displacement rather than acceleration might mean focusing on large scale, average properties that should be easier to model.

Appendix

The procedure described here aims at the estimate of the probability $H(\Delta t, I_0)$ which represents the degree of belief in the hypothesis that within a future exposure time Δt at the site of interest at least one seismic event will occur which will be characterized by local effects corresponding at least to the degree I_0 of the adopted macroseismic scale. Since $H(\Delta t, I_0) = 1 - R_{\Delta t}$, the probability H can be used to estimate the survivor function $R_{\Delta t}$ considered in eq. (4.1).

In order to estimate H, available information about seismic effects of past earthquakes at the site under study (local seismic history), on the reliability of the available catalogue at the site ('completeness') and on possible seismogenic conditions responsible for earthquakes at the site have to be considered. Each of these pieces of information is characterized in terms of an appropriate probability function, which expresses the respective degree of uncertainty.

Uncertainty about seismic effects of past earthquakes is expressed in terms of a probability $P_l(I_0)$ which represent the degree of belief in the hypothesis that the l_{th} seismic event has actually shaken the considered site with effects of degree at least equal to I_0. Values of $P_l(I_0)$, which represent the seismic history for the site under study, constitute the keystone of the whole procedure. Expert judgments of historians could allow direct estimates of P_l. When direct documentation about felt effects is not available, P_l values can be estimated from epicentral information. In general, the following relationship is considered

$$P_l(I_0) = \sum_{I=I_{min}}^{I_{max}} p_e(I)R(I_0|E),$$ (4.4)

where p_e represents the probability that the maximum macroseismic intensity for the considered earthquake was equal to I. R is the probability that the local effects have been of intensity I_0 given that the relevant earthquake has been characterized by a set of epicentral parameters E (including I). A possible parametrization of R relative to the Italian region is reported by *Magri et al.* (1994).

From P_l values it is possible to estimate, for each specific time interval Δt_j of duration equal to the exposure time Δt, the probability Q_j that at least one of these events at the site has been actually felt with an intensity not less than I_0. If N_j is the total number of events occurred during Δt_j it follows that

$$Q_j = Q(\Delta t, I_0|\Delta t_j) = 1 - \prod_{l=1}^{N_j}[1 - P_l(I_0)].$$ (4.5)

The probabilities Q_j computed by equation (4.5) can be used to compute H. To this purpose, a specific segment of the local seismic history ΔT_i (e.g., that comprised between T_i and the present time), which is believed to be representative of actual ('complete') seismicity, is considered. A number K_i of distinct time intervals, each spanning Δt, can be individuated which cover (with partial overlapping allowed) the time interval ΔT_i. One could imagine that in each specific interval Δt_j, particular conditions occurred in terms of seismotectonic situations and building vulnerability. In some cases these conditions resulted in the occurrence of at least one earthquake that produced at the site effects corresponding at least to the degree I_0 of interest. In order to forecast the future occurrence of seismic effects at least equal to I_0 during any future time interval of duration Δt, we should know if conditions in that future time interval will correspond to those which occurred during that specific past interval Δt_j. Due to our incomplete knowledge about present and past conditions we can only define a probability $g(\Delta t_j)$ that represents our degree of belief in the hypothesis that during any future interval Δt the same conditions active during the specific past interval of duration Δt_j will occur again. In the assumption that the seismogenic process is stationary and that considered seismic history is sufficient to characterize (in a probabilistic sense) future seismicity, the probability H_i (conditioned at the specific choice of ΔT_i) can be computed in the form

$$H_i = H(\Delta t, I_0|\Delta t_i) = \sum_{j=1}^{K_i} g(\Delta t_j) Q(\Delta t, I_0|\Delta t_j). \qquad (4.6)$$

This relation presumes that the joint occurrences of at least one earthquake with intensity not lower than I_0 during Δt_j and of analogous geodynamic conditions during the future time span Δt constitute a set of mutually exclusive events. In the common case that information on future and past seismotectonic conditions is not sufficient to constrain $g(\Delta t_j)$ one can assume that $g(\Delta t_j) = 1/K_i$.

In general, different opinions could exist about the interval ΔT_i for which the available seismic history can be considered as fully representative of actual seismicity. To take this problem into account the probability density function probability! density function $r(\Delta T_i)$ is defined which represents the degree of belief in the hypothesis that for the specific time span ΔT_i the catalogue is actually complete. Several possible procedures can be considered to assess $r(\Delta T_i)$ spanning form expert judgments to statistical approaches (see, e.g., *Albarello et al.*, 2001). In order to take the problem of completeness into account, equation (4.6) can be generalized in the form

$$H(\Delta t, I_0) = \sum_{i=1}^{L} r(\Delta T_i) H(\Delta t, I_0|\Delta T_i), \qquad (4.7)$$

where L is the number of possible choices for sub-catalogues to be considered for seismic hazard analysis.

4.2   USGS and partners: approaches to estimating earthquake probabilities

[By A. J. Michael, N. Field, A. Frankel, J. Gomberg, and K. Shedlock]

To make informed decisions about how to reduce the risks from earthquakes, individuals, companies, and governments must know what hazards they will face and how often these hazards will occur. For earthquakes this information is most often given as a Probabilistic Seismic Hazard Assessment (PSHA) which gives the probability of different levels of shaking occurring over a given time period (*Cornell*, 1968). This section reviews current work being done by the United States Geological Survey (USGS) and its partners on estimating the probabilities of earthquake occurrence, which is one of the items required to make such hazard estimates. This is a large body of work for two reasons. First, the United States is a large geographic region with diverse tectonic settings ranging from stable craton to multiple plate boundaries; techniques appropriate to one area are often inapplicable to others.

The second reason for the breadth of this work is the variety of time scales that are considered. The estimation of earthquake probabilities over long time-periods is critical for mitigation of the risks earthquakes pose to society, because it allows for the development of appropriate building codes, seismic retrofitting plans, bridge designs, loss estimation, insurance policies, and regulations for siting important facilities, all of which can take years to decades to carry out. The identification of short periods of time with increased hazards can also play a beneficial in risk mitigation. For instance, warnings of increased hazard provide an impetus for organizations to review emergency plans and supplies, and could be used by hospitals to increase stocks of medical supplies. Such actions have little cost and little impact on the overall economy, and may be beneficial even if a damaging earthquake does not occur during the duration of the warning. This is true because the warning can stimulate earthquake preparedness and mitigation measures which outlive the specific warning period (*Michael et al.*, 1995). Knowing the probability of damaging aftershocks can also help determine when it is safe to reoccupy partially damaged buildings during an earthquake sequence (*Gallagher et al.*, 1999).

As a federal agency mandated to provide information to reduce earthquake risk throughout the country, the work of the USGS must range from application of existing methods on a national scale to regional studies of specific urban areas, from long-term to short-term time scales, and must include the development of new methods to improve future estimates. All of these efforts take place with a variety of partners ranging from other national governments, US federal agencies, state governments, and researchers at both universities and in private industry. Space limitations preclude acknowledging all of our partners here.

Given the wide range of this work and the need for brevity, only basic summaries of the work and key issues for future development can be provided here. We start with some basic principles and then proceed from national estimates of earthquake probabilities to regional studies, and from long-term to short-term estimates. Comments on new research directions can be found in section 5.1.

4.2.1   Basic principles

Before computing earthquake probabilities, one must estimate the rate at which earthquakes of a given size will occur in a region or on a specific fault. One approach to this problem assumes that the future will resemble the past. Observations of past earthquakes are provided by seismographic recordings, historical records, and paleoseismic investigations (section 2.1.1). To obtain a record of independent events, foreshocks and aftershocks are generally removed from catalogues using a variety of algorithms (see works cited by *Shedlock*, 1999). Given sufficient data this may show the rate at which earthquakes of different sizes occur. However, usually the seismicity rate as a function of magnitude is parameterized by the Gutenberg-Richter (G-R) relation (section 2.3). Another approach to estimating earthquake rates is to balance the deformation caused by the motion of tectonic plates and the slip on faults, which occurs primarily during earthquakes. These two quantities must balance over long periods of time because the motion of tectonic plates is the source of energy for the earthquakes. This approach is called moment-balancing or conservation of moment, because the deformation during an earthquake is quantified by its seismic moment. Seismic moment is the average slip across the fault multiplied by the area of the fault multiplied by the rigidity of the medium surrounding the fault. The amount of deformation due to tectonic plate motions can be estimated from studies of plate motions over geologic time or geodetic surveying over a period of years to decades. GPS data can also be used. The distribution of earthquake sizes required to account for the total deformation can come from either the G-R relation or a physical model that predicts earthquake sizes for the particular region.

Once we have an estimate of the rate at which earthquakes of a given magnitude occur in a region, we can estimate the probability of future earthquakes of this size under the assumption that the earthquakes are independent random events that are equally likely at any time. This assumption is mathematically described as a Poisson process and leads to time-independent earthquake probabilities, meaning that the probabilities are constant as a function of time.

Sometimes, time-dependent earthquake probabilities are computed by assuming that earthquakes are a cyclic or semi-periodic process. In this case, we need to consider individual earthquake sources and must know not only the rate at which these sources activate but also the time when the last event occurred and a statistical distribution that describes how earthquakes recur as a function of time. For instance, if an earthquake occurs on average every 100 years we need to know the probability that it will recur after only 90 years. A variety of statistical distributions have been used to describe this process including log-normal, Weibull, and the Brownian Passage Time distributions. Time-dependent earthquake probabilities tend to be more controversial due to debates over whether earthquakes recur as a function of time in the way assumed by the model and also due to the extra information required. However, advocates of this approach point out that specific large earthquakes do not recur after very short time intervals as would be allowed by the Poisson distribution.

4.2.2 Earthquake recurrence rates for national and international seismic hazard maps

The USGS national seismic hazard maps are based on time-independent estimates of earthquake probabilities. Although different methods were used, time-independent estimates were also determined for the Seismic hazard map of North and Central America and the Caribbean (*Shedlock*, 1999) as part of the Global Seismic Hazard Assessment Program (*Shedlock et al.*, 2000). In order to estimate the earthquake probabilities under these assumptions one needs to determine the rate at which earthquakes of different sizes occur in a given region or on a specific fault. Two different methods were used to select the regions for which the seismicity rates based on historical catalogues were computed: the spatially-smoothed gridded seismicity method and the delineation of background or seismic source zones.

The historic parametric method determines seismicity rates (based on the G-R relation) for each cell of a grid (*Frankel*, 1995). This approach assumes that future large earthquakes will occur in the general vicinity of past earthquakes, even smaller ones. There is much evidence supporting this approach (e.g. *Kafka and Walcott*, 1998), although there are some large earthquakes that have occurred in areas of previous seismic quiescence in the historic catalogue. The implicit assumption in this approach to historical seismicity is that the geologically short catalogues are adequate indicators of long-term, and hence future, seismicity.

To apply this method the region is divided into a finely spaced grid of cells. An important issue that must be dealt with for all historical seismicity catalogues, is estimating the time at which the catalogue is complete for a given magnitude threshold. Using incomplete catalogues will lead to underestimating earthquake recurrence rates and the resulting probabilities and hazard assessments. To deal with this problem, the recurrence rates for each grid cell are determined by the maximum likelihood method of *Weichert* (1980), which allows for the possibility that the earthquake catalogue becomes complete at different times for different magnitude events. The seismicity rate as a function of magnitude is parameterized with the G-R relation with the *b*-values determined for large regions and only the constant *a*-values are determined for each cell. These seismicity rates are spatially smoothed with a Gaussian function.

As mentioned above, some large earthquakes have occurred in areas with no known historic seismicity. In this case, the short sampling time of the historic catalogues may be unrepresentative of the longer-term seismicity patterns. To address this possibility, the seismicity can be averaged over larger areas called either background or source zones. This approach distributes the hazard throughout each zone even where there have been no past earthquakes. The justification for such averaging is the assumption that earthquakes within each zone are of the same (or very similar) type of faulting and have common driving mechanisms. The delineation of seismic source zones involves specifying the geographical coordinates of an area (polygonal) or fault (linear/planar) source. The hazard is assumed to be uniform within each polygon or along each fault segment and may be described by the rate of seismicity using the G-R relation. This component quantifies the hazard for areas that have low rates of historic seismicity, but still have the potential for generating moderate and large earthquakes.

The background zones for the USGS national seismic hazard maps were based on broad geologic criteria. For instance, in the western U.S. there is a large background zone that includes the Basin and Range Province. In the Basin and Range there are a very large number of faults responsible for the formation of this large province. However, geologic studies of individual earthquakes suggest that the largest individual events recur only on the order of thousands to tens of thousands of years. Thus, the historic record will not include an adequate spatial representation of possible earthquakes and the background zone method is a more reasonable approach to estimating earthquake probabilities.

In essence, the background zones use a larger and subjectively defined averaging zone for the historical seismicity while the spatially-smoothed seismicity is based on smaller, objectively defined averaging areas.

For some areas, the historic time period may be too short to accurately estimate seismicity rates even when averaged over a large background zone. This can be diagnosed by a misfit between deformation rates observed geodetically and those estimated from known earthquakes. In such areas, expected seismicity rates can be estimated by computing the rate of seismic moment release required to balance the observed deformation. This moment release rate is then converted into the rates of earthquakes versus magnitude using a G-R distribution. For the national hazard maps VLBI (Very Long Baseline Interferometry) and/or GPS data were used to estimate shear rates for source zones in western Nevada, northeastern California, and the Puget Lowland in the Pacific Northwest.

Another approach used in these maps is estimating the hazard from specific faults. The hazard from about 450 Quaternary faults, mostly in the western U.S., was included in the 1996 maps. The faults selected had either published geologic slip rates or chronologies of past earthquakes derived from paleoseismic methods (e.g., trenching, paleoliquefaction studies, or coastal uplift studies). Faults were divided into two classes. Class A faults have been sufficiently studied so that they can be broken into segments that may rupture independently of the rest of the fault. Class B faults are those without existing segmentation models (see *Petersen et al.*, 1996). For class A faults, earthquakes were assumed to recur in a characteristic manner, such that earthquakes rupture an entire fault segment or, in some cases, multiple segments. The class A faults are the San Andreas, Hayward-Rogers Creek, San Jacinto, Elsinore faults in California and the Wasatch fault in Utah. The intervals between earthquakes on the San Andreas and Wasatch faults were constrained by paleoseismic chronologies of large earthquakes determined from trenching studies.

The vast majority of faults were treated as class B faults for the 1996 version of the national maps. For these faults, both characteristic and truncated G-R models of recurrence were used. For the characteristic model, the earthquake magnitude is determined from either the fault area or length of the surface trace (*Wells and Coppersmith*, 1994). This can be converted into expected seismic moment for a single event. The long-term moment release rate is determined using the same fault area and rigidity but replacing the slip in an event with the geologically observed slip rate. Then the average rate at which earthquakes must occur can be determined so that the seismic moment released in earthquakes equals the amount required to keep the fault slipping at the geologically observed rate. However, not all earthquakes are characteristic ruptures of an entire fault. For instance, the $m = 6.7$ Northridge, California, earthquake of 1994 is a possible example of an earthquake rupturing only a portion of a longer thrust fault. Thus, the characteristic model may overestimate both the time between earthquakes and the size of earthquakes. While larger earthquakes are more devastating as individual events, more frequent but smaller events are more important when calculating the expected hazard over geologically short time periods such as a few decades. To include these smaller but more frequent events caused by rupture of only part of a fault, the truncated G-R model uses a minimum magnitude of 6.5 and a maximum magnitude corresponding to the characteristic rupture of the entire fault. Between these limits the relative number of events is set by the b-value in the G-R relation and the total rate of events is set by adjusting the a-value so that the total rate of moment released in all earthquakes equals the geologically observed moment release rate. The results from these different methods were combined by using a logic tree which produces a weighted average of the results. The 1996 maps assign equal weights to the characteristic and truncated G-R models for class-B faults.

For some well studied faults, the intervals between earthquakes are estimated directly from paleoseismic chronologies of large earthquakes. For the San Andreas and Wasatch faults there were sufficient fault-crossing trenches to constrain the history of past, prehistoric events. For other regions such as the central US and the Pacific Northwest indirect paleoseismic evidence was used.

A sequence of three earthquakes occurred near New Madrid in 1811–1812. For the national hazard maps the magnitude of these earthquakes was estimated between 7.5 and 8.0 based on isoseismal data. This value was adopted for the National Hazard maps based on consensus at a meeting of experts in 2000. The recurrence rate of these earthquakes is controlled by paleoliquefaction evidence which has clearly demonstrated the repeated occurrence of similar shaking levels over the past few thousand years in the New Madrid area (e.g. *Johnston and Schweig*, 1996). The magnitude of the 1886 earthquake near Charleston, SC is estimated at 7.3 from the isoseismals (*Johnston*, 1996). Again, paleoliquefaction evidence attests to the repeated recurrence of earthquakes with similar sizes over the past few thousand years in this region (e.g. *Amick and Gelinas*, 1991).

Paleoseismic evidence of drowned coastlines and tsunamis form the basis of the recurrence times used in the national maps for great earthquakes on the Cascadia subduction zone. This evidence points to sudden ground deformation events, consistent with earthquakes, occurring about every 500 years in many places along the coast (*Atwater and Hemphill-Haley*, 1996). The magnitude of these earthquakes is more uncertain and therefore it is unclear whether this paleoseismic evidence points to one $m = 9.0$ earthquake (such as the January 1700 event, *Satake et al.*, 1996) that ruptures the entire subduction zone each 500 years or multiple smaller events that occur more frequently and are spread out along the subduction zone. The 1996 maps used a logic tree with a $m = 9.0$ occurring every 500 years and multiple $m = 8.3$ events completely rupturing the subduction zone every 500 years. The relative weighting of these scenarios is a key issue for updates of the maps.

To produce the earthquake hazard maps, the contributions of earthquake occurrence under the three components must be combined. This is done by producing hazard curves (the probability of shaking at different levels at a given site), based on the probability of earthquakes under each component and a variety of ground-motion prediction relations with variability (see *Frankel*, 1995). Three hazard curves (one each for the historical parametric, background zone, and specific fault approaches) are computed for each grid point. These curves can then be combined by using a logic tree and thus provide a natural means of combining the effects of many sources and different approaches. Probabilistic ground motions, the annual probability of exceeding the specified ground motions, are computed for each grid point (see *Cornell*, 1968). These are contoured to produce a hazard map.

There are many important issues on earthquake recurrence that need further work to reduce the uncertainties in the hazard calculations. The current national maps utilize time-independent estimates of recurrence rates. But, time-dependence should be considered for some of the major faults that have semi-periodic behavior and for which sufficient information on previous earthquakes is available. However, it is important to properly include the variability of recurrence rates caused by multiple-segment ruptures that change from event to event and due to fault interaction. Thus, multiple-segment ruptures should be refined for the San Andreas and Hayward faults and quantified for other A-class faults.

There is evidence for temporal and spatial clustering of large earthquakes in some regions. While such clustering does not change the average rate of earthquake occurrence over long periods of time, it can change the rate over specific shorter periods for which the maps are computed. Thus, clustering changes the variability in the occurrence of earthquakes rather than the mean rate of occurrence and therefore affects the uncertainty of the hazard estimate. Currently, clustering can be incorporated into the maps if there is direct evidence of changes in the slip rate of a fault by using a logic tree with branches for different slip rates.

Earthquake catalogues, seismicity-rate grids, and fault parameters used to make the national maps are available through the website for the USGS National Seismic Hazard Mapping Project[1]. The California portion of the hazard maps was produced jointly by the USGS and the California Division of Mines and Geology. The input and methodology of the national maps were discussed and modified at several regional workshops of geoscientists and users. The methodology of the 1996 national maps is described in detail by *Frankel et al.* (1996, 2000). The maps will be periodically updated to include new methods and data.

In addition to the national hazards maps, the USGS has worked with other countries to produce a seismic hazard map of North and Central America and the Caribbean (*Shedlock*, 1999) as part of the Global Seismic Hazard Assessment Program (*Shedlock et al.*, 2000). Greenland and Canada/Mexico/United States comprise the principal landmasses of the North American plate. The seven countries of Central America are the principal landmass of the Caribbean plate, while the smaller Caribbean Island nations serve as the above-sea-level expression of the rest of the plate. Plate boundary processes dominate the seismicity and seismic hazard in North and Central America and the Caribbean. The plate boundaries are either subduction (i.e. the Cocos/Caribbean, eastern Caribbean, Pacific/North American, Juan de Fuca/North American) or transform (i.e. Pacific/North American along the San Andreas fault and Gulf of California, the north and south

---

[1]http://earthquake.usgs.gov/

Caribbean plate boundaries). The largest earthquakes recorded in the region have been plate boundary earthquakes, including nine earthquakes over magnitude 8 in this century.

One of the differences with the US national hazard maps is that outside of the United States, no mapped fault parameters (length, slip rate) were used. Two factors are responsible for not using mapped faults when calculating earthquake occurrence rates outside of the US: (1) fewer detailed mapping and paleoseismic studies have been done outside of the US and (2) the dominance of subduction zones with their hidden, often submarine, faults that make carrying out such studies either much more difficult or impossible. Another difference is that for areas outside of the US, the historic parametric method of *McQueen* (1997) was used instead of the spatially-smoothed gridded seismicity method used within the US.

4.2.3   San Francisco Bay region

The heavily urbanized San Francisco Bay region straddles a segment of the North American-Pacific plate boundary. Its tectonics is dominated by horizontal motion on several nearly parallel faults including the San Andreas, Hayward, Calaveras, Rodgers Creek, Concord-Green Valley, and San Gregorio faults. While the geometry of these intersecting faults is complex, the overall motion is well understood. Over 90 per cent of the geologically and geodetically observed motions take place on these six faults. The rest of the slip takes place on a variety of minor strike-slip faults and a series of thrust faults that take up compression across the region.

To provide new estimates of earthquake probabilities for this region in time for the tenth anniversary of the 1989 $m_w = 6.9$ Loma Prieta earthquake, the USGS-led Working Group on California Earthquake Probabilities (WGCEP) was expanded to include over 100 scientists from federal and state government, consulting geologists, and academia. This incarnation of WGCEP was known as Working Group 99 (WG99) and while probabilities were released in October, 1999, WG99 has continued to work on various aspects of the problem. The following describes the work released in 1999 (*WGCEP*, 1999, and http://quake.usgs.gov/study/wg99) and some of the developments since then.

To produce accurate probability estimates over a period of 30 years, the group decided to calculate time-dependent probabilities. Two factors support this decision. First, there was a 4-fold change in the rate of $m \geq 6$ earthquakes after the $m_w = 7.8$ 1906 earthquake with 12 such events from 1850 to 1906 and only 5 from 1907 to 2002. Estimating the earthquake rate as an average over the entire time-period from 1850 to the present would thus not be a good representation of the rate during subsets of the period, as a good model of this region must take into account the time-dependent changes of rate caused by the occurrence of the great 1906 earthquake. Second, a combination of historical and paleoseismological data provides information on recurrence rates and dates of previous events for many of the faults making it possible to calculate time-dependent probabilities for these events.

The WG99 process can be broken into three basic steps. First, rather than imposing a G-R distribution for earthquake sizes on the region, a regional earthquake source model was constructed that provides long-term estimates of earthquake recurrence that are consistent with the rate of regional strain accumulation. The earthquake source model is based on dividing the principal faults into segments based on fault geometry, along fault changes in long-term slip rates at fault intersections, lithologic changes, and the historic earthquake record. Neighboring segments may combine into longer, and therefore larger, earthquakes. The scaling relation between earthquake source areas and size as measured by moment is a critical part of this process. However, the frequently used relations (*Wells and Coppersmith*, 1994) overestimate the amount of slip for a given fault length for strike-slip earthquakes in California. Thus, new scaling relations were developed for use in the WG99 model (*Hanks and Bakun*, 2002). Fully modeling smaller events would require many short segments that would be difficult to constrain with existing data and segmentation concepts. So, the earthquake source model also included randomly distributed floating $m = 6.2$ or $m = 6.9$ (depending on the fault) earthquakes and a G-R distribution of smaller events. The rate of the characterized segmented earthquakes, the floating earthquakes, and the G-R (also referred to as the 'exponential tail') events were adjusted so that the total moment per year balances the geodetic deformation data. Developing the earthquake fault model relied heavily on the use of expert opinion and many of the decisions (including the concept of fault segments itself) are controversial. However, it is worth noting that the resulting regional distribution of earthquakes fits a G-R distribution. Thus, abandoning this approach and adopting a G-R distribution would likely have had little effect on the regional probability estimates.

Accounting for slow aseismic slip is important in the San Francisco Bay region because it explains for up to 80 per cent of the slip on segments of the Hayward, Calaveras, San Andreas, and Concord-Green Valley faults. Because the sum of seismic and aseismic slip during earthquakes must balance the plate motions, the presence of aseismic slip reduces the amount of slip that must take place during earthquakes. On the Calaveras fault, *Oppenheimer et al.* (1990) suggest that the fault is a mix of patches that slip only in earthquakes and other patches that are completely aseismic. Under this model, the amount of aseismic slip reduces the total fault area. WG99 includes this in the calculation by reducing the depth extent of the creeping faults. Thus, the aseismic slip reduces the fault area that slips during earthquakes. This, in turn, reduces the magnitude of the events and the amount of slip in each event on the creeping segments. However, to balance the plate motions these smaller events must occur more often than if there was no creep on the faults. So, the effect of fault creep in the WG99 model is to create smaller but more frequent earthquakes.

The regional earthquake source model provides earthquake recurrence rates that are used as the input to the second step of the process: computing earthquake probabilities. Time-independent probabilities were computed using a Poisson model while time-dependent probabilities were computed using a new model which applies the Brownian Passage Time (BPT) function to earthquake probabilities (*Ellsworth et al.*, 1998). The BPT model was derived by assuming that earthquakes occur when a failure state is reached. The approach to this failure state is a combination of a linear increase in a state variable (a theoretical quantity which could represent stress or a combination of physical parameters) which represents the slow, continuous plate motions, and Brownian noise which represents intrinsic variability in the earthquake generation process and the effects of fault interaction. After an earthquake, the state variable returns to zero on that fault segment. The probabilities that result from this model are similar to those obtained from previous models such as the log-normal distribution. However, the BPT model places the statistical model on a more physical footing. In addition, the BPT model has the ability to determine time-dependent probabilities even when the time of the last event on a fault segment must have occurred before a given date but the actual date is unknown. The San Francisco Bay region earthquake history is unknown before settlement by Europeans in 1776, so for faults with unknown last earthquakes that year was used as a bound. Time-dependent probabilities were also computed with a time-predictable model in which the amount of slip in the previous earthquake is important. This could only be done for the San Andreas fault because there are estimates for the slip distribution during the 1906 earthquake.

In addition to estimates which ignore interactions between different faults, the change in earthquake rate after 1906 was modeled by considering the elastic stress transfer effects of the 1906 earthquake on the other fault segments. Because the 1906 earthquake ruptured along the entire region, all of the other faults are in its 'stress shadow' (*Harris and Simpson*, 1998). The stress changes imposed by the 1906 earthquake on the other faults were converted to clock delays in the BPT model. This process creates lower probabilities after the 1906 earthquake but does not seem to completely capture the magnitude of the change in seismicity rate. Viscoelastic models may improve this fit, but were not yet sufficiently well understood for use in this project. Also, the observed seismicity rate change is simply one realization of a stochastic process and may represent a larger change than the mean of many realizations. To provide another estimate of the effects of the 1906 earthquake, the rates used to calculate time-independent probabilities were modified by an empirical fit to the change in rate of smaller earthquakes (*Reasenberg et al.*, 2003). This assumes that the change in rate is equal for both small and large earthquakes and does not require a physical model of the interactions between the 1906 earthquake and the other faults.

The third and final step of the process was combining the probabilities calculated from the separate probability models into a single set of fault-specific and regional estimates. This was done by using a decision tree and expert opinion voting to weight the various choices.

4.2.4 Earthquake likelihood models in Southern California

Southern California is even more populated than the San Francisco Bay area but occupies a region of greater tectonic complexity. This leads to differing approaches to estimating earthquake likelihood based on the many geophysical and geological observations which can help constrain where and when earthquakes are likely to occur, including the historic earthquake record, geological observations, paleoseismic data, and crustal deformation estimates from geodetic measurements. The Phase II report of the Southern California Earthquake Center (SCEC) (*Working Group on California Earthquake Probabilities, WGCEP*, 1995) represented the first working-group effort to integrate such a wide variety of information into a 'consensus' earthquake-forecast model, using conservation of seismic moment rate as a guiding principle. One innovation was allowing individual segments of larger 'type A' faults, such as the San Andreas, to sometimes rupture simultaneously in multiple segment 'cascade' earthquakes. Another innovation was accounting for earthquakes on not only lesser faults (e.g., the Sierra Madre), but on unidentified faults as well (distributed seismicity). As such, the first complete model of $m \geq 6$ seismicity was constructed for southern California.

By far the most contentious issue with the *WGCEP* (1995) model was that it predicted twice the rate of $m = 6$ to $m = 7$ earthquakes that has been observed historically - a fact inherited somewhat from the 'official' model that forms the basis of today's building-codes (*Petersen et al.*, 1996). A debate ensued over whether this apparent deficit was real, an artifact of excluding 'huge' earthquakes from the model, or simply a reflection of present uncertainties (*Hough*, 1996; *Jackson*, 1996; *Schwartz*, 1996; *Stirling and Wesnousky*, 1997; *Stein and Hanks*, 1998; *Field et al.*, 1999). This debate highlighted several important issues that might not have been recognized had the model fit the historical record. Although an alternative 'mutually consistent' earthquake-forecast model has since been developed (*Field et al.*, 1999), several issues remain unresolved and the research community is far from reaching consensus on how to construct such models.

Given this lack of consensus on how to proceed with earthquake probability models for southern California, a new working group has been formed for the development of Regional Earthquake Likelihood Models in southern California

(RELM, http://www.relm.org). RELM is a collaboration between SCEC and the USGS and is currently lead by N. Field of the USGS. In contrast to previous working groups, a fundamentally different approach is being taken in that we are developing and evaluating a range of viable models, rather than constructing one 'consensus model'. The goal is to create a working environment and infrastructure that not only accommodates, but encourages alternative models that might otherwise get filtered out in a consensus-forcing mode. WG99 also developed a range of viable models which were then combined in a logic tree to produce a single estimate. One difference is that WG99 was led by a oversight committee that decided which models would be included. RELM will be more inclusive and will allow anyone to submit a model as part of the project. The types of models under development include those based on geologic fault data (with various possible spatial and temporal dependencies); historical seismicity (smoothed in various ways); geodetic observations; alternative, internally-consistent fault-system representations (e.g., different perspectives on the accommodation of shortening across the LA basin); stress interaction between faults (static and/or dynamic); spatial and temporal aftershock statistics; and hybrids of the above models.

Not only will these models be tested, to the extent possible, for consistency with existing geophysical data (e.g., historical seismicity), but RELM will also design and document conclusive tests with respect to future observations. Perhaps most important, however, is that the hazard implications of each model will be examined immediately. This will not only help to quantify existing uncertainties in seismic hazard analysis, but will identify influential parameters and research topics needed to reduce these uncertainties. It will also exploit southern California as a natural laboratory for identifying which types of models are exportable to other regions where the options are fewer. Our approach will provide agencies charged with establishing official seismic-hazard estimates a menu of viable options (branches of a logic tree in the parlance of PSHA). These agencies can then weight the branches of this logic tree and add other models to produce the official seismic-hazard estimates for use in developing building codes and other risk mitigation efforts. RELM itself will not publish its own consensus model, nor will it use a logic tree to provide a single estimate from the multiple models, to avoid conflicts with the official hazard estimates.

Developing and evaluating such a wide variety of models in a fair and consistent manner requires some standardization with respect to the data constraints. For example, resolution of the earthquake deficit problem raised by *WGCEP* (1995) was exacerbated by the fact that virtually every study used a slightly different earthquake catalogue. To avoid such problems when evaluating RELM models, we are establishing various 'community' databases and models. This will also avoid duplication of efforts by different researchers and make it easier for people to apply their models to southern California. Two examples are a community earthquake catalogue and a community fault-activity database. This is not to say that we are forcing consensus on, say, the exact magnitude of the 1857 Fort Tejon earthquake, but that we want the various perspectives represented in one, web-accessible location rather than distributed among various appendices. Note that the distinction between data and model becomes fuzzy in that, for example, the slip-rate assigned to a fault ultimately assumes a model. Therefore, it is important that our community models/databases provide an electronic paper trail of where such values ultimately derive from. Another RELM-related development is new Java-based PSHA code that will not only standardize the hazard calculations for the wide-variety of models, but will also enable on-line, interactive publication of results with a graphical user interface (see http://www.relm.org/ for more information on these developments).

4.2.5 The New Madrid Seismic Zone

At probability levels where large infrequent earthquakes dominate the earthquake hazard, consensus estimates show the hazard to be comparable in the New Madrid seismic zone (NMSZ) with that in the San Francisco Bay region. The three New Madrid earthquakes of 1811-1812 affected an area much larger than the $m = 7.8$ 1906 California earthquake. However, estimating the earthquake probabilities in this region poses some unique challenges due to controversies over the magnitudes of the past earthquakes, the slow rate of seismic activity, and a poor understanding of how this intraplate region is loaded. Currently earthquake probabilities for this region are estimated as described in the National Hazard Maps described earlier. In this section we describe the challenges this intraplate environment poses to estimating earthquake probabilities both from known past seismicity and from moment balancing.

Magnitude estimates of the 1811 and 1812 New Madrid events differ by half a magnitude unit, because they rely on uncertain and potentially biased conversions of observations of their effect (*Johnston*, 1996) into magnitudes. Two lines of reasoning suggest that the magnitudes are lower than the earlier estimates according to which they were $m = 8$ events. First, early settlers tended to live along rivers where sediments amplified seismic shaking. Accounting for this and revisiting the conversion of written accounts of the events into Modified Mercalli Intensity values reduces the magnitude of these events (*Hough et al.*, 2000). However, the extent to which magnitudes are biased is still debated, in part because *Hough et al.* (2000) used relations to convert from isoseimal areas to magnitudes in which the isoseismal areas may be similarly biased. Second, a new method that simultaneously inverts the individual intensity observations together with a new estimate of the regional attenuation of seismic shaking also suggests that the events are smaller than originally surmised (*Bakun and McGarr*, 2002; *Bakun and Hopper*,

2003).

Nonetheless, from the perspective of assessing earthquake hazard, this uncertainty in magnitude may not be so important because it is the earthquake effect that determines the hazard and we have direct observations of this effect. Paleoliquefaction and geologic studies indicate that earthquakes with effects comparable to those of 1811–1812 have occurred at least twice before, in approximately 1450 and 900, corresponding to a median recurrence time of 267 to 644 years (*Tuttle et al.*, 2000). These largest NMSZ earthquakes appear to be events with 'characteristic' sizes and repeat times. This conclusion is based on the observations that earthquakes of moderate size are missing from the geologic and historic records, and extrapolation from these records and the rate of small instrumentally recorded earthquakes (assuming a G-R relation) predict inter-event times for major earthquakes that grossly exceed those observed. The paleoliquefaction record also suggests that the clustering of earthquakes that occurred in 1811–1812 also occurred in prior events. Accurate estimates of earthquake probabilities based on the past seismic record require accounting for the clustering in these events and on understanding whether or not the apparent lack of moderate earthquakes is real or due to the short record of observations.

Alternatively, we can assess seismic hazard in this region by balancing the seismic moment against the tectonic loading of the faults. Doing this in an intraplate environment provides greater challenges than at a plate boundary. The fault system known to be active in the NMSZ is probably less than 200 km long, and recent geodetic observations are notable for the lack of relative motion measured across the system (*Newman et al.*, 1999). These observations are inconsistent with a plate-boundary model in which relative far-field displacements 'drive' the deformation across an essentially infinite-length fault zone. The absence of topographic relief in the NMSZ and a variety of subsurface geologic data imply that the current rate of large New Madrid earthquakes applies only to very recent geologic times (*Schweig and Ellis*, 1994). Thus, any viable tectonic model of the NMSZ must be consistent with a fault system of finite length, recurrence of events comparable to those in 1811–1812 about every 500 years, almost negligible contemporary surface strain rates, and pre-Holocene deformation rates that were orders of magnitude lower than today. A few models satisfying these criteria have been proposed. They assume a uniform stress field and no recharge of strain energy as in plate tectonic models, and some sort of heterogeneity in the lower crust that focuses deformation. For example (*Kenner and Segall*, 2000) propose that there is an elongated zone of low viscosity in the lower crust and that viscoelastic relaxation of this zone transfers stress into the overlying brittle crust, causing earthquakes, which in turn partially reload the viscoelastic zone. Surface deformation rates decrease through the interseismic period to values similar to those observed today, 188 years after 1812. The transfer of stress repeats, but loosing energy until the whole process eventually ceases. In another model, *Pollitz* (2001) suggests that a sinking mafic body could be responsible for concentrating stress in this area.

Perhaps the greatest enigma of the NMSZ is the question of what might have caused repeating major earthquakes to begin, particularly so geologically recently? Changes in loading or fault state that might possibly explain a recent increase in seismicity are related to Holocene deglaciation and removal of a large load of ice as close as 100 km from the NMSZ (*Grollimund and Zoback*, 2001). Also at that time, the Ohio River captured the Mississippi River, radically altering the hydrologic system directly above the NMSZ. The connection between these events and seismicity is uncertain. Designing ways to turn these ideas into testable hypotheses is a challenge. The amount of data available to test such hypotheses is small due to the low level of seismicity and (until recently) the paucity of recording instrumentation. Additionally surficial geologic data is reduced by the effects of the Mississippi River which erodes any tectonic deformation and has deposited up to 1 km of unconsolidated sediments thereby masking deeper evidence of faulting and source processes.

4.2.6 Foreshocks and aftershocks

In California, short-term warnings of increased earthquake probabilities are issued by the State of California but are generally based on advice from the USGS and the two groups act in concert. Some short-term warnings, such as aftershock probabilities, are most useful when released rapidly after a mainshock. To allow for quick release of these warnings, some standard situations are pre-approved after negotiations between the two groups and then can be programmed into the automatic earthquake response and information systems. In this section we describe a number of studies done by USGS scientists, some in collaboration with others, that currently provide the scientific foundation of these warnings.

To date, the only verified form of short-term earthquake forecasting is the use of earthquake clustering in the form of aftershock and foreshock advisories. The existence of aftershock sequences readily demonstrates that earthquakes cluster. *Reasenberg and Jones* (1989) produced a generic statistical model for California that allows us to estimate the probability of future aftershocks while a sequence is in progress. This was done by combining two well known laws that describe the behavior of seismicity: the G-R relation that describes the distribution of seismicity in terms of the magnitude and the modified Omori law that describes the rate of aftershock occurrence as a function of time after a mainshock. Applications of this method include estimating aftershock probabilities with the generic parameter values immediately after a mainshock. This is now done within minutes of $m \geq 5$ earthquakes in northern California and these estimates are released through the USGS web site http://quake.usgs.gov/. Then, the generic values are replaced with sequence specific parameters once enough aftershocks have been recorded.

Foreshocks are another part of earthquake clustering and two models have been used to produce probabilistic warnings. *Reasenberg and Jones* (1989) extended their model to aftershocks that are larger than the initial mainshock. This allows for a generic description of foreshock-mainshock behavior that can be quickly applied after an earthquake. However, this approach has two limitations. First, because it is generic it does not take into account the different clustering characteristics of strike-slip and thrust earthquakes, both of which occur in California. Second, also because it is generic, it does not take into account the effect of proximity of a potential foreshock to a major fault considered capable of producing large events. Due to the use of average parameters for a region, all earthquakes are equally likely to be followed by a larger event regardless of whether or not they occurred near a major fault that has been judged capable of producing large events. *Agnew and Jones* (1991) developed a model of foreshock behavior that accurately describes the average behavior and takes into account variations of mainshock probabilities and background seismicity as a function of location. For instance, an earthquake that occurs far from a major fault in a region with many background earthquakes has a low probability of being a foreshock, while a rare event near a major fault has a higher probability of being one. One problem with this model comes up when attempting to apply it to a region with great uncertainties in the long-term mainshock probabilities, background earthquake rates, and/or the rate at which mainshocks are preceded by foreshocks (*Michael and Jones*, 1998).

Warnings based on foreshocks generally produce mainshock probabilities on the order of a few per cent over a week. Aftershock warnings carry similarly low probabilities for events worse than the mainshock that has already occurred. Such low probabilities, coupled with efforts to educate the public about appropriate actions, produce reasonable societal actions and do not produce the types of harmful behavior that some have worried about (*Michael et al.*, 1995).

Future short-term warnings might someday also be based on fault creep or crustal deformation anomalies or other possible earthquake precursors, but, to date, none has been verified for such use. To provide data for future developments, the USGS and collaborators operate a variety of geophysical monitoring networks at Parkfield, California (*Bakun et al.*, 1987) and around the San Francisco Bay area.

4.2.7 Conclusions

Unlike an individual researcher, the USGS and its partners cannot limit their activities to research on seismogenic processes and earthquake probabilities in a limited area. Given its responsibility to provide the best possible hazard estimates over a broad area, the range of approaches must be diverse. We are fundamentally limited by the lack of data on large earthquakes and thus our progress toward better estimates of earthquake probabilities must rely heavily on understanding the physics of the earthquake process. As we do this it will be critical to keep estimates of uncertainties in the probabilities in mind in considering whether these more advanced models provide better estimates for societal purposes.

4.3   References

Agnew, D. C., and Jones, L. M., 1991. Prediction probabilities from foreshocks, *J. Geophys. Res.*, **96**, 11959–11971.

Albarello, D., 2001. Detection of spurious maxima in the site amplification characteristics estimated by the HVSR technique, *Bull. Seism. Soc. Am.*, **91**, 718–724.

Albarello, D., and Mucciarelli, M., 2002. Seismic hazard estimates using ill-defined macroseismic data at site, *Pure Appl. Geophys.*, **159**, 1289–1304.

Albarello, D., Camassi, R., and Rebez, A., 2001. Detection of space and time heterogeneity in the completeness of a seismic catalog by a statistical approach: an application to the Italian area, *Bull. Seism. Soc. Am.*, **91**, 1694–1703.

Amick, D., and Gelinas, R., 1991. The search for evidence of large prehistoric earthquakes along the Atlantic Seaboard, *Science*, **251**, 655–658.

Atwater, B. F., and Hemphill-Haley, E., 1996. Preliminary estimates of recurrence intervals for great earthquakes of the past 3500 years at northeastern Willapa Bay, Washington, *U.S. Geological Survey Open-fi le Report*, 96-101.

Bakun, W. H., and Hopper, M. G., 2003. Magnitudes and locations of the 1811-1812 New Madrid, Missouri, and the 1886 Charleston, South Carolina, earthquakes, *Bull. Seism. Soc. Am.*, in press.

Bakun, W. H., and McGarr, A., 2002. Differences in attenuation among the stable continental regions, *Geophys. Res. Lett.*, **29** (23), 2121, doi: 10.1029/2002GL015457.

Bakun, W. H., Breckenridge, K. S., Bredehoeft, J., Burford, R. O., Ellsworth, W. L., Johnston, M. J. S., Jones, L., Lindh, A. G., Mortensen, C., Mueller, R. J., Poley, C. M., Roeloffs, E., Schultz, S., Segall, P., and Thatcher, W., 1987. Parkfield, California, Earthquake prediction scenarios and response plans, U.S. Geological Survey Open-File Report, 87–192.

Bender, B., and Perkins, D. M., 1987. Seisrisk III: a computer program for seismic hazard estimation, *U.S. Geological Survey*, **1772**, 1–48.

Bernreuter, D. L., Savy, J. B., and Mensing, R. W., 1986. The LLNL approach to seismic hazard estimation in an environment of uncertainty, in *Proc. Conf. XXXIV workshop on "Probabilistic Earthquake hazard assessment"*, Walter H. Hays (ed.), U.S. Geol.Surv., Reston, Va., 314–352.

Boschi, E., Guidoboni, E., Ferrari, G., Mariotti, D., Valensise, G., and Gasperini, P. (eds.), 2000. Catalogue of strong Italian earthquakes from 461 B.C. to 1997, *Ann. Geofi s.*, **43**, 609–868.

Cornell, C. A., 1968. Engineering seismic risk analysis, *Bull. Seism. Soc. Am.*, **58**, 1583–1606.

Eaton, M. L., and Freedman, D. A., 2003. Dutch book against objective' priors. Technical Report 642, Department of Statistics, University of California, Berkeley. http://www.stat.berkeley.edu/ census/642.pdf.

Egozcue, J. J., and Ruttener, E., 1997. Bayesian techniques for seismic hazard assessment using imprecise data, *Nat. Haz.*, **14**, 91–112.

Electric Power Research Institute, 1986. Seismic hazard methodology for the Central and Eastern United States, EPRI NP-4726.

Ellsworth, W. L., Matthews, M. V., Nadeau, R. M., Nishenko, S. P., Reasenberg, P. A., and Simpson, R. W., 1998. A physically-based earthquake recurrence model for estimation of long-term earthquake probabilities, *Proceedings of the 2nd Joint Meeting of the UJNR Panel on Earthquake Research*, Geographical Survey Institute, Japan.

Field, E. H., Jackson, D. D., and Dolan, J. F., 1999. A mutually consistent seismic-hazard source model for Southern California, *Bull. Seism. Soc. Am.*, **89**, 559–578.

Frankel, A., 1995. Mapping seismic hazard in the Central and Eastern United States, *Seism. Res. Lett.*, **66**, 8–21.

Frankel, A. D., Mueller, C. S., Barnhard, T. P., Leyendecker, E. V., Wesson, R. L., Harmsen, S. C., Klein, F. W., Perkins, D. M., Dickman, N. C., Hanson, S. L., and Hopper, M. G., 2000. USGS national seismic hazard maps, *Earthquake Spectra*, **16**, 1–19.

Frankel, A., Mueller, C., Barnhard, T., Perkins, D., Leyendecker, E., Dickman, N., Hanson, S., and Hopper, M., 1996. National seismic-hazard maps: documentation June 1996, *U.S. Geological Survey Open-fi le Report*, 96-532.

Gallagher, R. P., Reasenberg, P. A., and Poland, C. D., 1999. *Earthquake Aftershocks-Entering Damaged Buildings*, Applied Technology Council, Redwood City.

Gallipoli, M. R., Mucciarelli, M., Ponzo, F., and Dolce, M., 2003. Seismic waves generated by oscillating buildings: analysis of a snap-back experiment, *Soil Dyn. and Earthq. Eng.*, **23**, 255–262.

Giardini, D., and Basham, P. (eds.), 1993. Global seismic hazard assessment program, *Ann. Geofi s.*, **36**, 3–14.

Grandori, G., Guagenti, E., and Tagliani, A., 1998. A proposal for comparing the reliabilities of alternative seismic hazard estimates, *J. Seism.*, **2**, 27–35.

Grollimund, B., and Zoback, M. D., 2001. Did deglaciation trigger intraplate seismicity in the New Madrid seismic zone?, *Geology*, **29**, 175–178.

Gueguen, P., Bard, P. Y., and Oliveira, C. S., 2000. Experimental and numerical analysis of soil motion caused by free vibration of a building model, *Bull. Seism. Soc. Am.*, **90**, 1464–1479.

Hanks, T. C., and Bakun, W. H., 2002. A bilinear source-scaling model for $M - \log A$ observations of continental earthquakes, *Bull. Seism. Soc. Am.*, **92**, 1841–1846.

Harris, R. A., and Simpson, R. W., 1998. Suppression of large earthquakes by stress shadows: a comparison of Coulomb and rate-and-state failure, *J. Geophys. Res.*, **103**, 24439–24451.

Hough, S. E., 1996. The case against huge earthquakes, *Seism. Res. Lett.*, **67**(3), 3–4.

Hough, S. E., Armbruster, J. G., Seeber, L., and Hough, J. F., 2000. On the modified Mercalli intensities of the 1811-1812 New Madrd, central United States earthquakes, *J. Geophys. Res.*, **105**, 23839–23864.

Jackson, D. D., 1996. The case for huge earthquakes, *Seism. Res. Lett.*, **67**(1), 3–5.

Johnston, A. C., 1996. Seismic moment assessment of stable continental earthquake, Part 3, 1811-182 New Madrid, 1886 Charleston, and 1755 Lisbon, *Geophys. J. Int.*, **126**, 314–344.

Johnston, A. C., and Schweig, E. S., 1996. The enigma of the New Madrid earthquakes of 1811-1812, *Ann. Rev. Earth Planet. Sci.*, **24**, 339–384.

Kafka, A. L., and Walcott, J. R., 1998. How well does the spatial distribution of smaller earthquakes forecast the locations of larger earthquakes in the northeastern United States?, *Seism. Res. Lett.*, **69**, 428–440.

Kagan, Y. Y., and Jackson, D. D., 1991. Long term earthquake clustering, *Geophys. J. Int.*, **104**, 117–133.

Kagan, Y. Y., and Jackson, D. D., 2000. Probabilistic forecasting of earthquakes, *Geophys. J. Int.*, **143**, 438–453.

Kenner, S. J., and Segall, P., 2000. A mechanical model for intraplate earthquakes: application to the New Madrid Seismic Zone, *Science*, **289**, 2329–2332.

Kijko, A., Lasoki, S., and Graham, G., 2001. Non-parameric seismic hazard of mines, *Pure Appl. Geophys.*, **158**, 1655–1675.

Krinitzky, E. L., 1993. Earthquake probability in engineering - part I: the use and misuse of expert opinion, *Eng. Geol.*, **33**, 257–288.

Krinitzky, E. L., 1995. Deterministic versus probabilistic seismic hazard analysis for critical structures, *Eng. Geol.*, **40**, 1–7.

Lomnitz, C., 1996. Search for a worldwide catalogue for earthquakes triggered at intermediate distances, *Bull. Seism. Soc. Am.*, **86**, 293–298.

Magri, L., Mucciarelli, M., and Albarello, D., 1994. Estimates of site seismicity rates using ill-defined macroseismic data, *Pure Appl. Geophys.*, **143**, 617–632.

Main, I., 1996. Statistical physics, seismogenesis and seismic hazard, *Rev. Geophys.*, **34**, 433–462.

McGuire, R. K., 1979. Adequacy of simple probability models for calculating felt shaking hazard using the Chinese earthquake catalog, *Bull. Seism. Soc. Am.*, **69**, 877–892.

McGuire, R. K., 1993a. Computations of seismic hazard, in *Global seismic hazard assessment program*, Giardini, D., and Basham, P. (eds.) ILP publication 209, *Ann. Geofi s.*, **36**, 181–200.

McGuire, R. K. (ed.), 1993b. *The Practice of Earthquake Hazard Assessment*, IASPEI, Denver, 1–284.

McGuire, R. K., and Barnhard, T. P., 1981. Effects of temporal variations in seismicity on seismic hazard, *Bull. Seism. Soc. Am.*, **71**, 321–334.

McQueen, C. M., 1997. *An Evaluation of Seismic Hazard in the Caribbean and Central America Using Three Probabilistic Methods*, Ph. D. thesis, University of Lancaster.

Michael, A. J., and Jones, L. M., 1998. Seismicity alert probabilities at Parkfield, California, revisited, *Bull. Seism. Soc. Am.*, **88**, 117–130.

Michael, A. J., Reasenberg, P. H., Stauffer, P. H., and Hendley, J. W., 1995. II, Quake forecasting - an emerging capability, USGS Fact Sheet 242–295.

Monachesi, G., and Stucchi, M., 1997. DOM. 4.1 a database of macroseismic observations of Italian earthquakes above the damage threshold (in Italian), http://emidius.itim.mi.cnr.it/GNDT/.

Mucciarelli, M., 1998a. A different $M = A + BI_o$ relationships due to a statistical bias, *Ann. Geofi s.*, **41**, 17–26.

Mucciarelli, M., 1998b. Reliability and applicability of Nakamura's technique using microtremors: an experimental approach, *J. Earthq. Engin.*, **2**, 1–14.

Mucciarelli, M., Magri, L., and Albarello, D., 1992. For an adequate use of intensity data in site hazard estimates: mixing theoretical and observed intensities, in *Proc. X World Conference on Earthquake Engineering*, Balkema, Rotterdam, 345–350.

Mucciarelli, M., Peruzza, L., and Caroli, P., 2000. Calibration of seismic hazard estimates by means of observed site intensities, *J. Earthq. Eng.*, **4**, 141–159.

Musson, R. M. W., 1999. Determination of design earthquakes in seismic hazard analysis through Monte-Carlo simulations, *J. Earthq. Eng.*, **3**, 463–474.

Musson, R. M. W., 2000. The use of Monte Carlo simulations for seismic hazard assessment in the U.K., *Ann. Geofi s.*, **43**, 1–9.

Nakamura, Y., 1989. A method for dynamic characteristics estimation of subsurface using microtremor on the ground surface, *Quarterly Report of RTRI*, **30**, 25–33.

Nakamura, Y., 2000. Clear identification of fundamental idea of Nakamura's technique and its applications, *Proceedings of 12th World Conference on Earthquake Engineering*, New Zeland.

Newman, A., Stein, S., Weber, J., Engeln, J., Mao, A., and Dixon, T., 1999. Slow deformation and lower seismic hazard at the New Madrid seismic zone, *Science*, **284**, 619–621.

Oppenheimer, D. H., Bakun, W. H., and Lindh, A. G., 1990. Slip partitioning of the Calaveras Fault, California, and prospects for future earthquakes, *J. Geophys. Res.*, **95**, 8483–8498.

Petersen, M., Bryant, W., Cramer, C., Cao, T., Reichle, M., Frankel, A., Lienkaemper, J., McCrory, P., and Schwartz, D., 1996. Probabilistic seismic hazard assessment for the state of California, *U.S. Geological Survey Open-fi le Report*, 96–706.

Petersen, M. D., Cramer, C. H., Reichle, M. S., Franckel, A. D., and Hanks, T. C., 2000. Discrepancy between earthquake rates implied by historical earthquakes and consensus geological source model for California, *Bull. Seism. Soc. Am.*, **90**, 1117–1132.

Pollitz, F. F., Kellogg, L., and Burgmann., R., 2001. Sinking of mafic body in a reactivated lower crust: a mechanism for stress concentration in the New Madrid Seismic Zone, *Bull. Seism. Soc. Am.*, **91**, 1882–1897.

Rabinowitz, N., and Steinberg, D. M., 1991. Seismic hazard sensitivity analysis: a multiparameter approach, *Bull. Seism. Soc. Am.*, **81**, 796–817.

Reasenberg, P. A., and Jones, L. M., 1989. Earthquake hazard after a mainshock in California, *Science*, **243**, 1173–1176.

Reasenberg, P. A., Hanks T. C., and Bakun, W. H., 2003. An empirical model for earthquake probabilities in the San Francisco bay region, California, 2002-2031, *Bull. Seism. Soc. Am.*, **93**, 1–13.

Reiter, L., 1990. *Earthquakes Hazard Analysis*, Columbia University Press, New York.

Romeo, R., and Pugliese, A., 2000. Seismicity, seismotectonics and seismic hazard in Italy, *Eng. Geol.*, **55**, 241–266.

Satake, K., Shimizaki, K., Tsuji, Y., and Ueda, K., 1996. Time and size of a giant earthquake in Cascadia inferred from Japanese tsunami records of January 1700, *Nature*, **379**, 246–249.

Schwartz, D. P., 1996. The case against huge earthquakes (Opinion), *Seism. Res. Lett.*, **67**(3), 3–5.

Schweig, E. S., and Ellis, M. A., 1994. Reconciling short recurrence intervals with minor deformation in the New Madrid seismic zone, *Science*, **264**, 1308–1311.

Shedlock, K. M., 1999. Seismic hazard map of North and Central America and the Caribbean, *Ann. Geofi s.*, **42**, 977–997.

Shedlock, K. M., Giardini, D., Grunthal, G., and Zhang, P., 2000. The GSHAP Global Seismic Hazard Map, *Seism. Res. Lett.*, **71**, 679–686.

Slejko, D., Peruzza, L., and Rebez, A., 1998. Seismic hazard maps of Italy, *Ann. Geofi s.*, **41**, 183–214.

Stein, R. S., and Hanks, T. C., 1998. $M6$ earthquakes in southern California during the twentieth century: no evidence for a seismicity or moment deficit, *Bull. Seism. Soc. Am.*, **88**, 635–652.

Stirling, M. W., and Wesnousky, S. G., 1997. Do historical rates of seismicity in southern California require the occurrence of earthquake magnitudes greater than would be predicted from fault length?, *Bull. Seism. Soc. Am.*, **87**, 1662–1666.

Tuttle, M. P., Sims, J. D., Dyer-Williams, K., Lafferty, I. R. H., and Schweig, E. S., 2000. *Dating of Liquefaction Features in the New Madrid Seismic Zone*, U.S. Nuclear Regulatory Commission, NUREG/GR-0018.

Veneziano, D., Cornell, C. A., and O'Hara, T., 1984. Historical method of seismic hazard analysis, *Elect. Power Res. Inst.*, Rep. NP-3438, Palo Alto.

Ward, S., 1994. A multidisciplinary approach to seismic hazard in Southern California, *Bull. Seism. Soc. Am.*, **84**, 1293–1309.

Weichert, D. H., 1980. Estimation of earthquake recurrence parameters for unequal observation periods for different magnitudes, *Bull. Seism. Soc. Am.*, **70**, 1337–1356.

Wells, D. L., and Coppersmith, K. J., 1994. New empirical relationships among magnitude, rupture length, rupture width, rupture area, and surface displacement, *Bull. Seism. Soc. Am.*, **84**, 974–1002.

WG95. Working Group on California Earthquake Probabilities, 1995. Seismic hazards in southern California: probable earthquakes, 1994 to 2024, *Bull. Seism. Soc. Am.*, **85**, 379–439.

WG99. Working Group on California Earthquake Probabilities, 1999. Earthquake probabilities in the San Francisco Bay region: 2000 to 2030, a summary of findings, *U.S. Geological Survey Open-fi le Report*, 99–517.

Wirgin, A., and Bard, P. Y., 1996. Effects of building on the duration and amplitude of ground motion in Mexico City, *Bull. Seism. Soc. Am.*, **86**, 914–920.

Wu, S. C., Cornell, A. C., and Winterstein, S. R., 1995. A hybrid recurrence model and its implication on seismic hazard results, *Bull. Seism. Soc. Am.*, **85**, 1–16.

Chapter 5

Time-dependent hazard estimates and forecasts, and their uncertainties

Editors' introduction. This chapter discusses time-dependent hazard estimation and forecasting. Work in this area is based on the hope that by using information on past earthquakes together with a model better estimates of hazards can be obtained than by time-independent estimates. Section 5.1 presents some approaches to this issue, and section 5.2 presents a data-based approach to estimating earthquake occurrence probabilities based on the history of past seismic activity. Section 5.3 looks at hazard estimation from a statistical point of view and strongly questions the significance of some of the approaches now being used. These disparate points of view reflect an ongoing controversy within the scientific community; the presentations in this chapter make it clear what the issues are. Note that all three sections of this chapter were written independently, and that none of the authors saw the other sections in the course of writing or editing their own contribution.

5.1 USGS and partners: research on earthquake probabilities

[By A. J. Michael, N. Field, A. Frankel, J. Gomberg, and K. Shedlock]

In section 4.2 we discussed current USGS efforts to estimate both long-term and short-term earthquake probabilities. Here we discuss a number of research topics that may help to improve these probability estimates. While many other topics could be discussed, these are representative of current work at the USGS. All of the work discussed in this section is by USGS authors and their collaborators. This section is not intended as a general review, because a great deal of work done outside the USGS is not covered.

Many of the probability estimates discussed in section 4.2 are obtained using empirical models and as such are heavily dependent on having a good record of past earthquakes. Much work continues on improving this record through reanalysis of historical documents and seismograms, obtaining new information such as paleoseismic data, and improving geodetic data. Despite these efforts we will always be fundamentally limited by the lack of data on large earthquakes and thus our progress toward better estimates of earthquake probabilities will also rely heavily on understanding the physics of the earthquake source process. This part of the problem is discussed in this section.

5.1.1   Physics, recurrence, and probabilities

Simple, statistical models used to estimate earthquake probabilities can be used to make earthquake hazards assessments without any understanding of the physical processes behind earthquakes. For instance, the seismicity rate methods discussed in section 4.2 assume only that the future will have the same number and sizes of earthquakes as the past. One way to refine earthquake probabilities is to move toward time-dependent models as exemplified by the Working Group on California Earthquake Probabilities (WG99) report on the San Francisco Bay Region (see section 4.2.3). However, knowing which model to use will depend on increasing our knowledge of the basic physics of the earthquake generation process.

The Brownian Passage Time (BPT) model (*Ellsworth et al.*, 1998) can be used to frame a few of these issues currently being addressed by research at the USGS. These questions are not unique to the BPT model (many are shared by any point model of recurring earthquakes) but it provides a framework for this discussion.

First, there is the question of the value of the state variable after a large earthquake. Does it drop identically to 0 after all events despite the great spatial complexity observed in moment released on the mainshock fault plane? For instance, the Loma Prieta mainshock had most of its slip on two separate patches of fault. Will a future earthquake repeat this pattern or fill in the gap between these two patches? If geologic structure controls the rupture pattern (e.g., *Michael and Eberhart-Phillips*, 1991) then the position of large slip patches may remain static. However, the number of asperities (or segments) that are linked together into a single event will likely vary with time. Alternatively, if the slip distribution is largely controlled by the preseismic stress distribution and not by structure, then the slip patterns and the locations of asperities could be very different from event to event. If either the position of the asperities or the number of asperities linked together in an individual event varies from earthquake to earthquake then models based on recurrence of identical events will be flawed.

Improving our understanding of the different possibilities requires improving our understanding of fault zone structure and processes. This is done by imaging fault zones with precise earthquake relocations (e.g., *Waldhauser and Ellsworth*, 2000), potential field studies (e.g., *Jachens and Griscom*, 2003), seismic imaging (e.g. *Michael and Eberhart-Phillips*, 1991), and geologic mapping. As the structure of fault zones becomes better understood, laboratory studies of friction and fluid processes may improve computational models of fault zone behavior (e.g., *Beeler et al.*, 2000). These models can then be used to test if asperities are fixed or move as a function of time and whether or not segmentation models are valid representations of the earthquake process. Surficial observations of fault zones are limited in their resolution and so borehole observations of fault zone properties are enticing. Current work toward drilling along the Parkfield segment of the San Andreas fault zone (*Hickman et al.*, 1994 and http://www.icdp-online.de/html/sites/sanandreas/index) should further our knowledge of fault zone materials and processes at the shallow edge of the seismogenic zone. This will be done both by analyzing cores and cuttings from the drill hole and by creating an observatory at depth in the fault zone. Seismometers and strainmeters close to the fault will have a better chance of constraining proposed models of earthquake physics than instruments at the surface and thus relatively far from the earthquakes.

Another part of the BPT model is the temporal evolution of the state variable. Currently the model superimposes Brownian noise on a linear trend. This linear trend is an analog of the slow motion of tectonic plates, which should constantly add strain to seismogenic faults. However, viscoelastic processes could modify this linear trend and affect the form of the earthquake probabilities. For instance, postseismic viscoelastic processes that transfer stress from the lower crust to the seismogenic faults could result in a relatively rapid, partial reloading of the faults. This could then result in greater variability in earthquake repeat times. Time-varying strain accumulation may also be a factor in intraplate regions, such as New Madrid, where earthquakes are, presumably, not generated directly by tectonic plate motions. Modeling these processes depends on accurate deformation observations currently being made with GPS and InSAR and augmented with borehole strainmeters (e.g. *Pollitz*, 2002). The expansion of these networks as part of the proposed Plate Boundary Observatory should help constrain our models of these viscoelastic processes and their role in the earthquake generation process.

The final part of the BPT model is a constant state level at which failure occurs. In reality, failure is a more complicated process that may be better modeled using rate- and state-dependent (*Dieterich*, 1972) or other realistic friction laws. Work in USGS rock physics labs aims at better understanding the behavior of geologic materials so that future models can use more realistic failure criteria.

5.1.2 Earthquake triggering

Earthquake clustering has been observed at a variety of spatial and temporal scales, from aftershock sequences to the clustering of large earthquakes over long periods of time. Modeling the deviations from the random behavior found in a Poisson distribution, or using time-dependent models based on independent events, may provide an important refinement in future earthquake hazard assessments. However, the methods for making these calculations are still under development.

For example, static stress changes with rate- and state-dependent friction models have been used to explain aftershock sequences and other short-term clusters composed of many events (for a review see *Harris*, 1998). These models have also been used to estimate the probability of future large earthquakes along the North Anatolian fault after the 1999 Koaceli earthquake (*Parsons et al.*, 2000). Probability calculations which include rate- and state-dependent friction depend on both the amount of stress change applied to the fault and on how close the fault was to failure at the time of the stress change. When considering a large number of faults, the stress state on these faults can be modeled with a statistical distribution. This statistical approach has been used to model aftershock sequences (*Dieterich*, 1994). However, when considering the small number of large faults in a region (e.g., *Parsons et al.*, 2000) the probabilities may critically depend on knowing the average recurrence interval and the date of the last event on each fault. There are often significant uncertainties associated with the above parameters and even the underlying concepts of recurrence, repeating events, and fault interaction are controversial. For example, the reduced seismicity rate after the 1906 San Francisco earthquake is obvious. However, WG99 found it difficult to reproduce this pattern with a physically based probability model using elastic models of the static stress changes imposed by the 1906 mainshock on faults in the region. These models fit the seismicity well soon after the 1906 mainshock but do not predict the reduced seismicity rate throughout the rest of the 20*th* century. Another problem when dealing with large earthquakes in a region is that they are infrequent and we rarely have sufficient data to understand the variability of behavior in large earthquakes and their interactions. While the effects of the 1906 earthquake on the regional seismicity are dramatic, the past century of earthquakes is only one realization of this stochastic process and may not represent average behavior. Overcoming this lack of data in any given region will require global tests of stress triggering hypotheses as they are developed.

Finally, there are multiple stress triggering hypotheses currently being developed and tested. While models of static stress changes have received the most attention, some work supports even greater importance for the larger, but temporary, dynamic stress changes caused by the seismic wavefield (*Gomberg et al.*, 2001; *Parsons*, 2002). Both static and dynamic stress transfer are likely to have an effect on future seismicity and we need to determine which is the most important or if they need to be combined into a single model for accurate applications.

5.1.3 Conclusions

Improved, and especially time-dependent, earthquake probabilities will require new models of earthquake behavior. As we develop a better understanding of the basic processes involved we will need to create physically-based statistical distributions of earthquake recurrence on a regional network of faults. This will be far more complex than the simple point processes currently used. Succeeding at this task will require not only a better understanding of the basic processes but also the ability to produce a variety of numerical models of earthquake behavior on a regional scale and the data needed to validate or reject these models. These goals are just starting to become realistic as computational power increases and new data is collected.

5.2 Probabilistic forecasting of seismicity

[By Y. Y. Kagan, Y. F. Rong, and D. D. Jackson]

Predicting individual earthquakes is not possible now, but long-term probabilistic forecasts can be validated and provide useful information for managing earthquake risk. Short-term probability estimates are important for emergency and scientific response, but they are considerably more difficult to construct and test than long-term forecasts. Here we present three different forecast models, along with some quantitative tests of their effectiveness. The three models are for long-term and short-term forecasts based on seismicity, and for long-term forecasts based on geodetically observed strain rate.

Our long-term seismicity model is described in *Kagan and Jackson* (1994) and *Jackson and Kagan* (2000). We made specific forecasts of earthquakes over magnitude 5.8 for two regions of the Pacific Rim at the start of year 1999, and we present here a true prospective test of those forecasts. The short-term model is much harder to test formally, but we show here some reasons to be optimistic. We also present a 'pseudo-prospective' test (that is, using data collected after 1993 to test a model based on pre-1993 data) of the geodetic strain model for southern California.

We assume that there are two processes causing earthquakes: a steady state process for which earthquake occurrence is independent of past events, and a clustering process in which each earthquake depends on previous 'parent' events (*Vere-Jones*, 1970; *Kagan*, 1991; *Ogata*, 1998; *Jackson and Kagan*, 1999). Seismicity is approximated by a Poisson cluster process, in which clusters or sequences of earthquakes are statistically independent with constant rate although individual earthquakes in the cluster are dependent events. We refer to earthquakes with no obvious parents as 'independent' events, and their progeny as 'dependent' events. Dependent events need not to be smaller than their parents. In practice, we need not to identify specific earthquakes as independent or dependent, but the distinction is useful for modeling the separate functions that describe their rates.

5.2.1   Long-term seismic hazard estimates

We have developed a long-term model for independent events based on smoothed seismicity. The method assumes that the rate density (the probability per unit area, magnitude, and time) at a given location is proportional to the magnitude of nearby past earthquakes and approximately proportional to a negative power of the epicentral distance out to a few hundred kilometers. The method is described in more detail in *Kagan and Jackson* (1994, 2000), *Jackson and Kagan* (1999, 2000), and *Jackson et al.* (2001). We assume that the rate density does not depend on time, although our estimate of it changes as the earthquake catalogue evolves. Forecasts based on the model, and tests of the model, can be seen at http://scec.ess.ucla.edu/ykagan.html. An example of the rate density for earthquakes larger than moment-magnitude $m_w = 5.8$ for the Northwest Pacific region is shown in figure 5.1.

In all of our calculations, we describe earthquakes, and their probability densities, in terms of seismic moment, which is the directly observed quantity in the Harvard earthquake catalogue (*Dziewonski et al.*, 2001) we used. However, many readers will be more familiar with a description of earthquake size in terms of moment-magnitude, so we will often refer to magnitude in the discussion below. We use the notation M for the scalar seismic moment and m for the moment-magnitude, which we calculate using

$$m = \frac{2}{3}\log_{10}M - 6.0, \tag{5.1}$$

where M is measured in Nm. Readers should be aware that not all published definitions of moment-magnitude use the same constant 6.0 in equation (5.1). Some authors use values different by as much as 0.1. Because we base our calculations on seismic moment only, our results are self consistent, but the magnitudes for individual earthquakes, and the earthquake rates we report for a given magnitude, might differ slightly from those that would be obtained using a different constant.

We use a statistical model to fit the catalogue of earthquake times, locations, and seismic moments, and we base forecasts on this model. While most model components have been validated by extensive geophysical research, some require further investigation. We assume that the occurrence rate $\phi(\mathbf{x}, M)$ of independent events at location $\mathbf{x}$ with magnitude M may be written purely in terms of the marginal rates, i.e.,

$$\phi(\mathbf{x}, M) \propto \phi_{\mathbf{x}}(\mathbf{x}) \times \phi_M(M), \qquad (5.2)$$

where $\mathbf{x}$ are horizontal spatial coordinates, and $\phi_{\mathbf{x}}(\mathbf{x})$ and $\phi_M(M)$ are normalized probability densities of events in area and magnitude, respectively. Equation (5.2)

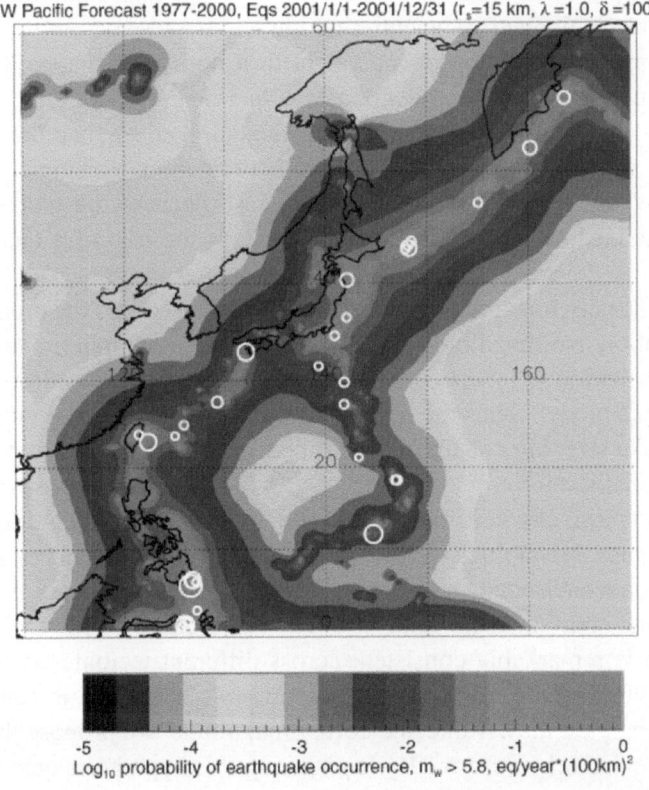

Figure 5.1: Color tones show the probability of earthquake occurrence calculated using the Harvard 1977-2000 catalogue; earthquakes 2001 are shown in white. Northwest Pacific long-term seismicity forecast: latitude limits from 0.25°S to 60.25°N, longitude limits from 109.75°E to 170.25°E.

signifies that the spatial and moment distributions of earthquake clusters are independent. The distribution $\phi_M(M)$ is defined by equation (5.3) below; a preliminary methodology for its use is described by *Kagan and Jackson* (1994), *Kagan and Jackson* (2000) and *Jackson and Kagan* (1999, 2000). The seismic moments are modeled as following the tapered Gutenberg-Richter distribution (*Kagan and Jackson*, 2000; *Kagan and Schönberg*, 2001; *Kagan*, 2002a):

$$1 - F_M(M) = \Phi_M(M) = M^{-\beta}\exp(-M/M_c) \quad \text{for} \quad M_t \leq M \leq \infty, \quad (5.3)$$

where $F_M(M)$ is a cumulative distribution function, M_t is a catalogue completeness threshold (cutoff), taken here to be $M_t = 10^{17.7}$ Nm ($m_t = 5.8$); and M_c is the parameter that controls the distribution in the upper ranges of M ('corner moment'). We assume that β and M_c are uniform over the entire area of study, and we estimate them by choosing the values that give the maximum likelihood fit to the earthquake catalogue for the entire region.

This moment distribution is equivalent to a normalized tapered Gutenberg-Richter magnitude distribution, defined by a 'b-value' and a 'corner magnitude'. The 'b-value' is 1.5 times the value β in the equation above, and the corner magnitude is the value obtained by converting M_c to magnitude using equation (5.1). The 'a-value' does not appear in equation (5.3), because the moment distribution is normalized, but the function $\phi(\mathbf{x}, M)$ plays the role of a locally variable 'a-value'.

The rate density function of equation (5.2) is factored into the product of functions of location and magnitude only, reflecting our assumption that the relative moment distribution is independent of location. Only the 'a-value' depends on location in this model. Thus the map of figure 5.1 represents the rate density for any earthquake size, after scaling the mapped values by a constant depending on the moment distribution.

The assumption of a uniform b-value and corner magnitude is somewhat controversial. Many seismologists have proposed that these parameters should depend on tectonic environment, geometry of faults, or other factors. However, our studies to date show that the ratio of small earthquakes to the overall tectonic moment rate is remarkably consistent across different tectonic environments in continental interiors and subduction zones (*Kagan*, 2002a, b). Because this ratio depends strongly on the b-value and corner magnitude, we propose that these parameters are relatively uniform. By formalizing this hypothesis quantitatively, as we have done here, we make it possible to test contrasting hypotheses against it using future large earthquakes. The alternative hypotheses for such a test must also be stated in terms of earthquake rate density.

We have begun a 'prospective' test of our model and earthquakes since the beginning of 1999 (*Kagan and Jackson*, 2000) are quite compatible with

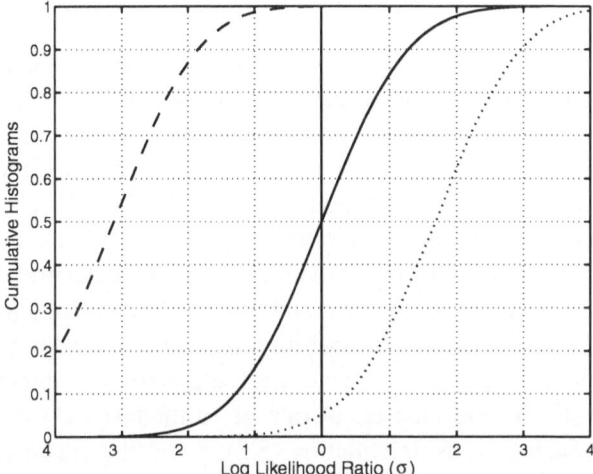

Figure 5.2: We use earthquakes from 1977 to 2000 as a control set. The solid line is the best Gaussian curve, having the same standard deviation as simulations. The dashed curve corresponds to simulation distributions for the NW-Pacifi c; the dotted curve to the SW-Pacifi c. Curves on the right from the Gaussian curve correspond to simulations worse than a real earthquake distribution; curves on the left correspond to simulations better than a real earthquake distribution.

it. To test the model we generated and scored a million synthetic earthquake catalogues each having the same number of events as the observed earthquake catalogue, and a spatial distribution consistent with the assumed $\phi_{\mathbf{x}}(\mathbf{x})$. The 'score' for each catalogue is the log-likelihood, computed by summing the log of the theoretical rate density $\phi_{\mathbf{x}}(\mathbf{x})$ at the location of each quake. We then computed the score for the actual catalogue, and compared it with the simulated values. Figure 5.2 shows cumulative histograms of the likelihood scores of synthetic catalogues relative to the observed ones for calendar year 2001. Figures for calendar years 1999 and 2000 were very similar (http://scec.ess.ucla.edu/~ykagan/tests_index.html). In most areas the observed likelihood score was within the central 95 per cent confidence limits based on the simulations. The exceptions were the SW Pacific in 1999, and the NW Pacific in 2001 (the latter is shown in figure 5.2). In both of the exceptional cases, the actual catalogue had a higher likelihood score than 97.5 per cent of the simulations.

When we defined spatial smoothing kernels $\phi_{\mathbf{x}}(\mathbf{x})$, we optimized them to forecast the latter half of a catalogue using the first catalogue half as a training set (*Kagan and Jackson*, 1994, 2000). Clearly, the kernels set to forecast 10–12 years in advance may not perform as well in yearly forecasts. More research is needed to develop appropriate spatial kernels.

We judge the model to represent the catalogue well when the likelihood of the observed catalogue falls in the middle range of synthetic values. When most of the synthetic catalogues plot to the right of the observed one, then the observed earthquakes are occurring in lower probability regions than expected, and the model fails. When most of the synthetic catalogues plot to left of the observed, then the observed earthquakes are falling in the high probability regions even more than expected, and the model fails by over-smoothing the earthquake probabilities.

The relationship between the forecast and observed earthquake locations can be more easily seen in displays of probability and earthquake location concentration, shown in figure 5.3a, b. To make these diagrams we divide the region into small cells; estimate the theoretical rate of earthquakes above the magnitude threshold for each cell, using equation (5.2); count the events that actually occurred in each cell; sort the cells in decreasing order of theoretical rate; and compute cumulative values of forecast and observed earthquake rates as plotted in the figures. We use two criteria to evaluate the forecasts using these figures. First, the most useful forecasts will have most of the probability concentrated in a small area; for such a forecast, the cumulative forecast rate will rise very steeply on the left side of the diagram and flatten out to the right. Second, for a successful forecast the observed earthquake distribution should mimic the theoretical forecast distribution. In fact the likelihood score for the actual catalogue can be computed directly from the table used to produce the concentrations diagrams like figure 5.3a, b.

In figure 5.3a (NW Pacific) *all* earthquakes are concentrated in the 'hottest' 30 per cent of the area, whereas the forecast predicts only about 93 per cent of epicenters to be in this part of the Pacific. The forecast for the NW Pacific over-smoothed the seismicity, and a more concentrated kernel function $\phi_x(x)$ would have provided a better fit. This explains why the model fails for having too high a likelihood, as shown in figure 5.2. Earthquakes in the SW Pacific were concentrated more nearly as forecasted, as shown in figure 5.3b. This explains why the forecast is within the 95 per cent confidence limits as shown in figure 5.2.

5.2.2   Short-term seismic hazard estimates

We are exploiting short-term trends in earthquake catalogues to provide daily estimates of the probabilities of strong earthquakes throughout the globe, and especially on the Pacific Rim. Short-term hazards may be viewed as temporary perturbations to the long-term earthquake potential. By short-term we mean earthquake hazard estimates on the order of a few days, weeks or even months. Our preliminary investigations indicate that following most mod-

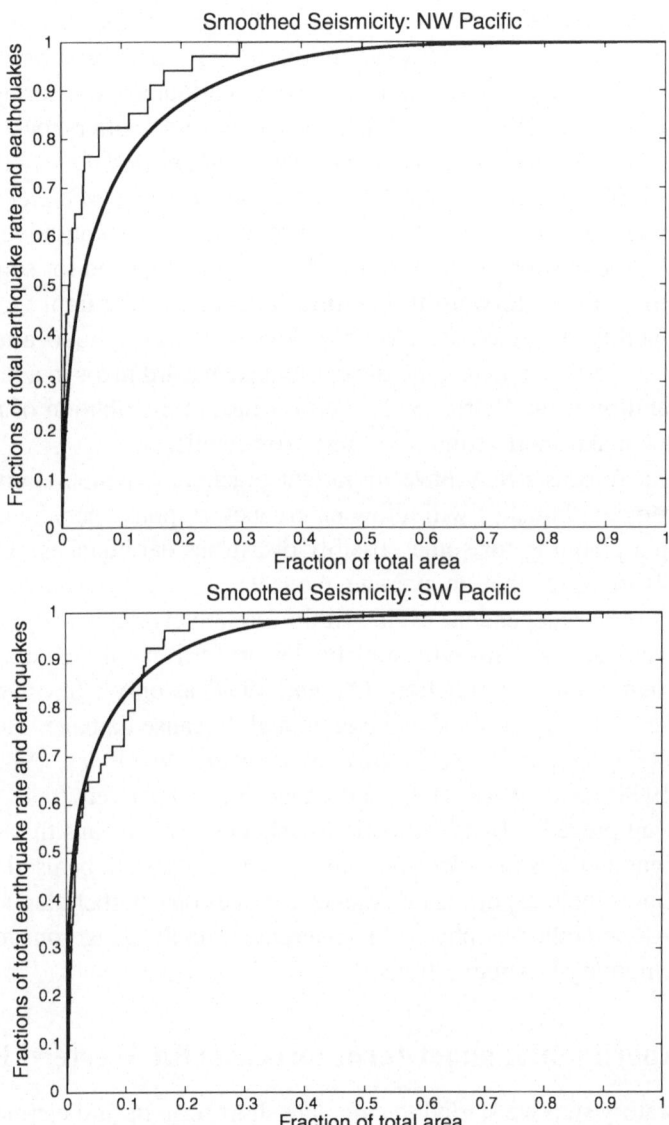

Figure 5.3: Concentration diagrams for forecast probability and earthquakes occurring in 2001. Solid line – probability distribution, thin line – earthquakes. (top) NW Pacifi c; (bottom) SW Pacifi c.

erate earthquakes (magnitude 6–7), seismic activity is dramatically increased for a few weeks, although the probability is measurably higher for a few years. Our forecasts are based on the most recent seismic record. They are and will be updated continuously and published on the World-Wide Web (http://scec.ess.ucla.edu/~ykagan/predictions_index.html).

Our short-term forecasts are largely based on earthquake clustering, the best-known manifestations of which are foreshock-mainshock-aftershock sequences. *Kagan and Knopoff* (1987), *Kagan* (1991), *Reasenberg and Jones* (1989), *Utsu and Ogata* (1997), *Ogata* (1988, 1998), *Console* (1998), *Michael and Jones* (1998), *Wiemer* (2000), and *Ebel et al.* (2000) described quantitative models of clustering. Although short-term clustering has been recognized for some time, its application to real time forward forecasting was not feasible until recently. The present availability of earthquake CMT solutions within a few hours after an event, the ability of fast computers to generate earthquake hazard maps in a few minutes, and the capability of the World-Wide Web for instant distribution of results to a wide audience make short-term forecasting worthwhile.

Short-term forecasts may have important practical benefits. Early alerts to probable strong earthquakes will allow emergency response personnel to antici-pate and begin planning for some possibly disastrous earthquakes. Our prelim-inary analysis indicates that on average about 9 per cent of earthquakes are fol-lowed by a stronger dependent event within a relatively short time (a few days or weeks). This fact explains why statistical short-term prediction does not meet the popular definition of 'prediction' (*Kagan*, 1997) as only a few alarms would be followed closely by earthquakes large enough to cause damage. However, in-vestigations (*Kagan and Knopoff*, 1987; *Reasenberg and Jones*, 1989; *Michael and Jones*, 1998; *Reasenberg*, 1999) show that from 20 per cent to 40 per cent of mainshocks are preceded by identifiable foreshocks. This means that some inex-pensive mitigation measures (*Molchan and Kagan*, 1992) can be used. Examples of such measures include putting emergency services on a higher alert status, step-ping up InSAR and other geophysical measurements in the dangerous regions, and deploying temporary seismic stations.

5.2.3 Experimental short-term forecasts for Western Pacific

As an exploratory step we started to calculate short-term hazard estimates for the western Pacific region. The forecasts are computed from the preliminary Har-vard CMT catalogue (*Dziewonski et al.*, 2001). We use email messages sent by the Harvard team to update our catalogue for all earthquakes $M \geq 10^{17.7}$ Nm ($m \geq 5.8$); the time delay between earthquake occurrence and update is on the order of a few hours, or sometimes one day. This catalogue is then used to es-timate both long- and short-term daily probabilities of future earthquake occur-rence. The short-term values are for one-day periods from the current midnight to the subsequent midnight, Los Angeles time. These forecasts are stored in a table and displayed in two figures, which are accessible via the World Wide Web

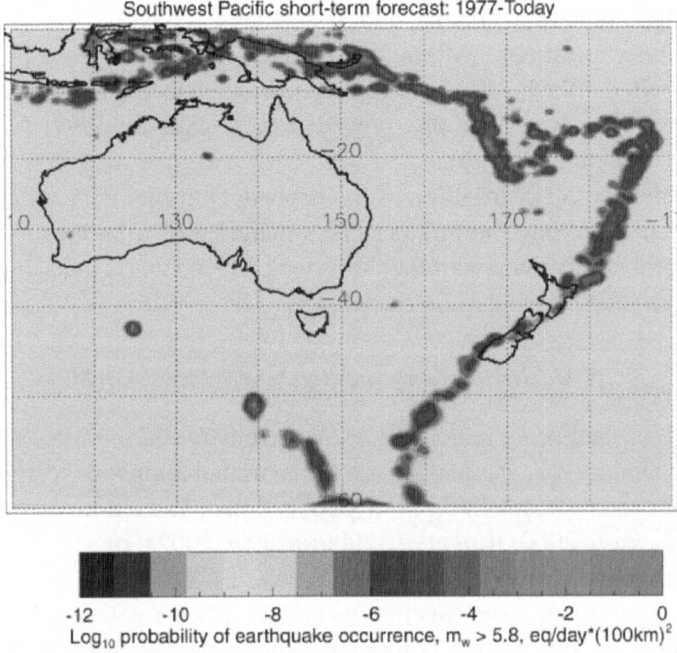

Figure 5.4: Short-term seismicity forecast for southwest Pacific. Color tones show the probability of earthquake occurrence calculated using the Harvard catalogue starting with January 1, 1977. This forecast as well as the forecast for the northwest Pacific is updated every day. The updated plots are available from http://scec.ess.ucla.edu/~ykagan/predictions_index.html/.

(http://scec.ess.ucla.edu/ykagan.html, see "FORECASTS FOR 1999-2002: TABLES AND FIGURES"). An example of such a forecast for the southwest Pacific region is shown in figure 5.4. The red glow in the New Ireland region shows the influence, as of February 13, 2002, caused by the sequence of strong earthquakes starting on November 16, 2000. After 15 months the earthquake occurrence rate is still much higher than the background rate. the recent sequence of four moderate events near Solomon islands (April 19, 2001, latitude 7.5° S, longitude 156° E) is also highlighted by its red color.

Each cluster is closely approximated by a stochastic space-time critical branching process. The space-time distribution of interrelated earthquake sources within a sequence is controlled by simple relations justified by analyzing the available statistical data on seismicity. Usually the first event in a sequence is the largest one and it is called a mainshock. Other dependent events are called aftershocks. If the first event in a sequence is smaller than subsequent shocks, it is called a foreshock. Retrospectively, it is relatively easy to subdivide an earthquake catalogue into fore-, main-, and aftershocks. However, in real time any event might be a foreshock. Although it is likely that dependent events will be smaller and called aftershocks, there is a significant chance that such earthquakes may be bigger than the events that preceded them (*Kagan*, 1991; *Michael and Jones*, 1998; *Reasenberg*, 1999).

We assume that the distribution of the dependent events *within* a cluster may also be broken down into the product of its marginal distributions. That is, the conditional probability density of the *j*-th shock dependent on the *i*-th mainshock ($j > i$) with seismic moment M_i is modeled as

$$\psi(\tau, \rho, M_j | M_i) = \psi_\tau(\tau) \times \psi_\rho(\rho) \times \psi_M(M_i) \times \phi_M(M_j), \qquad (5.4)$$

where $\tau = t_j - t_i$ and ρ is the horizontal distance between the *i*-th and *j*-th centroids. The functions ψ_τ, ψ_ρ, and ψ_μ are the marginal temporal, spatial, and moment densities, and are detailed in our publications (*Jackson and Kagan*, 1999; *Kagan and Jackson*, 2000; *Bird et al.*, 2000; *Kagan*, 2002a, b).

Quantitatively, the marginal densities behave as follows (*Kagan*, 1991; *Kagan and Jackson*, 2000). The function $\psi_\tau(\tau)$ describes a power-law decrease in earthquake rate *vs.* time, which is like the Omori law. We make an adjustment near $\tau = 0$ to prevent singularity of ψ_τ. The exponent in $\psi_\tau(\tau)$, estimated by the maximum likelihood (*Kagan*, 1991), differs substantially from the Omori exponent (section 2.3.2) because the earthquake rate density behaves on all previous events, not just on one previous mainshock. The spatial function $\psi_\rho(\rho)$ depends approximately as a Gaussian function of epicentral distance ρ (*Kagan*, 1991; *Kagan and Jackson*, 2000). The moment distribution $\psi_M(M_i)$ increases as a power law with the moment of previous events; $\phi_M(M_j)$ decreases as a negative power of the moment of later events. Thus large events increase the probability more than smaller events, and at any time large events are less probable than small ones. The density ϕ_M is proportional to the derivative of Φ_M in equation (5.3).

The earthquake rate density at a location is the sum of the values calculated from equation (5.4) for all past earthquakes. The individual terms from equation. (5.4) do not depend on whether the past earthquakes were foreshocks, mainshocks, or aftershocks, so this distinction becomes irrelevant for estimating earthquake potential. However, some other methods use a formula like equation (5.4), but the enhanced earthquake rate is presumed to stem from just one main event, so there is only one term in the sum. In such a case the choice of which event is called the mainshock could affect very strongly the estimated earthquake rate.

Although the focal mechanisms of future earthquakes might depend on the focal mechanism of previous events in the cluster (*Kagan and Jackson*, 1994), we presently assume that the distribution of focal mechanisms within a cluster is the same as the distribution governing long-term focal mechanism. Within an

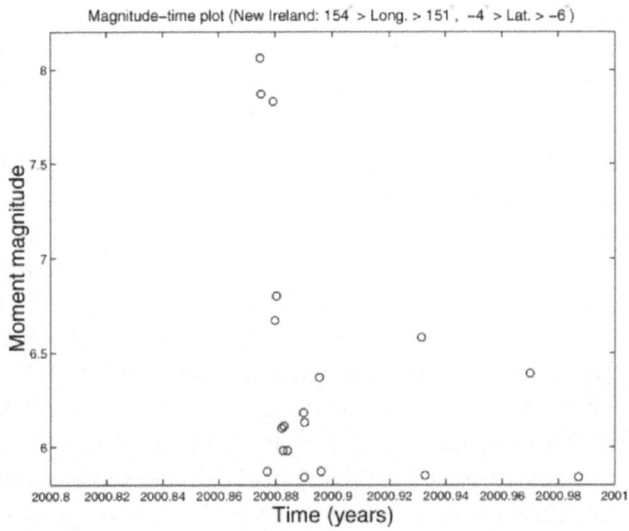

Figure 5.5: Time-magnitude plot for the New Ireland region.

earthquake sequence the focal mechanisms appear to be very tightly clustered, with the Cauchy rotational distribution describing the scatter of the 3D angle of rotation (*Kagan*, 2000).

Figures 5.5–5.7 show how the method would apply to the New Ireland sequence of large earthquakes which started on November 16, 2000. In figure 5.5 the time-magnitude sequence of events is shown. It is clear that the first three large earthquakes, which occurred in rapid succession, have very similar moment-magnitudes (8.06, 7.87, 7.83 in the final Harvard catalogue). The magnitude differences are comparable with the magnitude error in the catalogue (*Kagan*, 2002a). Their magnitudes in a preliminary catalogue are 8.03, 7.61, and 7.57, respectively. The mainshock would be the same in either the preliminary or final catalogue, but it is clear that in some cases the identity of the mainshock (i.e. the largest shock in a sequence) might change between the preliminary and final catalogue. Thus our method of summing over all events to estimate rate density has a strong advantage over methods based on using only a single event.

The closeness of magnitude values and their change from a preliminary to the final catalogue would present a serious difficulty for any forecasting procedure based on the Omori law and empirical rules (*Wiemer*, 2001). Such procedures depend on a real-time identification of foreshocks, mainshocks, and aftershocks. Probabilities are functions of time after the mainshock, so results are very sensitive to the choice of which event is called the mainshock. No such problem arises for our method. Since the model is based on a stochastic process formulation, no earthquake identification is necessary and minor adjustments in earthquake magnitude would only cause minor changes in the forecasted earthquake occurrence rate.

We constructed a hybrid earthquake forecast by taking weighted average of the long- and short-term forecasts (*Jackson and Kagan*, 1999; *Kagan and Jackson*, 2000). The total rate-density in a hybrid model is the short-term component plus 80 per cent of the long-term rate-density. In figure 5.6 we display the time history of the hybrid model, with the long-term model for comparison, at one point near the center of the November 16, 2000 sequence. The reference point is 5.0° S, 152.5° E, i.e., near the centroid of the sequence (see figure 5.7). Figure 5.6 shows that the estimated short-term rates increase by a few orders of magnitude for a very short time following each nearby event. Each of the three large earthquakes on November 16 and 17 spiked the theoretical earthquake rates, although the first two events were close enough in time that their effects merge in figure 5.6. The November 16 and 17 sequence and its aftershocks kept the short-term rate more than five times the long-term rate for a period of about a month and a half.

Figure 5.7 displays the spatial distribution of earthquakes in the New

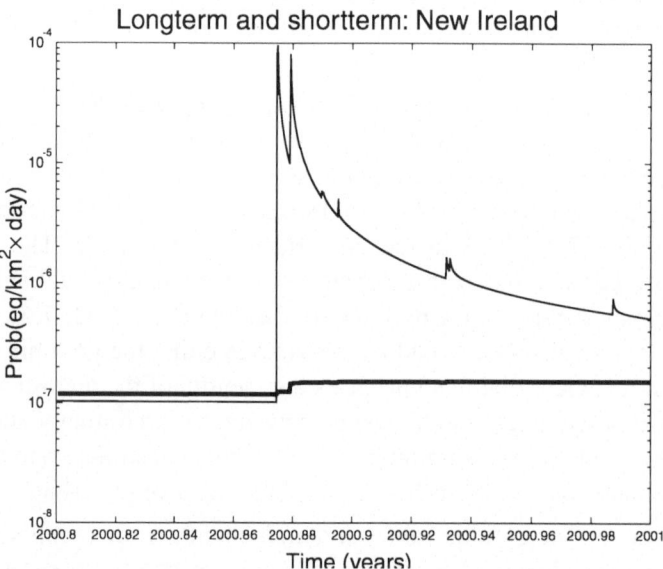

Figure 5.6: Time history of long-term and hybrid (short-term plus 0.8 × long-term) forecast for a point at 5.0° N, 152.5° E. Dark line is the long-term forecast; lighter line is the hybrid forecast.

Ireland region during 2000-2001. Earthquake positions are overlayed on the long-term forecast of seismic activity, which is based on the 1977–1999 earthquake history (*Jackson and Kagan*, 1999; *Kagan and Jackson*, 2000; *Jackson et al.*, 2001). Obviously, most of the earthquakes occurred in zones of a high forecasted activity level. A more quantitative measure of the forecasting efficiency (see FORECAST TEST FOR 2001: in http://scec.ess.ucla.edu/~ykagan/predictions_index.html) shows that our long-term forecast is quite satisfactory – the real catalogue falls within the 95 per cent confidence intervals of model prediction.

5.2.4   Experimental forecasts in Southern California

We also applied the technique described above to make a long-term seismicity forecast for southern California, although we did not put in on the web. The seismicity rate in California is too low for estimation of the smoothing kernel

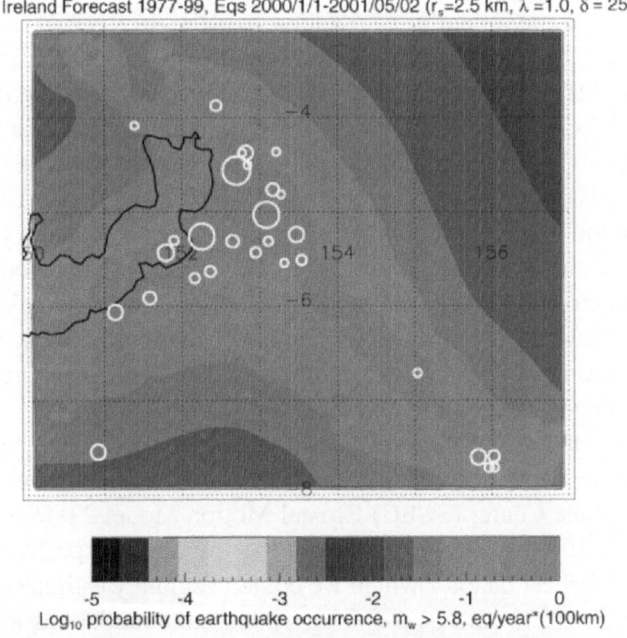

Figure 5.7: New Ireland region long-term seismicity forecast: color shows the rate-density of earthquake occurrence calculated using the Harvard 1977–1999 catalogue. Latitude limits are 3° to 8° South; longitude limits 150° to 157° West. Earthquakes in 2000 and 2001 (after the time of the forecast) are shown as white circles, with radius proportional to magnitude.

parameters by using one part of the catalogue to forecast a later part. Thus, we adopted the values of the kernel parameters from our analysis of global seismicity. However, in California we can extend our description of the past seismicity by representing earthquakes with magnitude 6.5 and greater as extended sources: i.e., as a sum of rectangular dislocation sources, and treating the separate dislocations as separate earthquakes. An example of such a representation of seismic sources is shown in figure 2 of *Kagan* (1994).

We calculate probabilities for earthquakes with $m > 5$ per unit area and time on a 5 km grid. For larger earthquakes the probabilities are lower according to the magnitude distribution, which we assume to be a tapered Gutenberg-Richter distribution with uniform b-value of 0.95 and a uniform corner magnitude of 8.5. The forecast model has just a few adjustable parameters: a normalizing constant, a smoothing distance, and an anisotropy factor. For California earthquakes we can almost always determine, using fault geology data, which of two planes in a focal mechanism solution is the fault plane. In a global catalogue such a decision is more difficult and is based on a statistical guess (*Kagan and Jackson*, 1994; *Kagan and Jackson*, 2000) which is correct only in about 75 per cent cases.

We have used the model in a 'pseudo-prospective' forecast in southern California (figure 5.8). We defined the probabilities using earthquakes before the beginning of 1993 and tested against later earthquakes, obtaining diagrams similar to figure 5.3a, b. The smoothed seismicity model predicts that 90 per cent of the earthquakes should lie in the 'hottest' 58 per cent of the area covered, while in fact all events after 1993 lay in the hottest 41 per cent. Thus, as in figure 5.3a, we smoothed too much. Optimizing the parameters with southern California data will probably result in a 'sharper' forecast. We can also use the model to construct random synthetic earthquake catalogues, including focal mechanisms. Since the likelihood score, a measure of compatibility of data and theory, was about the same for the real catalogue as it was for the simulated ones, the forecast was quite consistent with observed earthquakes.

We constructed an Earthquake Potential Model (figure 5.9) based on maximum horizontal shear strain rate, evaluated in 1993, based on the Southern California Earthquake Center (SCEC) Crustal Motion Model 2.0 (*Shen et al.*, 1996; *Jackson et al.*, 1997). We chose 1993 as the 'pseudo-prospective' test date because it is the earliest date for which we could accurately estimate the strain rate. Figure 5.9 also shows earthquakes after 1993. This model has a normalization constant and a smoothing parameter.

We set the normalization constant to fit the long-term earthquake rate, and we chose the smoothing to fit the observed GPS velocities. We assumed the same magnitude distribution as for the smoothed seismicity model. From a diagram similar to figure 5.3a, b, we infer that the geodetic model forecasted 90 per cent of the earthquakes in the fastest deforming 67 per cent of the area. In fact, all events

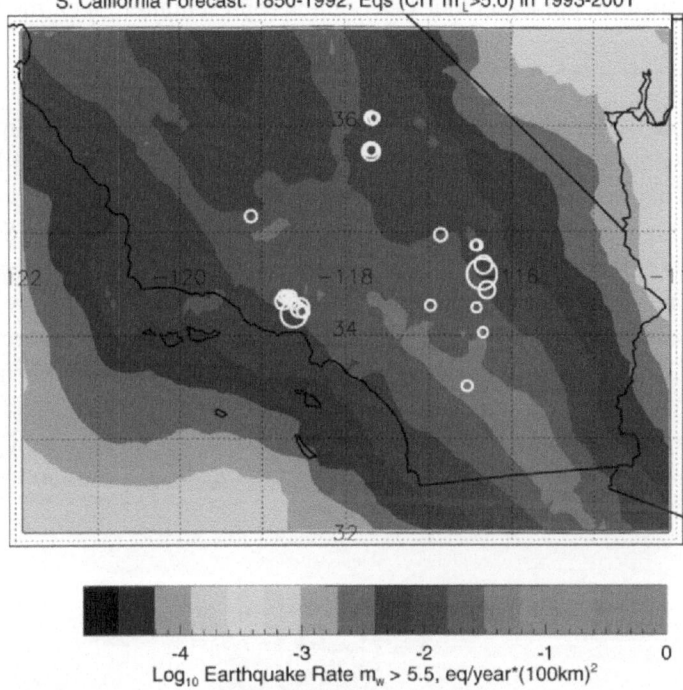

Figure 5.8: Long-term seismicity forecast for southern California: Latitude limits 32.0–37.0°N, longitude limits 114.0–122.0°W. Earthquakes after December 31, 1992 are shown in white. The size of circles is proportional to earthquake magnitude. Color scale shows the long-term probability of earthquake occurrence calculated with the historical and Harvard 1977–1992 catalogue.

after 1993 occurred in this area. In our calculations we excluded from consideration the parts of the southern California where strain data are not available (NE and SW parts of figure 5.9), thus probability values in our forecast are more uniform. The simulated earthquake catalogues looked very much like the observed one.

5.2.5 Conclusions

The best indicator of earthquake probability appears to be the occurrence of recent earthquakes. We regard the seismicity models as 'null hypotheses'. In other words, they are necessary first steps required to test more sophisticated models based on earthquake physics, active faults, or tectonic stresses. The long-term seismicity model assumes that earthquake probability does not vary with time, while the short-term model includes an explicit representation of earthquake clustering.

In general the long-term seismicity model performs well, although in two cases the model failed our two-sided likelihood test because the likelihood was too high. This occurred because the observed earthquakes were concentrated more in the higher probability regions than the model had predicted. Short-term probabilities seldom reach the high levels that would justify the label 'earthquake prediction', but they do vary enough to justify preliminary emergency activities in some circumstances. Aftershocks exemplify earthquakes made more probable by previous events, but earthquake clustering goes well beyond simple aftershock sequences. Many large earthquakes are preceded by smaller events, and a quantitative clustering model can forecast those larger events reasonably well.

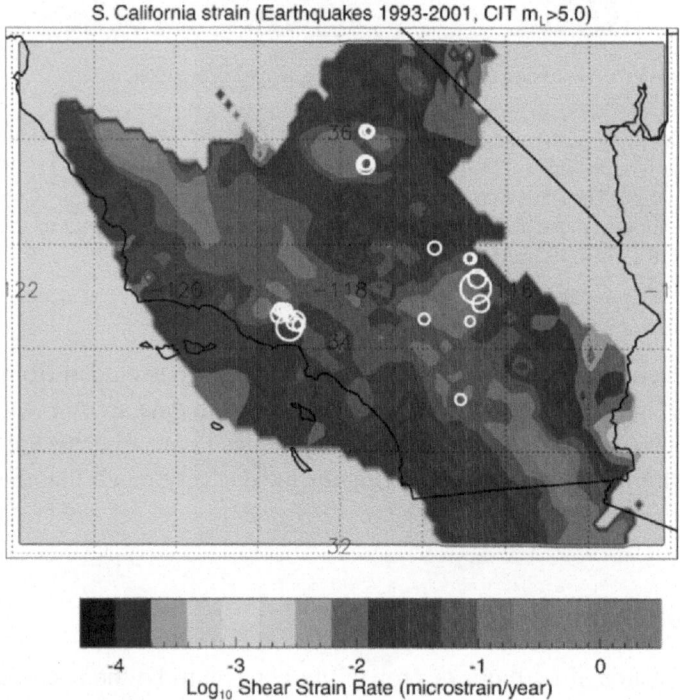

Figure 5.9: Same as fi gure 5.8 except that color scale tones show the long-term probability of earthquake occurrence calculated using strain map.

5.3 What is the chance of an earthquake?

[By P. B. Stark and D. A. Freedman]

The footnotes for this section are grouped at the end of this section.

What is the chance that an earthquake of magnitude 6.7 or greater will occur before the year 2030 in the San Francisco Bay Area? The U.S. Geological Survey estimated the chance to be 0.7 ± 0.1 (*WG99*, 1999). In this paper, we try to interpret such probabilities.

Making sense of earthquake forecasts is surprisingly difficult. In part, this is because the forecasts are based on a complicated mixture of geological maps, rules of thumb, expert opinion, physical models, stochastic models, numerical simulations, as well as geodetic, seismic, and paleoseismic data. Even the concept of probability is hard to define in this context. We examine the problems in applying standard definitions of probability to earthquakes, taking the USGS forecast (the product of a particularly careful and ambitious study) as our lead example. The issues are general, and concern the interpretation more than the numerical values. Despite the work involved in the USGS forecast, their probability estimate is shaky, as is the uncertainty estimate.

5.3.1 Interpreting probability

Probability has two aspects. There is a formal mathematical theory, axiomatized by *Kolmogorov* (1956). And there is an informal theory that connects the mathematics to the world, i.e., defines what 'probability' means when applied to real events. It helps to start by thinking about simple cases. For example, consider tossing a coin. What does it mean to say that the chance of heads is 1/2? In this section, we sketch some of the interpretations: symmetry, relative frequency, and strength of belief[1]. We examine whether the interpretation of weather forecasts can be adapted for earthquakes. Finally, we present Kolmogorov's axioms and discuss a model-based interpretation of probability, which seems the most promising.

Symmetry and equally likely outcomes

Perhaps the earliest interpretation of probability is in terms of 'equally likely outcomes', an approach that comes from the study of gambling. If the n possible outcomes of a chance experiment are judged equally likely (for instance, on the basis of symmetry) each must have probability $1/n$. For example, if a coin is tossed,

$n = 2$; the chance of heads is $1/2$, as is the chance of tails. Similarly, when a fair die is thrown, the six possible outcomes are equally likely. However, if the die is loaded, this argument does not apply. There are also more subtle difficulties. For example, if two dice are thrown, the total number of spots can be anything from 2 through 12—but these eleven outcomes are far from equally likely. In earthquake forecasting, there is no obvious symmetry to exploit. We therefore need a different theory of probability to make sense of earthquake forecasts.

The frequentist approach

The probability of an event is often defined as the limit of the relative frequency with which the event occurs in repeated trials under the same conditions. According to frequentists, if we toss a coin repeatedly under the same conditions[2], the fraction of tosses that result in heads will converge to $1/2$: that is why the chance of heads is $1/2$. The frequentist approach is inadequate for interpreting earthquake forecasts. Indeed, to interpret the USGS forecast for the Bay Area using the frequency theory, we would need to imagine repeating the years 2000–2030 over and over again, a tall order even for the most gifted imagination.

The Bayesian approach

According to Bayesians, probability means degree of belief. This is measured on a scale running from 0 to 1. An impossible event has probability 0; the probability of an event that is sure to happen equals 1. Different observers need not have the same beliefs, and differences among observers do not imply that anyone is wrong.

The Bayesian approach, despite its virtues, changes the topic. For Bayesians, probability is a summary of an opinion, not something inherent in the system being studied[3]. If the USGS says 'there is chance 0.7 of at least one earthquake with magnitude 6.7 or greater in the Bay Area between 2000 and 2030', the USGS is merely reporting its corporate state of mind, and may not be saying anything about tectonics and seismicity. More generally, it is not clear why one observer should care about the opinion of another. The Bayesian approach therefore seems to be inadequate for interpreting earthquake forecasts. For a more general discussion of the Bayesian and frequentist approaches, see *Freedman* (1995).

The principle of insufficient reason

Bayesians, and frequentists who should know better, often make probability assignments using Laplace's principle of insufficient reason (*Hartigan*, 1983, p. 2): if there is no reason to believe that outcomes are not equally likely, take them to be equally likely. However, not believed to be unequal is one thing; known to be equal is another. Moreover, all outcomes cannot be equally likely, so Laplace's prescription is ambiguous.

An example from thermodynamics illustrates the problem (*Feller*, 1968; *Reif*, 1965). Consider a gas that consists of n particles, each of which can be in any of r quantum states[4]. The state of the gas is defined by a 'state vector'. We describe three conventional models for such a gas, which differ only in the way the state vector is defined. Each model takes all possible values of the state vector—as defined in that model—to be equally likely.

1. Maxwell-Boltzmann. The state vector specifies the quantum state of each particle; there are

$$r^n$$

possible values of the state vector.

2. Bose-Einstein. The state vector specifies the number of particles in each quantum state. There are

$$\binom{n+r-1}{n}$$

possible values of the state vector[5].

3. Fermi-Dirac. As with Bose-Einstein statistics, the state vector specifies the number of particles in each quantum state, but no two particles can be in the same state. There are

$$\binom{r}{n}$$

possible values of the state vector[6].

Maxwell-Boltzmann statistics are widely applicable in probability theory[7] but describe no known gas. Bose-Einstein statistics describe the thermodynamic behavior of bosons, particles whose spin angular momentum is an integer multiple of $\hbar$, Planck's constant h divided by 2π. Photons and He^4 atoms are bosons. Fermi-Dirac statistics describe the behavior of fermions, particles whose spin angular momentum is a half-integer multiple of $\hbar$. Electrons and He^3 atoms are fermions[8].

Bose-Einstein condensates (very low temperature gases in which all the atoms are in the same quantum state) were first observed experimentally by *Anderson et al.* (1995). Such condensates occur for bosons, not fermions, compelling evidence for the difference in thermodynamic statistics. The principle of insufficient reason is not a sufficient basis for physics: it does not tell us when to use one model rather than another. Generally, the outcomes of an experiment can be defined in quite different ways, and it will seldom be clear a priori which set of outcomes, if any, obeys Laplace's dictum of equal likelihood.

Earthquake forecasts and weather forecasts

Earthquake forecasts look similar in many ways to weather forecasts, so we might look to meteorology for guidance. How do meteorologists interpret statements like "the chance of rain tomorrow is 0.7"? The standard interpretation applies frequentist ideas to forecasts. In this view, the chance of rain tomorrow is 0.7 means that 70 per cent of such forecasts are followed by rain the next day.

Whatever the merits of this view, meteorology differs from earthquake prediction in a critical respect. Large regional earthquakes are rare; they have recurrence times on the order of hundreds of years[9]. Weather forecasters have a much shorter time horizon. Therefore, weather prediction does not seem like a good analogue for earthquake prediction.

Mathematical probability: Kolmogorov's axioms

For most statisticians, Kolmogorov's axioms are the basis for probability theory, no matter how the probabilities are to be interpreted. Let Σ be a σ-algebra[10] of subsets of a set S. Let P be a real-valued function on Σ. Then P is a probability if it satisfies the following axioms:

- $P(A) \geq 0$ for every $A \in \Sigma$;

- $P(S) = 1$;

- if $A_j \in \Sigma$ for $j = 1, 2, \ldots$, and $A_j \cap A_k = \emptyset$ whenever $j \neq k$, then

$$P\left(\bigcup_{j=1}^{\infty} A_j\right) = \sum_{j=1}^{\infty} P(A_j).$$ (5.5)

The first axiom says that probability is nonnegative. The second defines the scale: probability 1 means certainty. The third says that if $A_1, A_2, \ldots$ are pairwise disjoint, the probability that at least one A_j occurs is the sum of their probabilities.

Probability models

Another interpretation of probability seems more useful for making sense of earthquake predictions: probability is just a property of a mathematical model intended to describe some features of the natural world. For the model to be useful, it must be shown to be in good correspondence with the system it describes. That is where the science comes in.

Here is a description of coin-tossing that illustrates the model-based approach. A coin will be tossed n times. There are 2^n possible sequences of heads and tails.

In the mathematical model, those sequences are taken to be equally likely: each has probability $1/2^n$, corresponding to probability $1/2$ of heads on each toss and independence among the tosses.

This model has observational consequences that can be used to test its validity. For example, the probability distribution of the total number X of heads in n tosses is binomial:

$$P(X = k) = \binom{n}{k} \frac{1}{2^n}.$$

If the model is correct, when n is at all large we should see around $n/2$ heads, with an error on the order of $\sqrt{n}$. Similarly, the model gives probability distributions for the number of runs, their lengths, and so forth, which can be checked against data. The model is very good, but imperfect: with many thousands of tosses, the difference between a real coin and the model coin is likely to be detectable. The probability of heads will not be exactly 1/2, and there may be some correlation between successive tosses.

The interpretation that probability is a property of a mathematical model and has meaning for the world only by analogy seems the most appropriate for earthquake prediction. To apply this interpretation, one posits a stochastic model for earthquakes in a given region, and interprets a number calculated from the model to be the probability of an earthquake in some time interval. The problem in earthquake forecasts is that the models, unlike the models for coin-tossing, have not been tested against relevant data. Indeed, the models cannot be tested on a human time scale, so there is little reason to believe the probability estimates. As we shall see in the next section, although some parts of the earthquake models are constrained by the laws of physics, many steps involve extrapolating rules of thumb far beyond the data they summarize; other steps rely on expert judgment separate from any data; still other steps rely on *ad hoc* decisions made as much for convenience as for scientific relevance.

5.3.2 The USGS earthquake forecast

We turn to the USGS forecast for the San Francisco Bay Area (*WG99*, 1999). The forecast was constructed in two stages. The first stage built a collection of 2,000 models for linked fault segments, consistent with regional tectonic slip constraints, in order to estimate seismicity rates. The models were drawn by Monte Carlo from a probability distribution defined using data and expert opinion[11]. We had trouble understanding the details, but believe that the models differed in the geometry and dimensions of fault segments, the fraction of slip released aseismically on each fault segment, the relative frequencies with which different combinations of fault segments rupture together, the relationship between fault area and earthquake size, and so forth.

Each model generated by the Monte Carlo was used to predict the regional rate of tectonic deformation; if the predicted deformation was not close enough to the measured rate of deformation, the model was discarded[12]. This was repeated until 2,000 models met the constraints. That set of models was used to estimate the long-term recurrence rate of earthquakes of different sizes, and to estimate the uncertainties of those rate estimates, for use in the second stage.

The second stage of the procedure created three generic stochastic models for fault segment ruptures, estimating parameters in those models from the long-term recurrence rates developed in the first stage. The stochastic models were then used to estimate the probability that there will be at least one magnitude 6.7 or greater earthquake by 2030.

We shall try to enumerate the major steps in the first stage (the construction of the 2,000 models) to indicate the complexity.

1. Determine regional constraints on aggregate fault motions from geodetic measurements.

2. Map faults and fault segments; identify fault segments with slip rates of at least 1 mm/yr. Estimate the slip on each fault segment principally from paleoseismic data, occasionally augmented by geodetic and other data. Determine (by expert opinion) for each segment a 'slip factor', the extent to which long-term slip on the segment is accommodated aseismically. Represent uncertainty in fault segment lengths, widths, and slip factors as independent Gaussian random variables with mean zero[13]. Draw a set of fault segment dimensions and slip factors at random from that probability distribution.

3. Identify (by expert opinion) ways in which segments of each fault can rupture separately and together[14]. Each such combination of segments is a 'seismic source'.

4. Determine (by expert opinion) the extent to which long-term fault slip is accommodated by rupture of each combination of segments for each fault.

5. Choose at random (with probabilities of 0.2, 0.2, and 0.6 respectively) one of three generic relationships between fault area and moment release to characterize magnitudes of events that each combination of fault segments supports. Represent the uncertainty in the generic relationship as Gaussian with zero mean and standard deviation 0.12, independent of fault area[15].

6. Using the chosen relationship and the assumed probability distribution for its parameters, determine a mean event magnitude for each seismic source by Monte Carlo simulation.

7. Combine seismic sources along each fault "in such a way as to honor their relative likelihood as specified by the expert groups" (*WG99*, 1999, p. 10); adjust the relative frequencies of events on each source so that every fault segment matches its geologic slip rate, as estimated previously from paleoseismic and geodetic data. Discard the combination of sources if it violates a regional slip constraint.

8. Repeat the previous steps until 2,000 regional models meet the slip constraint. Treat the 2,000 models as equally likely for the purpose of estimating magnitudes, rates, and uncertainties.

9. Steps 1-8 model events on seven identified fault systems, but there are background events not associated with those faults. Estimate the background rate of seismicity as follows. Use an (unspecified) Bayesian procedure to categorize historical events from three catalogues either as associated or not associated with the seven fault systems. Fit a generic Gutenberg-Richter magnitude-frequency relation $N(m) = 10^{a-bm}$ to the events deemed not to be associated with the seven fault systems. Model this background seismicity as a marked Poisson process. Extrapolate the Poisson model to $m \geq 6.7$, which gives a probability of 0.09 of at least one event[16].

This first stage in the USGS procedure generates 2,000 models and estimates long-term seismicity rates as a function of magnitude for each seismic source. We now describe the second stage, the earthquake forecast itself. Our description is sketchy because we had trouble understanding the details from the USGS report. The second stage fits three types of stochastic models for earthquake recurrence - Poisson, Brownian passage time (*Ellsworth et al.*, 1998), and 'time-predictable' - to the long-term seismicity rates estimated in the first stage[17]. Ultimately, those stochastic models are combined to estimate the probability of a large earthquake.

The Poisson and Brownian passage time models were used to estimate the probability that an earthquake will rupture each fault segment. Some parameters of the Brownian passage time model were fitted to the data, and some were set more arbitrarily; for example, aperiodicity (standard deviation of recurrence time, divided by expected recurrence time) was set to three different values, 0.3, 0.5, and 0.7. The Poisson model does not require an estimate of the date of last rupture of each segment, but the Brownian Passage Time model does; those dates were estimated from the historical record. Redistribution of stress by large earthquakes was modeled; predictions were made with and without adjustments for stress redistribution. Predictions for each segment were combined into predictions for each fault using expert opinion about the relative likelihoods of different rupture sources.

A 'time-predictable model' (stress from tectonic loading needs to reach the level at which the segment ruptured in the previous event for the segment to initiate a new event) was used to estimate the probability that an earthquake will originate on each fault segment. Estimating the state of stress before the last event requires knowing the date of the last event and the slip during the last event. Those data are available only for the 1906 earthquake on the San Andreas Fault and the 1868 earthquake on the southern segment of the Hayward Fault (*WG99*, 1999, p. 17), so the time-predictable model could not be used for many Bay Area fault segments.

The calculations also require estimating the loading of the fault over time, which in turn relies on viscoelastic models of regional geological structure. Stress drops and loading rates were modeled probabilistically (*WG99*, 1999, p. 17); the form of the probability models is not given. The loading of the San Andreas fault by the 1989 Loma Prieta earthquake and the loading of the Hayward fault by the 1906 earthquake were modeled. The probabilities estimated using the time-predictable model were converted into forecasts using expert opinion about the relative likelihoods that an event initiating on one segment will stop or will propagate to other segments. The outputs of the three types of stochastic models for each fault segment were weighted according to the opinions of a panel of fifteen experts. When results from the time-predictable model were not available, the weights on its output were in effect set to zero.

There is no straightforward interpretation of the USGS probability forecast. Many steps involve models that are largely untestable; modeling choices often seem arbitrary. Frequencies are equated with probabilities, fiducial distributions are used, outcomes are assumed to be equally likely, and subjective probabilities are used in ways that violate Bayes rule[18].

What does the uncertainty estimate mean?

The USGS forecast is 0.7 ± 0.1, where 0.1 is an uncertainty estimate (*WG99*, 1999). The 2,000 regional models produced in stage 1 give an estimate of the long-term seismicity rate for each source (linked fault segments), and an estimate of the uncertainty in each rate. By a process we do not understand, those uncertainties were propagated through stage 2 to estimate the uncertainty of the estimated probability of a large earthquake. If this view is correct, 0.1 is a gross underestimate of the uncertainty. Many sources of error have been overlooked, some of which are listed below.

1. Errors in the fault maps and the identification of fault segments[19].

2. Errors in geodetic measurements, in paleoseismic data, and in the viscoelastic models used to estimate fault loading and sub-surface slip from surface data.

3. Errors in the estimated fraction of stress relieved aseismically through creep in each fault segment and errors in the relative amount of slip assumed to be accommodated by each seismic source.

4. Errors in the estimated magnitudes, moments, and locations of historical earthquakes.

5. Errors in the relationships between fault area and seismic moment.

6. Errors in the models for fault loading.

7. Errors in the models for fault interactions.

8. Errors in the generic Gutenberg-Richter relationships, not only in the parameter values but also in the functional form.

9. Errors in the estimated probability of an earthquake not associated with any of the faults included in the model.

10. Errors in the form of the probability models for earthquake recurrence and in the estimated parameters of those models.

5.3.3 A view from the past

Littlewood (1953) wrote:

> "Mathematics (by which I shall mean pure mathematics) has no grip on the real world; if probability is to deal with the real world it must contain elements outside mathematics; the *meaning* of 'probability' must relate to the real world, and there must be one or more 'primitive' propositions about the real world, from which we can then proceed deductively (i.e. mathematically). We will suppose (as we may by lumping several primitive propositions together) that there is just one primitive proposition, the 'probability axiom', and we will call it *A* for short. Although it has got to be *true*, *A* is by the nature of the case incapable of deductive proof, for the sufficient reason that it is about the real world. [...]
>
> "There are two schools. One, which I will call mathematical, stays inside mathematics, with results that I shall consider later. We will begin with the other school, which I will call philosophical. This attacks directly the 'real' probability problem; what are the axiom *A* and the meaning of 'probability' to be, and how can we justify *A*? It will be instructive to consider the attempt called the 'frequency

theory'. It is natural to believe that if (with the natural reservations) an act like throwing a die is repeated n times the proportion of 6's will, *with certainty*, tend to a limit, p say, as $n \to \infty$. (Attempts are made to sublimate the limit into some Pickwickian sense—'limit' in inverted commas. But either you *mean* the ordinary limit, or else you have the problem of explaining how 'limit' behaves, and you are no further. You do not make an illegitimate conception legitimate by putting it into inverted commas). If we take this proposition as 'A' we can at least settle off-hand the other problem, of the *meaning* of probability; we define its measure for the event in question to be the number p. But for the rest this A takes us nowhere. Suppose we throw 1000 times and wish to know what to expect. Is 1000 large enough for the convergence to have got under way, and how far? A does not say. We have, then, to add to it something about the rate of convergence. Now an A cannot assert a *certainty* about a particular number n of throws, such as 'the proportion of 6's will *certainly* be within $p \pm \varepsilon$ for large enough n (the largeness depending on ε)'. It can only say 'the proportion will lie between $p \pm \varepsilon$ *with at least such and such probability (depending on ε and n_0) whenever $n > n_0$*'. The vicious circle is apparent. We have not merely failed to *justify* a workable A; we have failed even to *state* one which would work if its truth were granted. It is generally agreed that the frequency theory won't work. But whatever the theory it is clear that the vicious circle is very deep-seated: certainty being impossible, whatever A is made to state can be stated only in terms of probability".

5.3.4   Conclusions

Making sense of earthquake forecasts is difficult, in part because standard interpretations of probability are inadequate. A model-based interpretation is better, but lacks empirical justification. Furthermore, probability models are only part of the forecasting machinery. For example, the USGS San Francisco Bay Area forecast for 2000–2030 involves geological mapping, geodetic mapping, viscoelastic loading calculations, paleoseismic observations, extrapolating rules of thumb across geography and magnitude, simulation, and many appeals to expert opinion. Philosophical difficulties aside, the numerical probability values seem rather arbitrary.

Another large earthquake in the San Francisco Bay Area is inevitable, and imminent in geologic time. Probabilities are a distraction. Instead of making forecasts, the USGS could help to improve building codes and to plan the government's response to the next large earthquake. Bay Area residents should take reasonable precautions, including bracing and bolting their homes as well as securing water heaters, bookcases, and other heavy objects. They should keep first aid supplies, water, and food on hand. They should largely ignore the USGS probability forecast.

Notes

[1] See *Stigler* (1986) for history prior to 1900. Currently, the two main schools are the frequentists and the Bayesians. Frequentists, also called objectivists, define probability in terms of relative frequency. Bayesians, also called subjectivists, define probability as degree of belief. We do not discuss other theories, such as those associated with Fisher, Jeffreys, and Keynes, although we touch on Fisher's 'fiducial probabilities' in note 11.

[2] It is hard to specify precisely which conditions must be the same across trials, and, indeed, what 'the same' means. Within classical physics, for instance, if all the conditions were exactly the same, the outcome would be the same every time, which is not what we mean by randomness.

[3] A Bayesian will have a prior belief about nature. This prior is updated as the data come in, using Bayes rule: in essence, the prior is reweighted according to the likelihood of the data (*Hartigan*, 1983, pp. 29ff). A Bayesian who does not have a proper prior (that is, whose prior is not a probability distribution) or who does not use Bayes rule to update, is behaving irrationally according to the tenets of his own doctrine (*Freedman*, 1995). For example, the Jeffreys prior is generally improper, because it has infinite mass; a Bayesian using this prior is exposed to a money-pump (*Eaton and Sudderth*, 1999, p. 849; *Eaton and Freedman*, 2003). It is often said that the data swamp the prior: the effect of the prior is not important if there are enough observations (*Hartigan*, 1983, pp. 34ff). This may be true when there are many observations and few parameters. In earthquake prediction, by contrast, there are few observations and many parameters.

[4] The number of states depends on the temperature of the gas, among other things. In the models we describe, the particles are 'non-interacting'. For example, they do not bond with each other chemically.

[5] To define the binomial coefficients, consider m things. How many ways are there to choose k out of the m? The answer is given by the binomial coefficient

$$\binom{m}{k} = \binom{m}{m-k} = \frac{m!}{k!(m-k)!}$$

for $k = 0, 1, \ldots, m$. Let n and r be positive integers. How many sequences $(j_1, j_2, \ldots, j_r)$ of nonnegative integers are there with $j_1 + j_2 + \cdots + j_r = n$? The answer is

$$\binom{n+r-1}{n}.$$

For the argument, see *Feller* (1968). To make the connection with Bose-Einstein statistics, think of $\{j_1, j_2, \ldots, j_r\}$ as a possible value of the state vector, with j_i equal to the number of particles in quantum state i.

[6] That is the number of ways of selecting n of the r states to be occupied by one particle each.

[7] In probability theory, we might think of a Maxwell-Boltzman 'gas' that consists of $n = 2$ coins. Each coin can be in either of $r = 2$ quantum states - heads or tails. In Maxwell-Boltzman statistics, the state vector has two components, one for each coin. The components tell whether

the corresponding coin is heads or tails. There are

$$r^n = 2^2 = 4$$

possible values of the state vector: HH, HT, TH, and TT. These are equally likely.

To generalize this example, consider a box of r tickets, labeled $1, 2, \ldots, r$. We draw n tickets at random with replacement from the box. We can think of the n draws as the quantum states of n particles, each of which has r possible states. This is 'ticket-gas'. There are r^n possible outcomes, all equally likely, corresponding to Maxwell-Boltzman statistics. The case $r = 2$ corresponds to coin-gas; the case $r = 6$ is 'dice-gas', the standard model for rolling n dice.

Let $X = \{X_1, \ldots, X_r\}$ be the occupancy numbers for ticket-gas: in other words, X_i is the number of particles in state i. There are

$$\binom{n+r-1}{n}$$

possible values of X. If ticket-gas were Bose-Einstein, those values would be equally likely. With Maxwell-Boltzman statistics, they are not: instead, X has a multinomial distribution. Let $j_1, j_2, \ldots j_r$ be nonnegative integers that sum to n. Then

$$P(X_1 = j_1, X_2 = j_2, \ldots, X_r = j_r) = \frac{n!}{j_1! j_2! \cdots j_r!} \times \frac{1}{r^n}.$$

The principle of insufficient reason is not sufficient for probability theory, because there is no canonical way to define the set of outcomes which are to be taken as equally likely.

[8] The most common isotope of Helium is He^4; each atom consists of two protons, two neutrons, and two electrons. He^3 lacks one of the neutrons, which radically changes the thermodynamics.

[9] There is only about one earthquake of magnitude 8+ per year globally. In the San Francisco Bay Area, unless the rate of seismicity changes, it will take on the order of a century for a large earthquake to occur, which is not a relevant time scale for evaluating predictions.

[10] The collection Σ must contain S and must be closed under complementation and countable unions. That is, Σ must satisfy the following conditions: $S \in \Sigma$; if $A \in \Sigma$ then $A^c \in \Sigma$; and if $A_1, A_2, \ldots \in \Sigma$, then $\cup_{j=1}^{\infty} A_j \in \Sigma$.

[11] Some parameters were estimated from data. The Monte Carlo procedure treats such parameters as random variables whose expected values are the estimated values, and whose variability follows a given parametric form (Gaussian). This is 'fiducial inference' (*Lehmann*, 1986, pp. 229–230), which is neither frequentist nor Bayesian. There are also several competing theories for some aspects of the models, such as the relationship between fault area and earthquake magnitude. In such cases, the Monte Carlo procedure selects one of the competing theories at random, according to a probability distribution that reflects 'expert opinion as it evolved in the study'. Because the opinions were modified after analyzing the data, these were not prior probability distributions; nor were opinions updated using Bayes rule. See note 3.

[12] About 40 per cent of the randomly generated models were discarded for violating a constraint that the regional tectonic slip be between 36 mm/yr and 43 mm/yr.

[13] The standard deviations are zero - no uncertainty - in several cases where the slip is thought to be accommodated purely seismically; see table 2 of (*WG99*, 1999). Even the non-zero standard deviations seem to be arbitrary.

[14] It seems that the study intended to treat as equally likely all $2^n - 1$ ways in which at least one of n fault segments can rupture; however, the example on p. 9 of *WG99* (1999) refers to 6 possible ways a three-segment fault can rupture, rather than $2^3 - 1 = 7$, but then adds the possibility of a 'floating earthquake', which returns the total number of possible combinations to 7. Exactly what

the authors had in mind is not clear. Perhaps there is an implicit constraint: segments that rupture must be contiguous. If so, then for a three-segment fault where the segments are numbered in order from one end of the fault (segment 1) to the other (segment 3), the following six rupture scenarios would be possible: $\{1\}, \{2\}, \{3\}, \{1,2\}, \{2,3\}, \{1,2,3\}$; to those, the study adds the seventh 'fbating' earthquake.

[15] The relationships are all of the functional form $m_w = k + \log A$, where m_w is the moment-magnitude and A is the area of the fault. There are few relevant measurements in California to constrain the relationships (only seven 'well-documented' strike-slip earthquakes with $m_w \geq 7$, dating back as far as 1857), and there is evidence that California seismicity does not follow the generic model (*WG99*, 1999).

[16] This probability is added at the end of the analysis, and no uncertainty is associated with this number.

[17] Stage 1 produced estimates of rates for each source; apparently, these are disaggregated in stage 2 into information about fault segments by using expert opinion about the relative likelihoods of segments rupturing separately and together.

[18] See notes 3 and 11.

[19] For example, the Mount Diablo Thrust Fault, which slips at 3 mm/yr, was not recognized in 1990 but is included in the 1999 model (*WG99*, 1999, p. 8). Moreover, seismic sources might not be represented well as linked fault segments.

5.4   References

Anderson, M., Ensher, J. R., Matthews, M. R., Wieman, C. E., and Cornell, E. A., 1995. Observation of Bose-Einstein condensation in a dilute atomic vapor, *Science*, **269**, 198–201.

Beeler, N. M., Simpson, R. W., Lockner, D. A., and Hickman, S. H., 2000. Pore fluid pressure, apparent friction and Coulomb failure, *J. Geophys. Res.*, **105**, 25533–25542.

Bird, P., Kagan, Y. Y., Houston, H., and Jackson, D. D., 2000. Earthquake Potential Estimated from Tectonic Motion, *Eos, Trans. AGU*, **81**(48), Fall AGU Meet. Suppl., (abstract), F1226–1227.

Console, R., 1998. Computer algorithms for testing earthquake forecasting hypotheses, *Internal Report*, Eq. Res. Inst., Tokyo Univ.

Dieterich, J. H., 1972. Time-dependent friction in rocks, *J. Geophys. Res.*, **77**, 3690–3697.

Dieterich, J. H., 1994. A constitutive law for rate of earthquake production and its application to earthquake clustering, *J. Geophys. Res.*, **99**, 2601–2618.

Dziewonski, A. M., Ekstrom, G., and Maternovskaya, N. N., 2001. Centroid-moment tensor solutions for April-June 2000, *Phys. Earth Planet. Inter.*, **123**, 1–14.

Eaton, M. L., and Sudderth, W. D., 1999. Consistency and strong inconsistency of group-invariant predictive inferences, *Bernoulli*, **5**, 833–854.

Ebel, J. E., Bonjer, K. P., and Oncescu, M. C., 2000. Paleoseismicity: seismicity evidence for past large earthquakes, *Seism. Res. Lett.*, **71**, 283–294.

Ellsworth, W. L., Matthews, M. V., Nadeau, R. M., Nishenko, S. P., Reasenberg, P. A., and Simpson, R. W., 1998. A physically-based earthquake recurrence model for estimation of long-term earthquake probabilities, in *Proceedings of the 2nd Joint Meeting of the UJNR Panel on Earthquake Research*, Geographical Survey Institute, Japan, 135–149.

Feller, W., 1968. *An Introduction to Probability Theory and Its Applications*, vol. I., 3rd ed., John Wiley & Sons, Inc., New York.

Freedman, D., 1995. Some issues in the foundations of statistics, *Foundations of Science*, **1**, 19–39.

Gomberg, J., Reasenberg, P., Bodin, P., and Harris, R., 2001. Earthquake triggering by transient seismic waves following the Landers and Hector Mine, California earthquakes, *Nature*, **411**, 462–465.

Harris, R. A., 1998. Stress triggers, stress shadows, and implications for seismic hazard, introduction to the special issue, *J. Geophys. Res.*, **103**, 24347–24358.

Hartigan, J., 1983. *Bayes Theory*, Springer-Verlag, New York.

Hickman, S. H., Zoback, M. D., Younker, L., and Ellsworth, W. L., 1994. Deep scientific drilling in the San Andreas fault zone, *Eos, Trans. Am. Geophys. Union*, **75**, 137–142.

Jachens, R. C., and Griscom, A., 2003. Geologic and geophysical setting of the 1989 Loma Prieta earthquake, California, inferred from magnetic and gravity anomalies, in The Loma Prieta, California Earthquake of October 17, 1989 - Geologic Setting and Crustal Structure, Wells, R. (ed.), USGS.

Jackson, D. D., and Kagan, Y. Y., 1999. Testable earthquake forecasts for 1999, *Seism. Res. Lett.*, **70**, 393–403.

Jackson, D. D., and Kagan, Y. Y., 2000. Earthquake Potential Estimated from Seismic History, *Eos, Trans. AGU*, **81**(48), Fall AGU Meet. Suppl., (abstract), F1226.

Jackson, D. D., Kagan, Y. Y., Rong, Y. F., and Shen, Z., 2001. Prospective Tests of Earthquake Forecasts, *Eos, Trans. AGU*, **82**(47), Fall AGU Meet. Suppl., (abstract), S41C-04, F890–891.

Jackson, D. D., Shen, Z. K., Potter, D., Ge, B. X., and Sung, L. Y., 1997. Southern California deformation, *Science*, **277**, 1621–1622.

Kagan, Y. Y., 1991. Likelihood analysis of earthquake catalogs, *Geophys. J. Int.*, **106**, 135–148.

Kagan, Y. Y., 1994. Observational evidence for earthquakes as a nonlinear dynamic process, *Physica D*, **77**, 160–192.

Kagan, Y. Y., 1997. Are earthquakes predictable?, *Geophys. J. Int.*, **131**, 505–525.

Kagan, Y. Y., 2000. Temporal correlations of earthquake focal mechanisms, *Geophys. J. Int.*, **143**, 881–897.

Kagan, Y. Y., 2002a. Seismic moment distribution revisited: I. Statistical results, *Geophys. J. Int.*, **148**, 520–541.

Kagan, Y. Y., 2002b. Seismic moment distribution revisited: II. Moment conservation principle, *Geophys. J. Int.*, **149**, 731–754.

Kagan, Y. Y., and Jackson, D. D., 1994. Long-term probabilistic forecasting of earthquakes, *J. Geophys. Res.*, **99**, 13685–13700.

Kagan, Y. Y., and Jackson, D. D., 2000. Probabilistic forecasting of earthquakes, *Geophys. J. Int.*, **143**, 438–453.

Kagan, Y. Y., and Knopoff, L., 1987. Statistical short-term earthquake prediction, *Science*, **236**, 1563–1567.

Kagan, Y. Y., and Schoenberg, F., 2001. Estimation of the upper cutoff parameter for the tapered Pareto distribution, *J. Appl. Probab.*, **38A**, 158–175.

Kolmogorov, A., 1956. *Foundations of the Theory of Probability*, 2nd ed., Chelsea Publishing Co., New York.

Lehmann, E., 1986. *Testing Statistical Hypotheses*, 2nd ed., John Wiley and Sons, New York.

Littlewood, J., 1953. *A Mathematician's Miscellany*, Methuen & Co. Ltd., London.

Michael, A. J., and Eberhart-Phillips, D. M., 1991. Relations among fault behavior, subsurface geology, and three-dimensional velocity models, *Science*, **253**, 651–654.

Michael, A. J., and Jones, L. M., 1998. Seismicity alert probabilities at Parkfield, California, revisited, *Bull. Seism. Soc. Amer.*, **88**, 117–130.

Molchan, G. M., and Kagan, Y. Y., 1992. Earthquake prediction and its optimization, *J. Geophys. Res.*, **97**, 4823–4838.

Ogata, Y., 1988. Statistical models for earthquake occurrence and residual analysis for point processes, *J. Amer. Statist. Assoc.*, **83**, 9–27.

Ogata, Y., 1998. Space–time point-process models for earthquake occurrences, *Ann. Inst. Statist. Mech.*, **50**, 379–402.

Parsons, T., 2002. Global Omori law decay of triggered earthquakes: large aftershocks outside the classical aftershock zone, *J. Geophys. Res.*, **107** (B9), 2199, doi: 10.1029/2001JB000646.

Parsons, T., Toda, S., Stein, R. S., Barka, A., and Dieterich, J. H., 2000. Heightened odds of large earthquakes near Istanbul: an interaction-based probability calculation, *Science*, **288**, 661–665.

Pollitz, F. F., Sacks, I. S., 2002. Stress triggering of the 1999 Hector Mine earthquake by transient deformation following the 1992 Landers earthquake, *Bull. Seism. Soc. Am.*, **92**, 1487–1496.

Reasenberg, P. A., 1999. Foreshock occurrence before large earthquakes, *J. Geophys. Res.*, **104**, 4755–4768.

Reasenberg, P. A., and Jones, L. M., 1989. Earthquake hazard after a mainshock in California, *Science*, **243**, 1173–1176.

Reif, F., 1965. *Fundamentals of Statistical and Thermal Physics*, McGraw-Hill Book Publishing Co., New York.

Shen, Z. K., Jackson, D. D., and Ge, B. X., 1996. Crustal deformation across and beyond the Los Angeles basin from geodetic measurements, *J. Geophys. Res.*, **101**, 27957–27980.

Stigler, S., 1986. *The History of Statistics: the Measurement of Uncertainty before 1900*, Harvard University Press, Cambridge, MA.

Utsu, T., and Ogata, Y., 1997. Statistical analysis of seismicity, in *IASPEI Software Library*, **6**, Healy, J. H., Keilis-Borok, V. I., and Lee, W. H. K. (eds.), Int. Assoc. of Seismol. and Phys. of the Earth's Inter. and Seismol. Soc. Am., El Cerrito, CA, 13–94.

Vere-Jones, D., 1970. Stochastic models for earthquake occurrence (with discussion), *J. Roy. Stat. Soc.*, **B32**, 1–62.

Waldhauser, F., and Ellsworth, W. L., 2000. A double-difference earthquake location algorithm: method and application to the northern Hayward fault, California, *Bull. Seism. Soc. Am.*, **90**, 1353–1368.

WG99: Working Group on California Earthquake Probabilities, 1999. Earthquake Probabilities in the San Francisco Bay Region: 2000-2030–A Summary of Findings, *Technical Report Open-File Report*, 99-517, USGS, Menlo Park, CA.

Wiemer, S., 2000. Introducing probabilistic aftershock hazard mapping, *Geophys. Res. Lett.*, **27**, 3405–3408.

Wiemer, S., 2001. Adding time-dependent elements to earthquake hazard mapping, Second International Workshop on Statistical Seismology, 18–21 April 2001, held at Victoria University of Wellington, abstract.

Chapter 6

Gathering new data

Editors' introduction. We will now discuss how new data may help in solving some of the basic problems that both the classical as well as the PCS (Physics of Complex Systems) approaches leave open. We will analyze separately the time and spatial domains, since palaeoseismic and geodetic techniques allow to tackle them separately. The InSAR geodetic technique, however, is likely to allow to simultaneously resolve the detail in both the time and the spatial domains, at least with regard to the evolution on time scales ranging from a month to several years.

Since seismicity is produced by the stress distribution in the Earth's crust, the stress field should be ideally known everywhere in the crust. However, as direct measurements of stress are generally lacking and we must perforce rely on its proxy, strain, from which stress can be estimated through Hooke's law (cf. section 2.8). In general, geodetic methods allow us to measure the strain field only on a sparse grid of points both in time and space. In classical geodesy, the data points are typically spaced several kilometers apart in space and a few years in time in global networks. Denser sampling schemes can be afforded only in local networks, usually aimed at monitoring a particularly important structure. Satellite geodesy, and in particular INSAR, allows us to survey crustal strains over large areas with meter-size grids in space and monthly spacing in time.

Such dense sampling will hopefully allow us to establish the experimental basis for the detail of strain accumulation, propagation and release. It is fairly obvious that such detail, mostly disregarded in the classical seismological approach, is of crucial importance. For example, the few high resolution studies already available suggest that the classical seismological picture of single large faults is fully incompatible with the data, since an event like the Landers, 1992 earthquake, was composed of no less than 186 sub-events (*Wald and Heaton*, 1994).

As discussed below, the InSAR appears to be an ideal tool for studying the earthquake process, allowing us to monitor with millimeter accuracy preseismic, coseismic and postseismic dynamics. It will also allow to analyze the dynamics of strain propagation and its relation to epicenter migration and clustering. A closely linked problem is that of foreshocks: why do only some earthquakes have foreshocks? For example, *Reasenberg* (1999), using the Harvard catalogue found that for shallow events, 13.2 per cent of the $m \geq 6.0$ events had a foreshock of $m \geq 5$, with a foreshock rate density of approximately 12 per cent per magnitude unit. This value appears in good agreement with all similar estimates and means, for example, that 25 per cent of $m \geq 7$ events have $m \geq 5$ foreshocks, and so on.

The paleoseismology of coastlines provides spatially distributed sampling along shorelines. This complements classical palaeoseismological studies, which are based on trenches excavated on faults (*Sieh et al.*, 1989; *Yeats et al.*, 1997). As discussed below, coastline data might allow not only to identify and date the events, but also to gather some information on the detail of their dynamics.

6.1   Space geodesy

[By F. Rocca and F. Sans`o]

6.1.1   The observables of space geodesy

In 1957 the first artificial satellite of the Earth, Sputnik 1, was launched and fundamentally changed geophysical observations of the Earth, because for the first time geodesy had found a target visible simultaneously from points distant several thousands kilometers from one another. Indeed traditional astrogeodetic observations allowed one to place on the celestial sphere the direction of the vertical issued at any point on the Earth's surface at a certain time t, with respect to the system of 'fixed' stars (with radioscience providing an accurate bridge between different instants in time). With a model of a frozen Earth and with equations like[1]

$$\dot{\underline{V}}_P(t) = \omega(t)\underline{e}(t) \wedge \underline{V}_P(t) , \tag{6.1}$$

$$
\begin{aligned}
\omega &= \text{angular velocity} \\
\underline{e} &= \text{instantaneous rotation axis} \\
\wedge &= \text{vector cross-product operator} \\
\underline{V}_P &= \text{direction of the vertical in } P
\end{aligned}
$$

---

[1] In this section underlined symbols refer to vector variables.

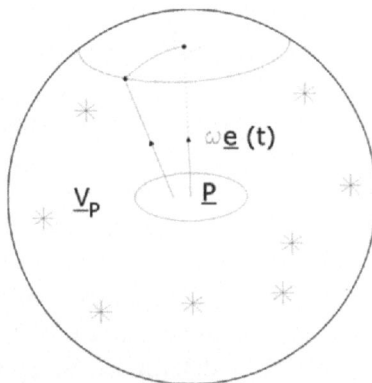

Figure 6.1: Rotation of a rigid Earth seen in a Celestial Reference System.

expressing that the vertical vector in P rotates daily around a rotation axis $\underline{e}(t)$ with angular velocity $\omega(t)$ (vectors referred to a Celestial Reference System linked to 'fixed' stars) one can retrieve, from data of many observatories, both ω and $\underline{e}$, assumed to be constant over a sensible time interval, e.g. one day.

This is what is basically done nowadays by a network of VLBI (Very Long Baseline Inferferometry) observatories, which together provide valuable information on the orientation of the Earth in an inertial reference system and in time.

Yet the model (6.1) is by far too simplistic for many reasons including the fact that the Earth is not a rigid body and its surface undergoes a significant deformations, i.e. there is no reference system in which points of the Earth's surface can keep their coordinates constant in time. Rather, every point is subject to several motions, from the short periodic tidal motions to long term motions, related to large scale tectonics. In addition, equation (6.1) depends on the local direction of the vertical which can change due to a pure re-adjustment of internal masses, without any geometrical deformation of the surface. Only from the point of view of spatial resolution, can a real description of such deformation patterns be achieved, even on a large scale (say of 10^3 km) by a network of several hundreds observatories, which in the case of VLBI seems rather unrealistic. Thus there is a need for new types of observations, and this is provided by artificial satellites.

Apart from a first attempt to determine the direction between an observer on ground and a satellite, which is too strongly affected by atmospheric refraction, it was soon understood that ranging was the key to get very accurate measurements in satellite tracking. Nowadays we observe the distance Earth station - satellite, by laser ranging ($\sigma \sim 1$ mm) or by electromagnetic sounding, for instance by GPS where phases are measured with errors of a few millimeters (but in this case the observable is the distance modulo an initial integer bias, the famous phase ambiguity at the time when observations start).

It seems interesting to notice that modeling such observations at the accuracy level of millimeters is indeed a difficult affair, yet in principle we understand that there are basically two different types of information they can provide: one geometric, the other on the gravity field. That there is a pure geometric content in ranging is easily understood if we imagine a case in which the same position of the satellite is simultaneously tracked by four ground stations; then, after eliminating in a purely geometrical mode the coordinates of the satellite, we get an equation linking the coordinates of ground stations only. Indeed this is a very poor approach to the analysis of tracking data, as no use is made of the dynamical model of the satellite. In fact satellites' motion is determined by dynamic equations linking their position to a force model, where the leading term is gravity.

It is because of this dynamical link that the global gravity field can be estimated from satellite tracking, at least up to some harmonic degree depending on the altitude of the satellite[2]. In this way it was possible to determine the gravity field up to a maximum degree 50. It is only recently that satellite missions have been designed to estimate the gravity field with much higher resolution.

This, for instance, can be done by introducing instruments (accelerometers, gradiometers) which could determine or eliminate the effect of the non-gravitational forces on the space craft (atmospheric drag, direct and indirect radiation pressure, etc.), thus enhancing enormously the gravimetric signal.

With spaceborne gradiometry (the GOCE Project) we reach the limit where the gravimetric measurement is almost completely decoupled from geometry and orbital dynamics, so that with an orbit at 300 km altitude, the global gravity field can be retrieved up to degree 200.

6.1.2   Reference system and deformation concepts

Let us go back now to geometry and stipulate, with a good level of abstraction, that we observe from many 'fixed' stations on the Earth the orbital evolution of a number of satellites in terms of ranging, i.e. we measure the distances

$$D_r^s(t) = |\underline{X}^s(t-\tau) - \underline{X}_r(t)| \tag{6.2}$$

from the receiver r at reception time t, to the satellite s at emission time $t - \tau$, for many r, s, t. The distances in equation (6.2) are by definition invariant under rotation or translation of the reference system with the only proviso that the transformation is the same for both vectors.

---

[2]Let us remember here that the gravity field, like all harmonic fields, decays with altitude with an exponent controlled by the harmonic degree; the higher the degree, the faster the decay.

One choice is a quasi-inertial frame with the origin at the barycenter of the Earth and an attitude with the Z axis following the direction of the instantaneous rotation axis as seen in the celestial system and as reconstructed from VLBI observations. In this way, with some approximation, Newton's law holds for the satellite S,

$$\ddot{\underline{X}}^s = R\underline{G}(R^+\underline{X}^s) + \underline{F} , \tag{6.3}$$

where $\underline{F}$ is a vector of perturbative forces which we consider here as completely known, while R is the rotation matrix connecting any 'conventional terrestrial system', i.e. a reference system where coordinates of points of the Earth's surface have 'small' variations in time, with the quasi-inertial system. If we call $\underline{x}$ any position vector expressed in the conventional terrestrial system, the connection between $\underline{X}$ and $\underline{x}$ is then given by

$$\underline{X} = R(\underline{x} - \underline{b}) , \tag{6.4}$$

where $\underline{b}$ is the position of the barycenter in the system $\underline{x}$ and R is the rotation between $\{\underline{b}\}$ and $\{\underline{x}\}$.

Let us explicitly observe that with the above definition R includes the main rotations between the terrestrial system and the quasi-inertial system, namely the 1-day cycle around the instantaneous rotation axis and the wandering of the rotation axis itself inside the physical body of the Earth (polar motion).

In (6.3) $\underline{G}$ can be assumed as a known function, but in terms of terrestrial coordinates; this is why we use R^+ (i.e. the inverse of R) in the argument of $\underline{G}$, while the vector $\underline{G}$ itself is then rotated by R to bring it back to the quasi-inertial system.

This hypothesis, which might be unjustified in general, is certainly true for very high satellites like those of the GPS constellation, because at these altitudes only the first degrees ($5 \sim 7$) of gravity harmonics have a dynamical relevance.

Once integrated (6.3) provides the satellites' ephemerides

$$\underline{X}^s \equiv \underline{X}^s(\underline{X}_0^s, \dot{\underline{X}}_0^s, R), \tag{6.5}$$

which are functions of the initial state as well as of the unknown rotation R. There is no need to specify that (6.5) is a very simplified orbital propagation model, where only variables of 'conceptual' importance are made explicit.

Using (6.4) and (6.5) in (6.2) yields

$$D_r^s(t) = |\underline{X}^s(\underline{X}_0^s, \dot{\underline{X}}_0^s, R) - R \cdot (\underline{x}_r - \underline{b})| . \tag{6.6}$$

This equation is repeated for all the epochs t when a receiver r tracks a satellite s. Solving these equations in a giant least squares computation is what we would call the 'general tracking problem'. Let us notice here that the unknowns in (6.6) are the initial states of satellites $(\underline{X}_0^s, \underline{\dot{X}}_0^s)$, the coordinates of the stations $\underline{x}_r$ as well as the system transformation $(R, \underline{b})$.

Indeed since the terrestrial system is not yet precisely defined, we see that R and $\underline{b}$ depend on how we choose $\{\underline{x}\}$ and *vice versa*, so that the system (6.6) (though there are many more equations than unknowns) is rank deficient. This is not new in geodesy, but exactly the same phenomenon happens when we analyse geodetic networks and we know that by defining the additional conditions that allow to eliminate (without over-constraining) this rank deficiency we are in fact defining the reference system too (*Sillard and Boucher*, 2001).

Following a traditional concept we could perform a two-step least squares approach, to arrive at a consistent definition of a terrestrial reference system. One can start e.g. by solving the 'general tracking' problem by fixing a minimal number of terrestrial coordinates $\{\underline{x}_r\}$ and subsequently find the terrestrial reference as the one which satisfies a 'minimal motion' condition over all the terrestrial stations participating to the measurement operations.

In doing so the transformation parameters $R(t)$, $\underline{b}(t)$ are modeled as smooth functions of time and in particular, on the basis of a general dynamical reasoning as well as of the empirical experience, we now put

$$R(t) = [I + dR(t)]R_0(t) \tag{6.7}$$

with $R_0(t)$ a uniform rotation around a fixed (in the celestial system) axis $\underline{e}_0$ and

$$dR(t) \sim (d\underline{\omega}\wedge) \tag{6.8}$$
$$d\underline{\omega}(t) = d\omega(t)\underline{e}_0 + \omega\underline{\varepsilon}(t)$$
$$\underline{e}_0 \cdot \underline{\varepsilon}(t) = 0 .$$

In equation (6.8) $d\omega(t)$ is the variation of the so-called *Length of the Day* and $\underline{\varepsilon}(t)$ is a vector collecting the effects of both the motion of the rotation axis in the celestial system (very well-known and predicted) and the polar motion, within the body of the Earth.

What is important is that now we have a clear concept splitting the motion of the stations (r) in space into a rigid motion of the reference system and a residual motion

$$\underline{u}_r(t, t') = \underline{x}_r(t') - \underline{x}_r(t) ; \tag{6.9}$$

expressed in the Conventional Terrestrial Reference System, which is related to an authentic deformation of the Earth's surface.

The field of motions $\underline{u}_r$ contains short periodic contributions that are explained in terms of tides and elastic direct and indirect response of the Earth's crust; by averaging out these effects one is left with long term motions which are typically modeled as linear trends in time. This is reflected in that the ground stations have average coordinates and velocities. It is by analyzing these velocities that one can realize that they can be mostly explained in terms of 'plates' in rigid motion (i.e. endowed with their own differential rotation) on the Earth's surface.

We recall that beyond the horizontal displacements of plates, we have also vertical movements which, for physical reasons, are particularly interesting when referred to a reference equipotential surface: the geoid.

Although we can detect geometric deformations of the order of few millimeters in the vertical direction by the same space methods, we do not have yet such a precise knowledge of the geoid on scales typical of the distance between satellite observatories. Nevertheless now satellite missions are coming (e.g. GRACE) where geoid variations of millimeters at such large scales are truly observable. At smaller scales, for the purpose of determining geoid variations, the painstaking work with ground gravity observations cannot be substituted for the moment by satellite observations.

6.1.3 The observing networks

Summarizing, the techniques of space observations available for the purpose of defining a worldwide unified terrestrial reference system and monitoring the deformation of the Earth's crust in time are

VLBI - Very Long Baseline Interferometry

SLR - Satellite Laser Ranging

GPS-GLONASS - Global Positioning System

DORIS - a French Doppler System

PRARE - a German system of range and range rate measurement

We shall concentrate on the first three techniques for which the International Association of Geodesy (IAG), in cooperation with other scientific organizations, is providing well organized services.

The first technique is different from the others in that it uses the signal of *quasars* to determine baselines (length and orientation in space) between radiotelescopes; as such the observations are not dependent on satellite dynamics and in particular they have no relation to the Earth's gravity field. Data are collected

Product	Available	Interval	Precision	
Satellite Orbits & clocks			*orbits*	*clocks*
Predicted	Real-time	15 min	0.50 m	30 ns
Rapid	17 hours	15 min	0.10 m	0.5 ns
Final	12 days	15 min	0.05 m	0.3 ns
IGS Combined (prelim.)			*Stat.Post*	*Velocities*
Weekly solutions	2-4 days	7 days	3-5 mm	1-3 mm/y
Earth Rotation Parameters			*ERPs*	*rates/LOD*
Rapid PM	17 hours	1 day	0.2 mas.	0.4 mas/d
Final PM	10 days	1 day	0.1 mas.	0.2 mas/d
Rapid UT/LOD	17 hours	1 day	0.1 ms.	0.06 mas/d
Final UT/LOD	17 days	1 day	0.05 ms.	0.03 mas/d
Troposheric ZPD	<4 weeks	2 hours	4 mm	
Ionospheric grid TEC	<4 weeks	2 hours	2 TEC unit ($\sim$ 0.20 m)	

Table 6.1: IGS Products.

and processed by the International VLBI Service, which includes 30 observatories. The accuracy of baselines is in the sub-millimetric level, although the most important contribution of VLBI is in providing a lot of information on the rotation of the terrestrial reference system with respect to the celestial reference system.

The Laser Ranging system is based on a space segment of the satellites LAGEOS I and II, Starlette, Stella, AJISAI and some others. The ground segment is formed by a network of 49 observatories and data are collected and analyzed by the International Laser Ranging Service (figure 6.2). This system provided the most accurate and reliable tracking data (millimeter accuracy) although its effectiveness is limited by the cost of the observing apparatus and by the fact that observations can be taken only in clear sky conditions.

GPS (GLONASS) observations from a network of permanent stations are collected, processed and distributed with variable delays by the International GPS Service (IGS). The number of the IGS primary stations is $\sim$ 200 but this number is rapidly increasing, because the ground apparatus is not very expensive and at the same time the presence of a permanent GPS station in an area allows to incorporate all local geophysical observations in a unified reference system. In addition GPS is an all weather operating system. The IGS products are summarized in table 6.1.

IGS is presently the one, among the existing services, providing the largest amount of data with continuity and homogeneous accuracy; these data are the most valuable source for the determination of the terrestrial reference system and at the same time they can provide a fairly detailed picture of the overall motion of deformation of the Earth's crust. In this respect it is important to underline

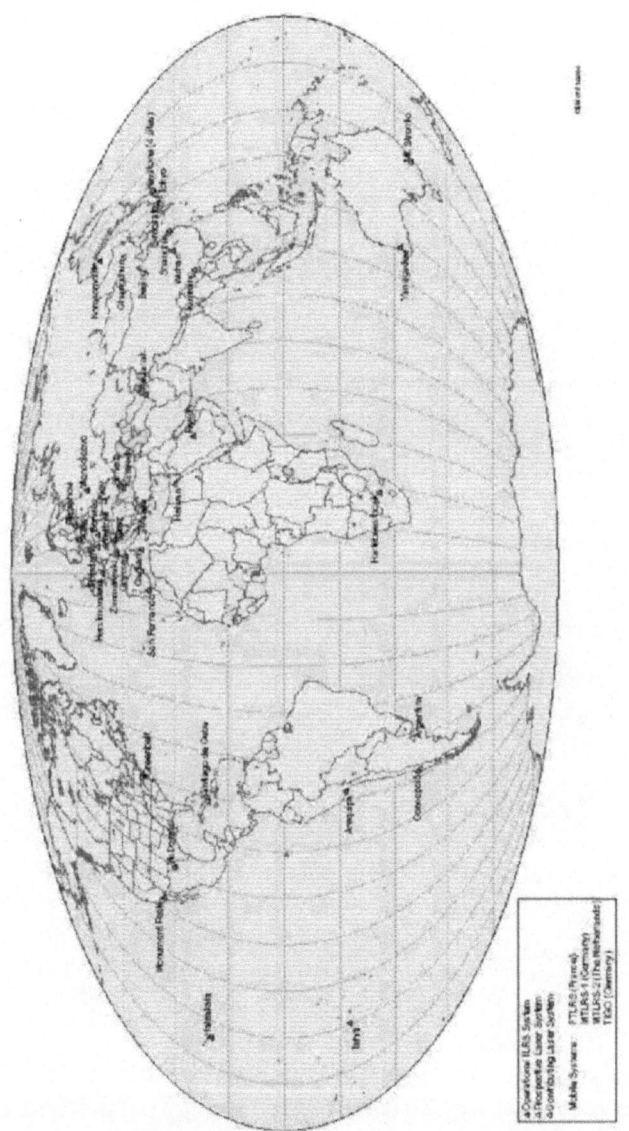

Figure 6.2: Worldwide ILRS sites.

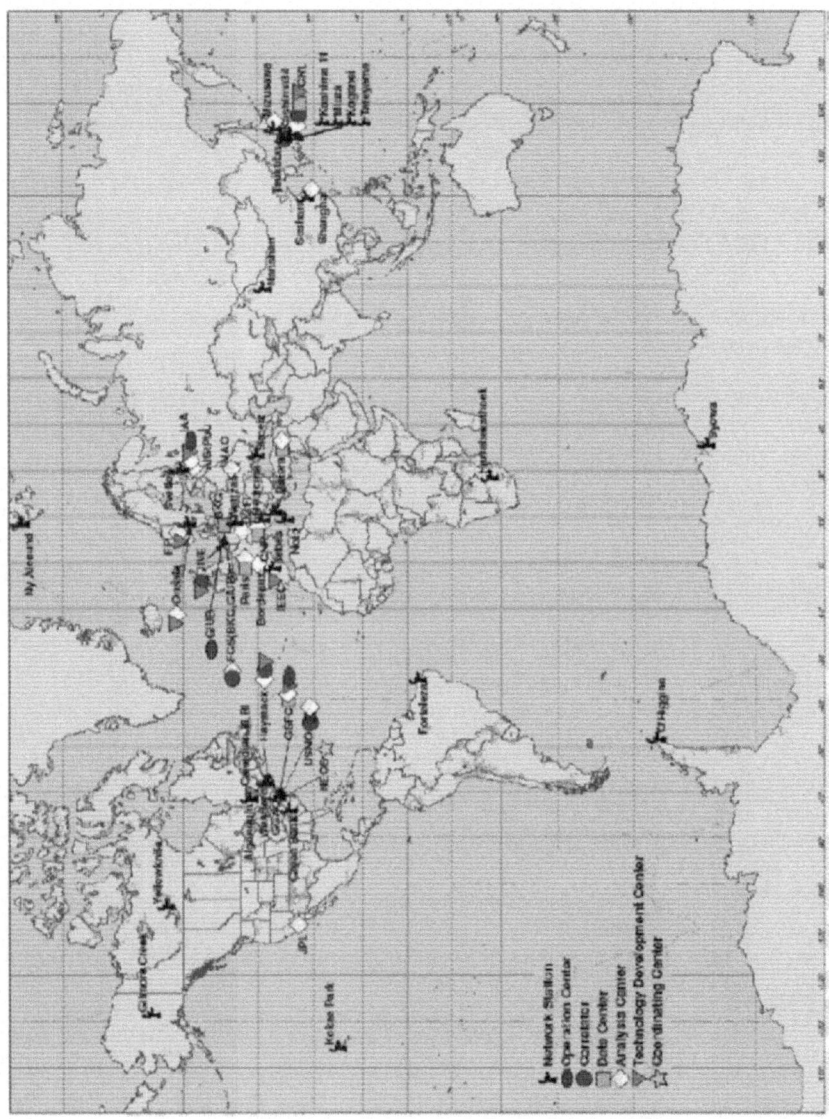

Figure 6.3: Worldwide IVS sites.

that when we want to reach the limit of 3 mm in coordinate determination, the GPS receiver cannot work for 1 or 2 days only on each point but it has to be a permanent station because there are phenomena that can affect the motion of the point in the range of some millimeters, which are related to seasonal or annual effects.

All the data from different systems (including those that we do not treat here) are then conveyed to the International Earth Rotation Service (IERS) which processes a combined solution (*Boucher et al.*, 1997).

A by-product of this solutions is the definition of the International Terrestrial Reference Frame (ITRF); this, as it was described in section 6.1.2, is in fact materialized by the specific stations that enter into the least squares solution and by the period in which data are taken. For that reason the realization of the terrestrial system is in fact called the terrestrial frame. At present (2001) we are still working with ITRF '97, though the new ITRF 2000 is ready to take over.

In the IERS combined solutions we can say (*Boucher et al.*, 1997) that VLBI provides the most important information on the rotation of the ITRF with respect to the celestial system while GPS tells us how the barycenter moves and how the rotation axis is materialized in the Earth body; SLR with its higher accuracy and reliability constitutes in fact the control system for GPS solutions. But GPS, as we said, is now developing its network of permanent stations at the continental and even national scales. This provides an accurate materialization of the global ITRF to a spatial resolution of say 100–200 km, at least in North America and Europe, with a uniform accuracy of a few millimeters. This constitutes the material reference and logical bridge to other techniques that can, with comparable accuracy, provide a monitoring of the crust deformation from such scales down to a very local level, say 0.1 km, like InSAR (Interferometric Synthetic Aperture Radar). Since this system is working on a satellite platform and therefore surveys the same area regularly in time, e.g. every month, we are close now to a concept of complete and unified surveying of the crust by satellite systems.

6.1.4 An introduction to SAR imaging and SAR interferometry

Synthetic Aperture Radar is a microwave imaging system of the Earth's surface (*Franceschetti and Lanari*, 1999) that has all weather capabilities. In its interferometric configuration, it allows accurate measurements of the radiation travel path. In turn, these measurements allow one to generate Digital Elevation Models (DEM) and to measure millimetric surface deformations of the terrain. A digital SAR image is a mosaic of picture elements that, in the case of the European Space Agency satellites ERS-1 and ERS-2, correspond to about 5 meters in azimuth (along track) and about 9.5 meters in slant-range (across track). The amplitude SAR image contains in each cell a measurement of the amplitude of the radiation backscattered toward the radar by the objects (scatterers) contained in the corresponding SAR resolution cell. This amplitude depends more on the roughness than on the chemical composition of the scatterers on the terrain. Typically, exposed rocks and urban areas show strong amplitudes whereas smooth flat surfaces (like quiet water basins) show low amplitudes, since the radiation is mainly mirrored away from the radar. Scatterers at different distances from the radar (different slant range) introduce a different delay between transmission and reception of the radiation. The SAR phase image is then proportional to the two way travel distance of the radiation divided by the transmitted wavelength.

Travel distances that differ by an integer multiple of the wavelength introduce exactly the same phase change. Due to the huge ratio between the resolution cell dimension and wavelength ($\simeq 5.6 \ 10^{-2}$ m for ERS), even discounting the randomness of the phases of the scatterers in each cell, the SAR phase image looks random passing from one pixel to another and it is of no practical utility.

A satellite based SAR can observe the same area from slightly different looking angles. It can be done simultaneously (two radars mounted on the same platform, as in the recent NASA/DLR/ASI survey SRTM) or at different times using repeated orbits (*Bamler and Hartl*, 1998; *Rosen et al.*, 2000; *Massonnet and Feigl*, 1998). The latter is the case of ERS-1 and ERS-2. In that case, time intervals between observations of 1, 35 or a multiple of 35 days are available. The distance between the two satellites in the plane perpendicular to the orbit is called the interferometer baseline and its projection perpendicular to the slant range is called the perpendicular baseline B_n. The interferometric phase is the phase difference between the two images: it depends only on the travel path difference since the phase of the scatterers is cancelled.

6.1.5   SAR and digital elevation models

The variation of the travel path difference Δr that results passing from one resolution cell to another has a simple expression that depends on a few geometric parameters as shown in figure 6.4: the perpendicular baseline B_n, the radar-target distance R, the displacement between the resolution cells along the perpendicular to the slant range q_s. The following approximate expression for Δr holds:

$$\Delta r = 2\frac{B_n q_s}{R} \ . \tag{6.10}$$

The interferometric phase variation thus has the following expression (where λ is the SAR wavelength):

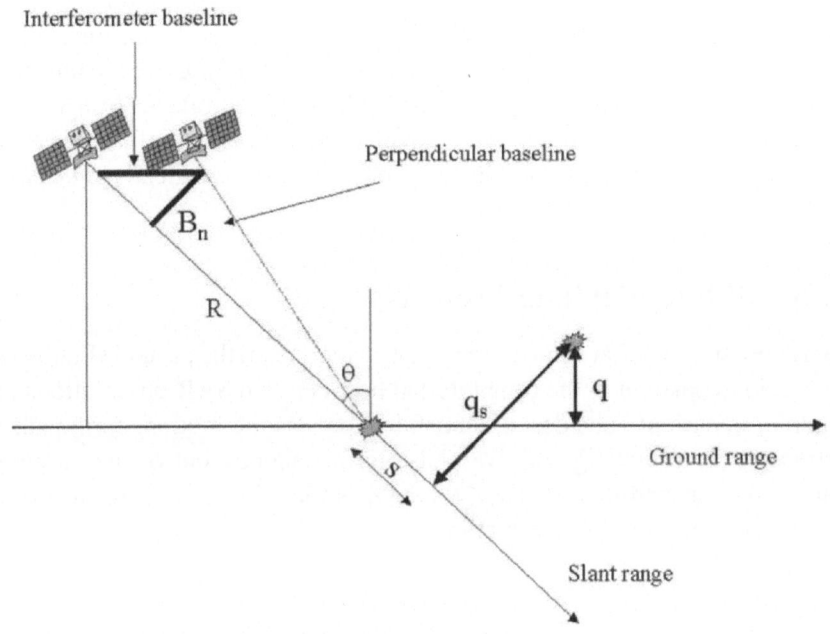

Figure 6.4: Geometric confi guration of SAR satellite.

$$\Delta\phi = \frac{2\pi\Delta r}{\lambda} = \frac{4\pi}{\lambda}\frac{B_n q_s}{R} . \qquad (6.11)$$

The altitude of ambiguity is defined as the altitude difference q_a that generates an interferometric phase change of 2π after removing the slope due to a planar Earth. The altitude of ambiguity is inversely proportional to B_n :

$$q_a = \frac{\lambda R \sin\theta}{2 B_n} . \qquad (6.12)$$

The higher the baseline, the more accurate is the altitude measurement, since the phase noise (see next section) is equivalent to a smaller altitude noise.

The flattened interferogram provides a measurement of the relative terrain altitude that is ambiguous. The phase variation between two points on the flattened interferogram provides a measurement of the actual altitude variation plus an integer number of altitude of ambiguity (equivalent to an integer number of 2π phase cycles). The process that allows to add to the interferometric fringes the correct number of altitude of ambiguity is called phase unwrapping. There are several well-known phase unwrapping techniques that are not discussed here. However, phase unwrapping does not have a unique solution and *a priori* information should be exploited to get the right solution (*Gabriel et al.*, 1989). Once the interferometric phases are unwrapped, an elevation map in SAR coordinates (range, azimuth) is obtained. If several interferograms taken with different baselines are available, then the phase unwrapping problem can be solved systematically (*Ferretti et al.*, 1999).

6.1.6   Differential interferometry

Let us now suppose that some of the point scatterers on the ground slightly change their relative position in the time interval between two SAR observations (as, for example, in case of subsidence, landslides, preseismic motion, earthquakes). In such cases the following additive phase term, independent of the baseline, appears in the interferometric phase, where d is the relative scatterer displacement projected on the slant range direction

$$\Delta\phi_d = \frac{4\pi}{\lambda}d \; . \tag{6.13}$$

It is thus evident that after interferogram flattening, the interferometric phase contains both altitude and motion components:

$$\Delta\phi = \frac{4\pi}{\lambda}\frac{B_n q}{R\sin\theta} + \frac{4\pi}{\lambda}d \; . \tag{6.14}$$

Moreover, if a DEM is available, the altitude contribution can be subtracted from the interferometric phase (generating the so-called differential interferogram) and the terrain motion component can be measured (*Gabriel et al.*, 1989; *Massonnet et al.*, 1993; *Massonnet et al.*, 1995). From this example it appears that the sensitivity of SAR interferometry to terrain motion is much larger than that to the altitude difference. A 2.810^{-2} m motion component in the slant range direction would generate a 2π interferometric phase variation. There are different ways to get a differential interferogram: (1) with a single interferometric pair (two SAR images) and baseline close to zero: then, the interferometric phase contains the motion contribution only; no other processing steps are required; (2) with a single interferometric pair (two SAR images) and a baseline different from zero: the interferometric phase contains both altitude and motion contributions. The altitude component has to be removed using *a priori* information or a DEM obtained from the data themselves.

When two interferometric SAR images are not simultaneous, the radiation travel path can be affected differently by the atmosphere. In particular, different humidity, temperature and pressure between the two takes can have a visible consequence on the interferometric phase (Atmospheric Phase Screen, APS) (*Ferretti et al.*, 2000; *Hanssen*, 1998; *Peltzer et al.*, 1999). This effect is usually confined within a 2π peak to peak interferometric phase change along the image with a smooth spatial variability (from a few hundreds meters to a few kilometers). The effect of such a contribution impacts both on altitude (especially in case of small baselines) and terrain deformation measurements. Other contributions to the phase noise should also be considered like that due to temporal change of the scatterers, to volume scattering, and to the different looking angle. The most important consequence of this last effect is that there exists a critical baseline over which the interferometric phase is pure noise. In the ERS case, the critical baseline for horizontal terrain is about 1,150 m.

The phase noise can be estimated from the interferometric SAR pair by means of the local coherence γ. The local coherence is the cross correlation coefficient of the SAR image pair estimated on a small window (a few pixels in range and azimuth), once all the deterministic phase components (mainly due to the terrain elevation) are compensated. The coherence map of the scene is then formed by computing its absolute value on a moving window that covers the whole SAR image. The coherence values ranges from 0 (pure noise) up to 1 (no noise).

6.1.7 Permanent scatterers

Stable natural reflectors (Permanent Scatterers, PS) can be identified from long temporal series of interferometric data (*Ferretti et al.*, 2000, 2001). They can be used in urban areas showing subsidence effects. One of the main difficulties encountered in differential SAR interferometry applications is due to temporal and geometric decorrelation. We have been able to identify single pixels (the PS's) coherent over years and for wide look-angle variations. This allows the use all ERS acquisitions relative to an area of interest. In fact when the dimension of the PS is smaller than the resolution cell, the coherence is good (the speckle is the same) even for image pairs taken with baselines larger than the decorrelation one. Then, on those pixels, sub-meter DEM accuracy and millimetric terrain motion detection can be achieved, even if the coherence is low in the surrounding areas. Reliable elevation and deformation measurements can then be obtained on this subset of image pixels that can be used as a 'natural' GPS network. Even though precise state vectors are available for ERS satellites (*Scharroo and Visser*, 1998), the impact of orbit indeterminations on the interferograms cannot be neglected (*Ferretti et al.*, 2001). The estimated APS that has to be removed is actually the sum of two contributions: atmospheric effects and orbital fringes due to baseline errors. However, the latter correspond to low-order phase polynomials and do not change the low-wavenumber character of the signal to be estimated on the sparse

PS grid.

At the PS, sub-meter elevation accuracy (due to the wide dispersion of the incidence angles available, usually 70 millidegrees with respect to the reference orbit) and millimetric terrain motion detection (due to the high phase coherence of these scatterers) can be achieved, once APS's are estimated and removed. In particular, relative target LOS (Line Of Sight) velocity can be estimated with unprecedented accuracy (often better than 0.1 mm/yr, due to the long time span). The higher the accuracy of the measurements, the more reliable the differentiations between models of the deformation process under study (*Gabriel et al.*, 1989), a key issue for risk assessment.

The final results of this multi-interferogram approach are the following: (1) a map of the PS identified in the image and their coordinates: latitude, longitude and precise elevation; (2) their average LOS velocity, and (3) the estimated motion component of each PS as a function of time. Common to all differential interferometry applications, the results are computed with respect to a Ground Control Point (GCP) of known elevation and motion.

6.1.8   Integration of GPS and SAR data: an example in Southern California

In figure 6.5 a perspective view of the LOS velocity field estimated from 55 ERS acquisitions from 1992 to 1999, over Southern California is reported. More than 400,000 PS were identified, with average density of as many as 150 PS/km^2. The reference point, supposed stable (zero LOS velocity), was chosen at Downey (in the center of the test site, 20 km SE of downtown Los Angeles), based on the data of a permanent GPS station of the Southern California Integrated GPS Network (SCIGN)[3].

Apart from subsiding areas due to oil and gas extraction[4] and water pumping, clearly visible in the picture, local maxima and minima of the velocity field gradient are strongly correlated to the map of known active faults in the area[5]. The position of fault's hanging walls can be inferred very precisely when such high density of accurate measurements is available (figure 6.6).

In particular, the velocity field shows local behavior that correlates well with blind thrust faults difficult to locate and map before occurrence of co-seismic displacements along them. The velocity field mapped in figure 6.5 presents local

---

[3] Web Site, http://www.scign.org/

[4] Department of Oil Properties, City Of Long Beach - Web Site, http://www.ci.long-beach.ca.us/oil/

[5] California Division Mines and Geology Fault Activity Map of California and Adjacent Areas (Calif. Dept. of Conservation, Sacramento, 1994)

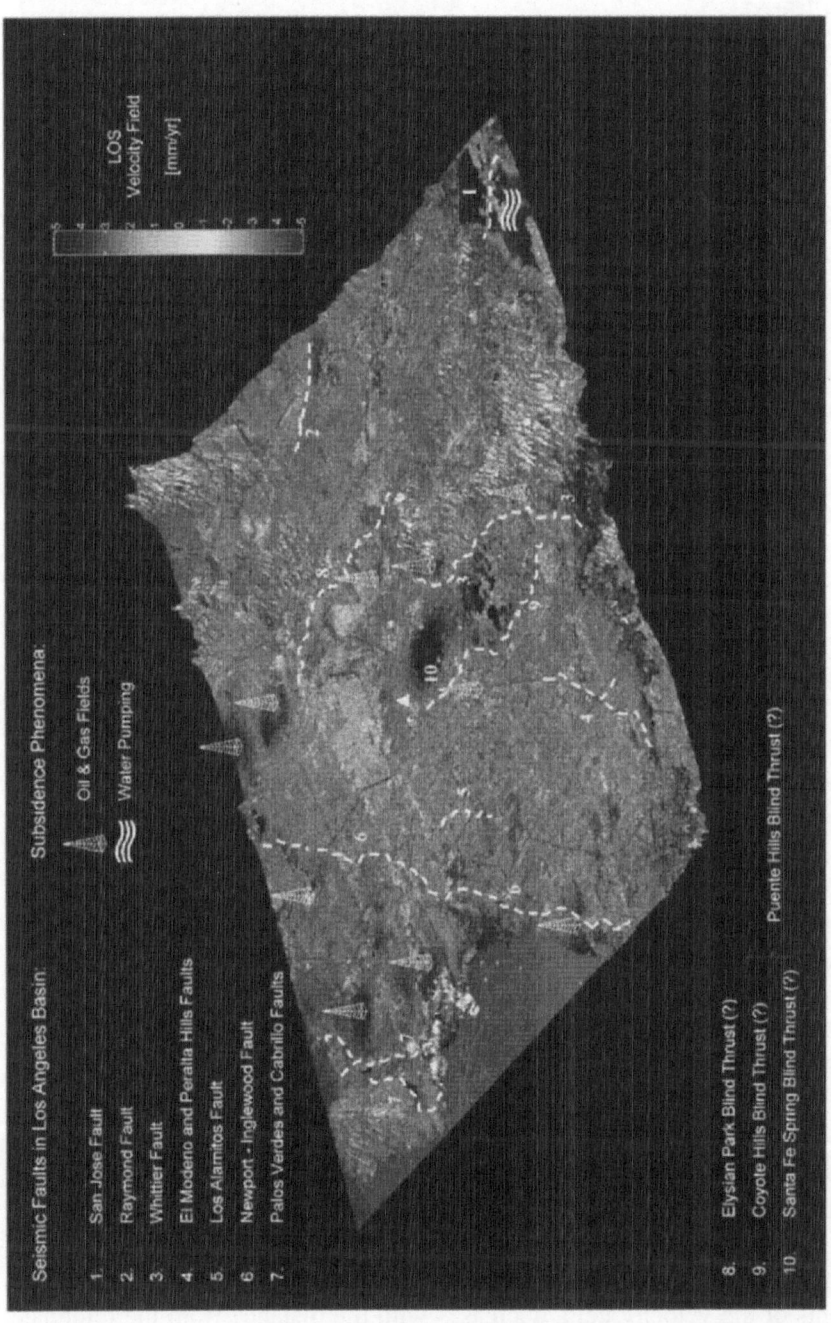

Figure 6.5: Seismic faults and subsidence in Southern California from PS data.

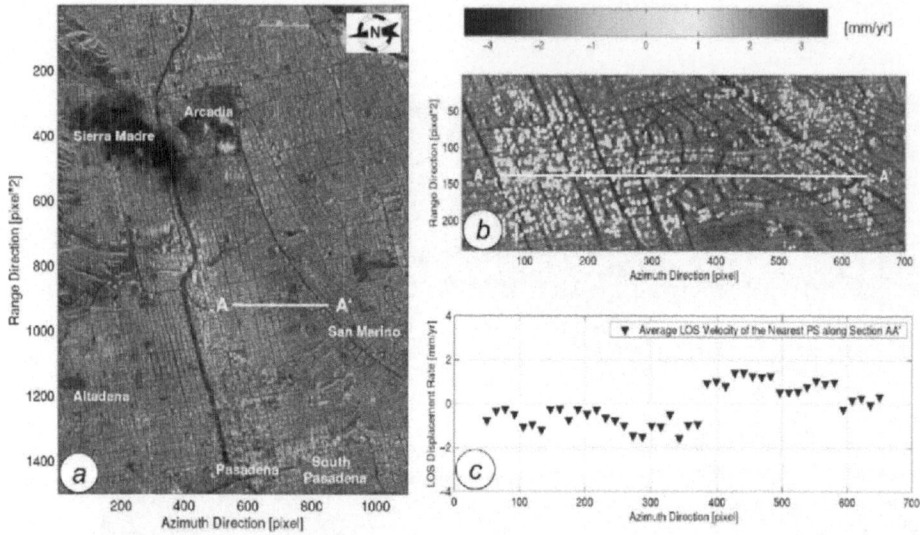

Figure 6.6: Velocity fi eld across a fault.

variations near central Los Angeles in very good agreement with the estimated location and slip-rate of the so-called Elysian Park blind thrust fault (*Oskin et al.*, 2000).

Comparison with eleven GPS stations of the SCIGN (figure 6.7), gathering data since at least 1996 (for a more reliable estimate of target velocity), has highlighted two main aspects of the results: the PS technique presents better accuracy than static GPS, and allows for much higher spatial sampling (hundreds of benchmarks per square kilometer, that can be revisited every 35 days). However, SAR displacement data are not 3D, and the temporal resolution of GPS and spaceborne SAR is not comparable: this calls for a synergistic use of SAR data (worldwide, with monthly frequency) and GPS data (where and when available). It is worth noting that combination of SAR data relative to both ascending and descending satellite overpasses could further improve the results, while reducing the time interval between two passes yielding East-West motion components, too.

The sensitivity of SAR and GPS are complementary each other to some extent. Indeed, SAR data are very sensitive to the vertical motion of the target, where GPS performs more poorly; on the other hand, North-South displacements (almost orthogonal to the range direction) can hardly be detected by means of SAR data only. Moreover, SAR data accuracy, in particular the low-wavenumber components of the velocity field, decreases with the distance from the reference point. Considering the Kolmogorov turbulence model for atmospheric artifacts

(*Hanssen*, 1998; *Williams et al.*, 1998) (with power 0.2/rad^2 1 km apart from the GCP: heavy turbulence conditions), the accuracy of target velocity estimate (with respect to the reference point) is theoretically better than 1 mm/yr within ca. 16 km from the GCP. Therefore, if the PS density is high, and the target motion strongly correlated in time, APS estimation and removal allow detecting relative LOS displacements with accuracy better than 1 mm (equivalent to a phase shift of only 13° at the operating frequency of the ERS-1/2 radars) on single acquisitions. Such accuracy is compatible with target displacements never measured hitherto as, for instance, thermal dilation of metallic-like targets. As a proof, figure 6.8 displays the PS time series in central Los Angeles versus air temperature records synchronous to ERS acquisitions[6]. Several high-reflectivity targets have been detected, showing sharp seasonal behavior (figure 6.7), providing a first example of

---

[6]U.S. National Weather Service Temperature Data available on the web.

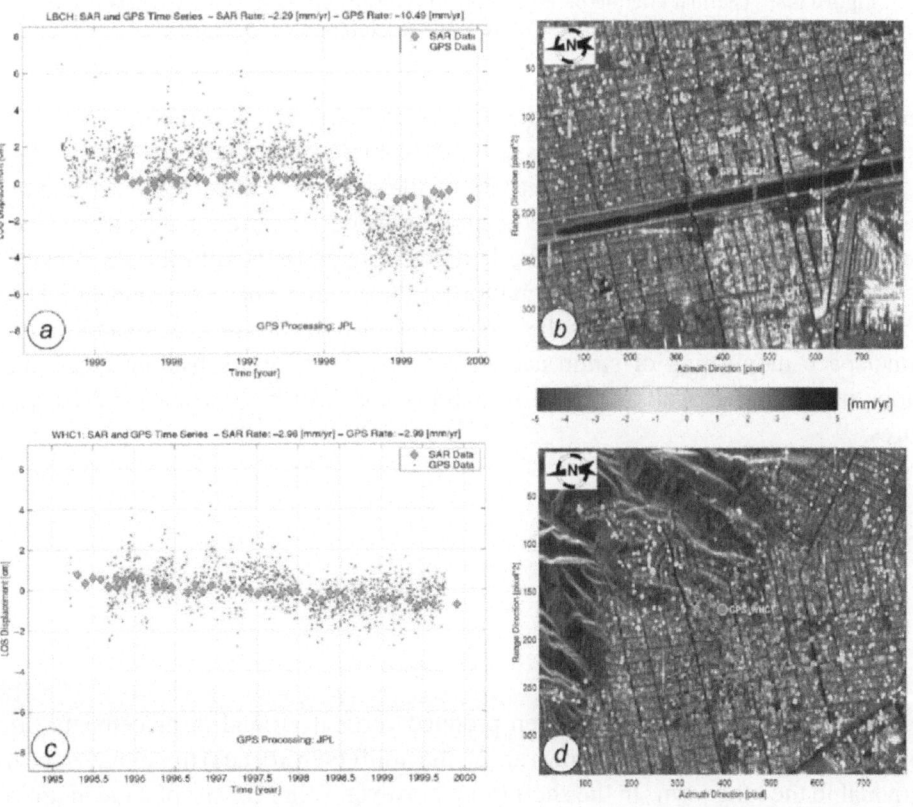

Figure 6.7: PS - GPS comparison in Los Angeles.

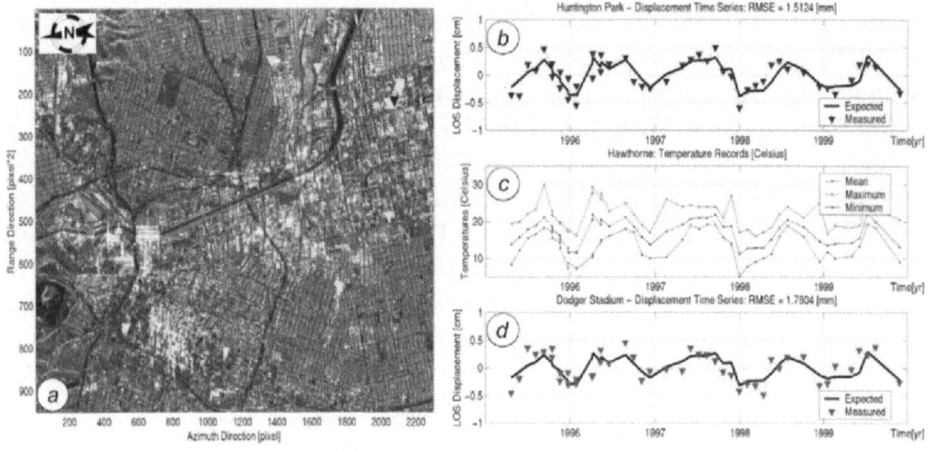

Figure 6.8: Thermal dilation of structures in Los Angeles: LOS motion versus maximum temperature of the day (the satellite is sun synchronous at 10:30 a.m.).

millimetric motion detection of individual natural targets (not corner reflectors) in full-resolution SAR interferograms, with no spatial smoothing.

Having demonstrated the sensitivity of the method, we expect that PS analysis may play a major role where accurate geodetic measurements are needed, especially in urban areas, where the building density (related to the PS density) is sufficient for a reliable discrimination of the motion phase component. This may open new possibilities for the monitoring of hazardous areas, including the time/space monitoring of strain accommodation on faults, subsiding areas and slope instability, as well as precision stability check of buildings and infrastructures.

6.2   Paleoseismic data

[By P. A. Pirazzoli]

Shallow great earthquakes often produce vertical ground displacements. In many tectonic uplifting (subsiding) areas, the uplift (subsidence) trend may appear gradual in the long term. In the short term, however, it can consist of sudden uplift (subsidence) movements, separated by more or less long periods of quiescence or of gradual subsidence (uplift) (figure 6.9). The uplifting (subsiding) movements usually take place at the time of large-magnitude earthquakes, which are often accompanied by surface faulting or folding and ground deformation.

Ground displacements occurring at the time of an earthquake are called co-seismic. Gradual displacements, often in opposition to the coseismic ones, may occur during an interval of time preceding (preseismic) or following (postseismic) the coseismic event. The duration of the time interval between two coseismic events (interseismic period) is not perfectly regular and depends on the variability of the local tectonic stress accumulation. It may tend however to be statistically repetitive. This recurrent time period can vary, according to the seismo-tectonic area considered, from a few centuries to over ten thousand years (*Pirazzoli et al.*, 1996b).

In coastal areas, where sea level is an obvious reference geodetic datum, records of past coseismic events may be preserved for a long time, making it possible to detect evidence of past earthquakes, reconstruct the vertical ground displacement which accompanied them, or even identify and date sequences of past seismic events. The accuracy of such reconstructions will depend on the origin (biological, geomorphologic, stratigraphic, archaeological) of past event markers, on the state of preservation of these markers, and on the limits of the dating methods used.

In coastal active seismic areas, investigation and dating of former shorelines different from the present ones have often been used to determine the age, distribution and succession of vertical displacements (*Wellman*, 1967; *Yonekura*, 1972, 1975; *Kaizuka et al.*, 1973; *Nakata et al.*, 1979; *Taylor et al.*, 1980; *Jouannic et al.*, 1982; *Ota*, 1985, 1991; *Atwater*, 1987; *Berryman et al.*, 1989; *Pirazzoli et al.*, 1989, 1994; *Darienzo and Peterson*, 1990; *Ota et al.*, 1990, 1991; *Shennan et al.*, 1999).

When the recurrence time of morphogenetic earthquakes is relatively short, it is even possible to localize possible areas of impending great-magnitude earthquakes and estimate approximately the vertical crustal movements which may take place in them. Such estimates, based on recurrence period considerations, have an undoubted statistical value over the long term, though they can hardly be used for accurate predictions over the short term.

A main difficulty in the identification of coseismic displacements is to distinguish them from other tectonic or eustatic displacements, of similar magnitude, which may have occurred gradually. This implies the recognition of appropriate coseismic coastal indicators.

6.2.1 Coastal indicators of coseismic vertical movements

On the shore, a period of relatively stable sea level will generally leave many depositional and erosion marks. Identification of coseismic movements is difficult in the case of subsequent subsidence: if a submergence movement occurs, marks of the former sea level will tend to be rapidly reworked by wave action, or eroded, or concealed by new sediments. In the case of an emergence movement, on the other hand, reworking and erosion processes will be very active in the mid-littoral zone (exposed to the tidal range and to average wave action), but preservation is possible above, in the supra-littoral zone, where marine erosion is much less active. When the rock type of the shore is disregarded, coastal zones relatively sheltered from wave action and affected by small or very small tidal range are therefore most favorable to the preservation of small uplift evidence. The fact that certain vertical displacements took place in a very short period of time, however, is not easy to demonstrate. Geomorphological indicators alone (stepped erosional benches, notches, or depositional platforms) may lead to ambiguous interpretations. To be fully convincing, the interpretation of a movement as coseismic should therefore include evidence (biological, stratigraphical, archaeological, historical, etc.), or at least a cluster of converging indications, suggesting that the displacement could have taken place only in a very brief space of time. For example, preservation of boring shells inside fossil burrows is generally indicative of a very rapid uplift (fall in the relative sea level) greater than the height of the local mid-littoral zone, because in the case of a slow uplift (of the order of mm/yr or even of cm/yr) these fragile remains would have been altered and destroyed within a few years

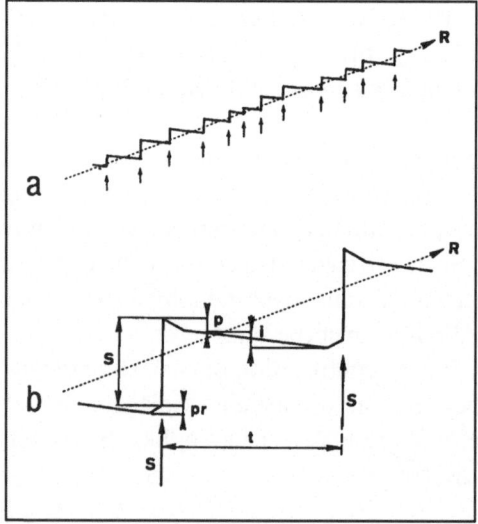

Figure 6.9: In seismic regions where uplift is predominant, the uplift trend, though apparently more or less gradual over the long term, may consist in the short term of sudden uplifting movements accompanying great earthquakes, separated by more or less long periods of seismic quiescence and gradual subsidence. S = coseismic, pr = preseismic, p = postseismic, i = interseismic, t = recurrence time.

by mid-littoral bio-erosion (*Laborel and Laborel-Deguen*, 1994).

A few case studies of coseismic displacements investigated by the present author in seismically active areas of the north-western Pacific and the eastern Mediterranean are summarized below.

6.2.2 Case studies

Okinawa Island, Japan

Emerged marine notches, hermatypic and autochthonous coral reefs, beachrocks and fossil intertidal barnacles collected in living position at the level of the notches have revealed that a sudden uplift movement, reaching approx. 2.5 m in central and south Okinawa, was caused by a major earthquake around 400 B.C. (figure 6.10). The magnitude of the earthquake is estimated at 7.4 from the size of the area uplifted and the amount of the vertical displacement. The recurrence interval of such major earthquakes in Okinawa is estimated at 7,000 to 10,000 years (*Kawana and Pirazzoli*, 1985).

Eastern coast of Taiwan

The eastern Coastal Range of Taiwan is an accreted prism of the Luzon arc, belonging to the Philippine Sea plate, which collided with the Eurasian plate margin. This area experiences lasting rates of tectonic uplift among the highest reported in the literature for coastal areas. In several cases uplift movements appear to have occurred by steps, spasmodically, with recurrence periods of less than 1,000 years (figure 6.11). Near Shih-yü-san (about 23°10'N, 121°23'E) evidence of at least nine stepped Holocene shorelines, reaching up to 37 m in altitude, have been reported. Four of these shorelines have been dated by radiocarbon (*Liew et al.*, 1993).

Çevlik, Hatay coast, Turkey

A similar tectonic history seems to have occurred in the Late Holocene along more than 250 km of the Levant coast of the Mediterranean, from Hatay to Syria and the Lebanon, in an area parallel to the boundary separating the African and the Arabian plates (figure 6.12). As shown by the identification and radiocarbon dating of raised shorelines, this history consists mainly of two uplift movements which took place roughly about 770 cal. B.C. and 550 cal. A.D., respectively (*Pirazzoli et al.*, 1996).

Cape Ladiko, Rhodes, Greece

The east coast of the island of Rhodes shows evidence of several crustal blocks affected by up and down movements, during the Holocene, with however a general trend towards an uplift increasing from south to north and reaching about 3.5 m in the northernmost part of the island (figure 6.13). There is a recurrent periodicity of coseismic displacements varying from a few hundreds to one or at most two thousand years (*Pirazzoli et al.*, 1989) that are associated by *Kontogianni et al.* (2002) with $m > 7.5$.

Western Crete, Greece

Marks of Holocene emergence are widespread in the western part of Crete (figure 6.14). Their altitude increases gradually towards the SW boundary of the island, where shoreline marks attain about +9 m. The uppermost marks of emer-

Figure 6.10: Raised notches near Gushikami (south coast of Okinawa) (~26°07'N, 127°45'E). Tidal range: 1.7 m. Water level at -0.65 m (below MSL). Elevation of the notch vertex: +2.9 m (a metre gives scale). Some fossil barnacles are preserved between +2.55 and +2.95 m on the notch undercut in limestone blocks projecting above a dead coral reef flat. In a nearby sea cave, at ~+2.9 m, similar barnacles (*Octomeris sulcata*) collected in growth position have been dated 2330±85 yr B.P. (Gak-5195) (*Machida et al.*, 1976) (photo 6604, March 1981).

Figure 6.11: Marks left by the last two fast uplift movements at Shih-yü-san. Tidal range: 1.45 m. At about +10 m, remnants of a notch and of an algal ridge (a), dated 1220±60 yr B.P. (Gif-8646), indicate the approximate time when the relative sea level fell to about +5.5 m, developing a continuous notch (c) and widening a sea cave. Oysters colonizing the walls of the cave died about 910±70 yr B.P. (GifA-91090), when a new uplift brought the relative sea level near its present position. The good state of preservation of oyster and barnacle shells inside the cave and of the base of the notch profi le indicate that the uplift movement was very fast, probably coseismic (photo B724, January, 1990).

gence have been dated (nine radiocarbon dates) between 261 and 425 cal. A.D. and their uplift ascribed tentatively to the earthquake of July 21, 365 A.D. The very rapid (coseismic) character of the uplift is shown, among other things, by (1) the absence of younger bioconstructions at intermediate levels between the uppermost Holocene shoreline and the present sea level, (2) the excellent state of preservation of very fragile sublittoral bioconstructions related to the uppermost shoreline but located at lower elevations, and (3) the confirmation by radiocarbon datings that the in situ sublittoral bioconstructions emerging at lower levels have the same age as the uppermost shoreline, i.e. that they were uplifted at the same time (*Pirazzoli et al.*, 1996).

Historical records do not mention uplift but indicate that 100 town were destroyed in A.D. 365 in Crete by an earthquake of unprecedented magnitude (*Stiros,*

Figure 6.12: A double raised shoreline at Çevlik (36°10'N, 35°52'E). Tidal range: negligible. According to four radiocarbon dates, the upper shoreline (an erosion notch at about +2.9 m) (N) corresponds to a relatively stable sea level position between about 4500 and, roughly, 770 B.C., when a 2.1 m uplift took place. The lower raised shoreline is here represented by spectacular cornices made by oysters and vermetids (c), which developed since about 770 B.C., until the occurrence of a coseismic uplift between 430 and 710 AD This displacement is tentatively ascribed to the A.D. 526 earthquake, which destroyed Antioch and Seleucia Pieria and caused a tsunami in the area. Several other earthquakes are known to have occurred during historical times in this region (*Guidoboni et al.*, 1994; *Ambraseys et al.*, 1994), but no other marks of vertical displacement could be observed (photo A719, May 1988).

2001). In one of these towns, Kisamos, archaeological excavations brought evidence of a prosperous period after A.D. 270, suddenly interrupted by a tremendous catastrophe, shortly after A.D. 355 and possibly after 361 (numismatic evidence), but before A.D. 400, with destruction of most houses; their roofs and walls collapsed, killing and burying people, or breaking and burying precious objects below a layer of roof tiles and debris from the fallen walls in a town-wide scale. Survivors of this catastrophe proved impotent to clear the debris, search for victims and offer them a proper burial, that was a fundamental ritual of the ancient Greek and Roman society (*Stiros and Papageorgiou*, 2001).

Figure 6.13: A series of notches dating from the last 6,000 years (four of which have been dated by radiocarbon) appears between sea level and +3.75 m on the limestone cliffs near Cape Ladiko (~36°19'N, 28°15'E). Tidal range: ~0.2 m. According to the notch profiles and to possible rates of notch undercutting by marine (bio)erosion, the sea must have remained relatively stable at the level of each notch for several centuries, whereas displacements of sea level from one notch to the other must have been rapid, probably coseismic (photo 4537C, May 1978).

Cephalonia, Greece

Marks of two Holocene shorelines differing from the present one have been found on Cephalonia Island (figure 6.15). The lower elevated shoreline, which corresponds to uplift occurred at the time of the A.D. 1953 earthquakes, reaches a maximum elevation of +0.7 m.

It was charted systematically in 1990 and compared to the estimations of coseimic uplift deduced from 1953 earthquake reports, showing that no postseismic displacements followed the 1953 coseismic uplift. On the other hand, the occurrence of preseismic subsidence movements shortly before 1953 has been demonstrated by the 1990 survey.

According to the morphology of notches and benches belonging to the upper, older shoreline, which reaches a maximal elevation of +1.2 m, also its uplift seems to have been relatively rapid. Five radiocarbon dates of samples belonging to

Figure 6.14: At Kisamos (now Kastelli) (~35°30'N, 28°38'E) (tidal range: negligible) the uplift ascribed to the A.D. 365 earthquake reaches about 6.5 m, as indicated by traces of marine erosion on limestone cliffs. A most prominent feature is the uplifted mole of rubble blocks of the Roman harbour, which rises several metres above sea level, delimiting a small basin now completely filled in with sand. Marks of in situ marine incrustations dating from the uplift time are still preserved on many blocks (photo G523, June 2000).

the upper shoreline have shown ages ranging between 3060–2570 cal. B.C. and cal. A.D. 350–710. The latter interval probably includes the time when the older shoreline was uplifted (*Pirazzoli et al.*, 1994).

6.2.3   Conclusions

With instrumental seismic records going rarely back for more than one century and historical records (usually far from complete) for at best 2,500 years in a few regions, the assessment of seismic risk still remains an uncertain task, especially in apparently 'aseismic' or 'low seismicity' regions, where the recurrence periods of great earthquakes may be relatively long.

Paleoseismic investigations can contribute with success, in some cases, to fill such gaps, by showing that the long-term trend can be different from what is shown by the short-term instrumental record.

Figure 6.15: A double bench, with steps at +0.6 and +1.2 m, respectively, around a mushroom rock in the harbour of Poros, Cephalonia (~38°09'N, 20°47'E) (tidal range: negligible). The lower step (L) was uplifted coseismically in 1953. The upper step (U) was probably also uplifted coseismically around A.D. 350–710 (photo B899, June 1990).

The case of western Crete is from this point of view exemplary: according to the official Seismotectonic Map of Greece (*IGME*, 1989), observed intensities of scale VI and VII, and only locally of scale VIII are reported to have occurred between 1700 and 1981. However, the earthquake which caused an up to 9 m coseismic uplift in the western part of the island (possibly the same which totally destroyed the Roman town of Kisamos) had a probable magnitude 8 or higher (*Stiros and Papageorgiou*, 2003), i.e. at least an order of magnitude greater than what has been observed during the instrumental period. In the same way, the seismic risk assessment in southern and central Okinawa would not be the same if the paleoseismic event of about 400 B.C. was not taken into account.

Paleoseismic records can therefore be very useful to complete, over the long term, seismic risk assessments that instrumental record can provide only for much shorter, recent periods.

6.3 References

Ambraseys, N. N., Melville, C. P., and Adams, R. D., 1994. *The Seismicity of Egypt, Arabia and the Red Sea: a Historical Review*, Cambridge University Press.

Atwater, B. F., 1987. Evidence for great Holocene earthquakes along the outer coast of Washington State, *Science*, **236**, 942–944.

Bamler, R., and Hartl, P., 1998. Synthetic aperture radar interferometry, *Inverse Problems*, **14**(4), R1–R54.

Berryman, K. R., Ota, Y., and Gull, A. G., 1989. Holocene paleoseismicity in the fold and thrust belt of the Hikurangi subduction zone, eastern North Island, New Zealand, *Tectonophysics*, **163**, 185–195.

Boucher, C., Altamini, L., and Sillard, P., 1997. The 1997 International Reference Frame (ITRF97), *IERS Technical Notes*, n. 27.

Darienzo, M. E., and Peterson, C. D., 1990. Episodic tectonic subsidence of late Holocene salt marshes, northern Oregon central Cascadia margin, *Tectonics*, **9**, 1–22.

Ferretti, A., Prati, C., and Rocca, F., 1999. Multibaseline InSAR DEM reconstruction: the wavelet approach, *IEEE Trans. Geosci. Remote Sensing*, **37**, 705–715.

Ferretti, A., Prati, C., and Rocca, F., 2000. Nonlinear subsidence rate estimation using permanent scatterers in differential SAR interferometry, *IEEE Trans. Geosci. Remote Sensing*, **38**, 2202–2212.

Ferretti, A., Prati, C., and Rocca, F., 2001. Permanent scatterers in SAR interferometry, *IEEE Trans. Geosci. Remote Sensing*, **39**, 8–20.

Franceschetti, G., and Lanari, R., 1999. *Synthetic Aperture Radar Processing*, CRC Press, Boca Raton.

Gabriel, A. K., Goldstein, R. M., and Zebker, H. A., 1989. Mapping small elevation changes over large areas, differential radar interferometry, *J. Geophys. Res.*, **94**, 9183.

Guidoboni, E., Comastri, A., and Traina, G., 1994. *Catalogue of Ancient Earthquakes in the Mediterranean Area up to the 10th Century*, Istituto Nazionale di Geofisica, Roma.

Hanssen, R. F., 1998. Atmospheric heterogeneities in ERS tandem SAR interferometry, Tech. Rep. No. 98.1, Delft University Press, Delft.

IGME (Institute of Geology and Mineral Exploration), 1989. Seismotectonic map of Greece. 1:500,000 scale. Athens.

Jouannic, C., Taylor, F. W., and Bloom, A., 1982. Uplift and deformation of a young arc: the New Hebrides arc (in French) in *Contribution à l'Étude Géodynamique du Sud-Ouest Pacifi que*, Travaux et Documents ORSTOM, Paris, **147**, 223–246.

Kaizuka, S., Matsuda, T., Nogami, M., and Yonekura, N., 1973. Quaternary tectonic and recent seismic crustal movements in the Arauco Peninsula and its environs, central Chile, *Geogr. Rep. Tokyo Metrop. Univ.*, **8**, 1–49.

Kawana, T., and Pirazzoli, P. A., 1985. Holocene coastline changes and seismic uplift in Okinawa Island, the Ryukyus, *Japan. Zeits. Geomorphol.*, Suppl. **57**, 11–31.

Kontogianni, V. A., Tsoulos, N., and Stiros, S. C., 2002. Coastal uplift, earthquakes and active faulting of Rhodes Island (Aegean Arc): modeling based on geodetic inversion, *Mar. Geol.*, **186**, 299–317.

Laborel, J., and Laborel-Deguen, F., 1994. Biological indicators of relative sea-level variations and of co-seismic displacements in the Mediterranean region, *J. Coastal Res.*, **10**, 395–415.

Liew, P. M., Pirazzoli, P. A., Hsieh, M. L., Arnold, M., Barusseau, J. P., Fontugne, M., and Giresse, P., 1993. Holocene tectonic uplift deduced from elevated shorelines, eastern Coastal Range of Taiwan, *Tectonophysics*, **222**, 55–68.

Machida, H., Nakagawa, H., and Pirazzoli, P. A., 1976. Preliminary study on the Holocene sea level in the central Ryukyu Islands, *Rev. Géomorphol. Dyn.*, **25**, 49–62.

Massonnet, D., and Feigl, K. L., 1998. Radar interferometry and its application to changes in the Earth's surface, *Rev. Geophys.*, **36**, 441–500.

Massonnet, D., Rossi, M., Carmona, C., Adragna, F., Peltzer, G., Feigl, K., and Rabaute, T., 1993. The displacement field of the Landers earthquake mapped by radar interferometry, *Nature*, **364**, 138–142.

Massonnet, D., Briole, P., and Arnaud, A. 1995. Deflation of the Mount Etna monitored by spaceborne radar interferometry, *Nature*, **375**, 567–570.

Nakata, T., Koba, M., Jo, W., Imaizumi, T., Matsumoto, H., and Suganuma, T., 1979. Holocene marine terraces and seismic crustal movements, *Sci. Rep. Tohoku Univ.*, 7th Ser. (Geogr.), **29**, 195–204.

Oskin, M., Sieh, K., Rockwell, T., Miller, G., Guptill, P., Curtis, M., McArdle, S., and Elliot, P., 2000. Active parasitic folds on the Elysian Park anticline: implications for seismic hazard in central Los Angeles, California, *Bull. Geol. Soc. Am*, **112**, 693–707.

Ota, Y., 1985. Marine terraces and active faults in Japan with special reference to co-seismic events, in *Tectonic Geomorphology*, Morisawa, M., and Hack, J. T. (eds.), Allen & Unwin, 345–366.

Ota, Y., 1991. Coseismic uplift in coastal zones of the western Pacific rim and its implications for coastal evolution, *Zeits. Geomorphol.*, Suppl. **81**, 163–179.

Ota, Y., Miyauchi, T., and Hull, A. G., 1990. Holocene marine terraces at Aramoana and Pourerere, eastern North Island, New Zealand, *N. Z. J. Geol. Geophys.*, **33**, 541–546.

Ota, Y., Hull, A. G., and Berryman, K. R., 1991. Coseismic uplift of Holocene marine terraces in the Pakarae River area, eastern North Island, *New Zealand Quat. Res.*, **35**, 331–346.

Peltzer, G., Crampe, F., and King, G., 1999. Evidence of nonlinear elasticity of the crust from the M_w 7.6 Manyi (Tibet) earthquake, *Science*, **286**, 272–276.

Pirazzoli, P. A., Laborel, J., and Stiros, S. C., 1996a. Earthquake clustering in the Eastern Mediterranean during historical times, *J. Geophys. Res.*, **101**, 6083–6097.

Pirazzoli, P. A., Laborel, J., and Stiros, S. C., 1996b. Coastal indicators of rapid uplift and subsidence: examples from Crete and other Mediterranean sites, *Zeits. Geomorphol.*, Suppl. **102**, 21–35.

Pirazzoli, P. A., Montaggioni, L. F., Sali`ege, J. F., Segonzac, G., Thommeret, Y., and Vergnaud-Grazzini C., 1989. Crustal block movement from Holocene shorelines: Rhodes Island (Greece), *Tectonophysics*, **170**, 89–114.

Pirazzoli, P. A., Stiros, S. C., Laborel, J., Laborel-Deguen, F., Arnold, M., Papageorgiou, S., and Morhange C., 1994. Late-Holocene shoreline changes related to palaeoseismic events in the Ionian Islands, Greece, *The Holocene*, **4**, 397–405.

Reasenberg, P. A., 1999. Foreshock occurrence rates before large earthquakes worldwide, *Pure Appl. Geophys.*, **155**, 355–379.

Rosen, P. A., Hensley, S., Joughin, I. R., Li, F. K., Madsen, S. N., Rodriguez, E., and Goldstein, R. M., 2000. Synthetic aperture radar interferometry, Invited paper, *Proc. IEEE*, **88**, 333–382.

Scharroo, R., and Visser, P., 1998. Precise orbit determination and gravity field improvement for the ERS satellites, *J. Geophys. Res.*, **103**, 8113–8127.

Shennan, I., Scott, D. B., Rutherford, M., and Zong, Y., 1999. Microfossil analysis of sediments representing the 1964 earthquake, exposed at Girwood Flats, Alaska, USA, *Quat. Int.*, **60**, 55–73.

Sieh, K., Stuiver, M., and Brillinger, D., 1989. A more precise chronology of earthquakes produced by the San Andreas fault in Southern California, *J. Geophys. Res.*, **94**, 603–623.

Sillard, P., and Boucher, C., 2001. A review of algebraic constraints in terrestrial reference frame datum definition, *J. Geodesy*, **75**, 63–74.

Stiros, S. C., 2001. The AD 365 Crete earthquake and possible seismic clustering during the fourth to sixth centuries AD in the Eastern Mediterranean: a review of historical and archaeological data, *J. Struct. Geol.*, **23**, 545–562.

Stiros, S. C., and Papageorgiou, S., 2001. Seismicity of Western Crete and the destruction of the town of Kisamos at AD 365: archaeological evidence, *J. Seismol.*, **5**, 381–397.

Taylor, F. W., Isacks, B. L., Jouannic, C., Bloom, A. L., and Dubois, J., 1980. Coseismic and Quaternary vertical tectonic movements, Santo and Malekula Islands, New Hebrides Island arc, *J. Geophys. Res.*, **85**, 5367–5381.

Wald, D. J., and Heaton, T. H., 1994. Spatial and temporal distribution of slip for the 1992 Landers, California, earthquake, *Bull. Seism. Soc. Am.*, **84**, 668–691.

Wellman, H. W., 1967. Tilted marine beach ridges at Cape Turakirae, New Zealand, *J. Geosci. Osaka City Univ.*, **10**, 123–129.

Williams, S., Bock, Y., and Pang, P., 1998. Integrated satellite interferometry: tropospheric noise, GPS estimates and implications for interferometric synthetic aperture radar products, *J. Geophys. Res.*, **103**, 27051–27067.

Yeats, R. S., Sieh, K., and Allen, C. R., 1997. *The Geology of Earthquakes*, Oxford University Press, New York.

Yonekura, N., 1972. A review on seismic crustal deformations in and near Japan, *Bull. Dep. Geogr. Univ. Tokyo*, **4**, 17–50.

Yonekura, N., 1975. Quaternary tectonic movements in the outer arc of southwest Japan with special reference to seismic crustal deformation, *Bull. Dep. Geogr. Univ. Tokyo*, **7**, 19–71.

Chapter 7

Seismic risk mitigation

7.1 Greek case study

[By S. A. Anagnostopoulos]

The irregularity and long time intervals between earthquakes are factors contributing to reduced awareness about earthquake risks among the public and government officials and hence to reduced allocation of resources for their mitigation. Moreover, it is not uncommon to see misallocation of resources by non-knowledgeable decision makers and politicians, especially when they act under fear of criticism and public opinion pressures in the aftermath of a catastrophic event. It is therefore up to scientists to help maintain an increased level of awareness about seismic hazards and also to help policy makers understand that seismic risk reduction requires continuous, long-term efforts with a multitude of activities covering all aspects of the problem.

This section discusses the seismic hazard and seismic risk of Greece in terms of expected magnitudes, intensities, building damage and numbers of lives lost. The activities required to mitigate seismic risk are then reviewed and, based on the pertinent Greek experience of the past twenty years, they are grouped into two lines of action: one aimed at increasing the seismic safety of future and existing construction, and another aimed at pre-disaster planning and preparedness for improving emergency response and rehabilitation. In the context of such activities, the pertinent contribution of the various areas of earthquake-related research is identified and policies are recommended for short and long term seismic risk mitigation.

The term seismic risk is used here to mean expected effects of earthquakes to humans, their activities and works (i.e. the built environment), in accordance with

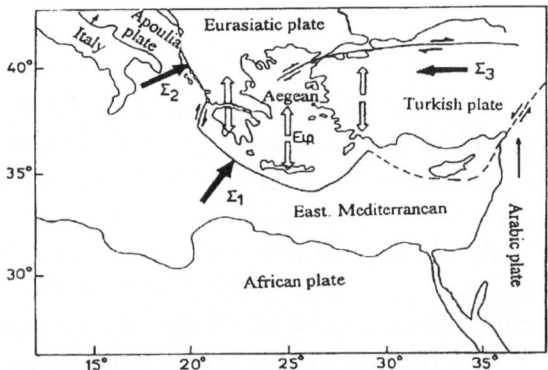

Figure 7.1: Tectonic setting of Greece.

the *UNDRO* (1979) definition. Quite often seismic risk is used as having the same meaning as seismic hazard, but, according to the UNDRO document, the latter should be used to express the probability of occurrence, within a specific period of time in a given area, of a potentially damaging earthquake.

7.1.1   The seismic risk in Greece

Greece, a small country of about 10 million people located at the African and Eurasiatic plate boundary, has the highest seismicity in Europe (figure 7.1). In its long history, it has suffered from many devastating earthquakes, which continue to threaten the lives of its people and to burden its economy. A quantitative description of the seismicity of Greece is given by the number N_t of earthquakes with magnitude $\geq m$ that occurred in the region during this century. A straight-line approximation for the logarithm of N_t, fitted to the data by the least squares method, shown in figure 7.2, is given by the following equation (*Papazachos and Papazachou*, 1989):

$$\log N_t = 8.24 - 1.00m \,. \tag{7.1}$$

On an annual basis, the logarithm of the number of earthquakes N with magnitude m or greater is given by:

$$\log N = 6.31 - 1.00m \,, \tag{7.2}$$

while the most probable maximum magnitude for any year is $\overline{m} = 6.3$.

Another indicator of the seismic hazard is the mean return period T_m of earthquakes having magnitude m or greater. This can be obtained from equation (7.2) and is given by:

$$T_m = 10^{m-6.31} \ . \tag{7.3}$$

Using equation (7.3), it is estimated that Greece and its surroundings are struck by earthquakes having magnitudes $m > 6, 7, 7.5$ and 8 on the average every 0.5, 5, 15 and 50 years, respectively. This places Greece quite high in the list of the most seismically active regions in the world.

Quantification of the seismic risk in Greece, requires indices related to the effects of earthquakes such as damage to structures and numbers of lives lost. Macroseismic intensities are among the most frequently used quantities to characterize seismic risk. In Greece, there is a wealth of macroseismic intensity assessments, derived from historic accounts that date back to the 6th century B.C.. However, only recently (since 1950) has the collection of such data been carried out in a systematic manner and thus only for the period thereafter are the intensity assessments uniformly reliable and complete. A list of the strongest Greek earthquakes for the period 1950–1996 and of their consequences is given in table 7.1. Using these data, the number of earthquakes N_t with maximum Modified Mercalli Intensity equal to or greater than I was found and its logarithm was plotted in figure 7.3, along with a straight line, least squares, approximation. The latter is given by the equation:

$$\log N_t = 6.5 - 0.615I \tag{7.4}$$

from which the corresponding expression for a one year period can be derived:

$$\log N = 4.70 - 0.60I \ . \tag{7.5}$$

The mean return period T_m (in years) of events giving MMI equal to I or greater can be computed from the following expression, derived from equation (7.5):

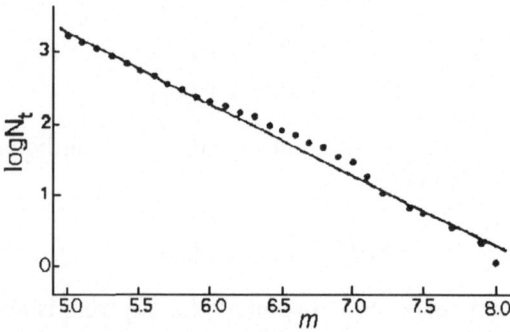

Figure 7.2: Logarithm of the number of earthquakes N_t in Greece during 1901–1985 with magnitude m (*Papazachos and Papazachou*, 1989).

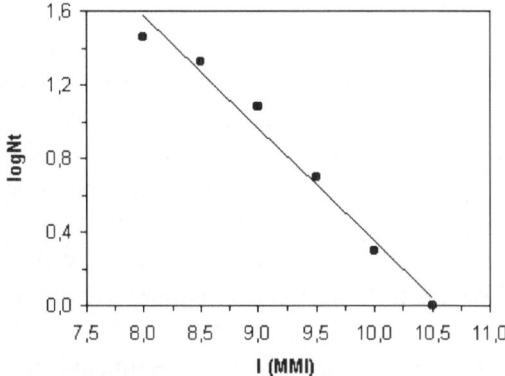

Figure 7.3: Logarithm of the number of earthquakes N_t in Greece during 1950–2000 with maximum Intensity (MMI) I.

$$T_m = 10^{0.60I-4.70} \tag{7.6}$$

It follows from equation (7.6) that the mean return periods for intensities $I >$ VIII, IX and X in Greece are equal to 1.25, 5.0 and 20 years, respectively.

Based on the same body of data for the period 1950–1996, it is estimated that the mean annual number of buildings totally destroyed by earthquakes in Greece is about 2,150. This number, however, has been estimated without any qualification as to the size of the building, construction material, engineered or non-engineered type etc. and hence it cannot be used for reliable predictions of annual economic losses. We must also note here that the epicenters of Greek earthquakes are often at sea and hence they do not cause much damage.

If a single number were to be used to characterize the disaster caused by an earthquake, this would undoubtedly be the casualty count. In Greece, the worst of all the known earthquakes in terms of human losses is most probably the earthquake that leveled the ancient city of Sparta in 464 B.C. killing over 20,000 people. In the last two centuries, the worst event was an earthquake of $m = 6.4$ that occurred on April 3, 1881 which destroyed most of the island of Chios in the eastern Aegean sea, killing 3,650 people and injuring about 7,000. During this century, the most catastrophic sequence of earthquakes was a series of shocks with $m_s = 6.4, 6.8, 7.2$ and 6.3 that almost completely destroyed the Ionian islands of Cephalonia, Zakynthos and Ithaca on August 9, 11 and 12, 1953 (the main shock with $m_s = 7.2$ occurred on Aug. 12), killing 476 people and injuring about 2,400. More recently, during 1978–2000 several strong earthquakes hit modern Greek cities, causing extensive damage and loss of life. The most damaging were the

Thessaloniki earthquake of 1978 (m_s = 6.5) that killed 48 people, most of them in one major collapse; the Alkyonides sequence of 1981 (m_s = 6.7, 6.4 and 6.3) that killed 20 people and caused widespread damage in towns around the Gulf of Corinth and in Athens; the Kalamata earthquake of 1986 (m_s = 6.2) that killed 20 people and destroyed many of the old houses in the city of Kalamata; the Aigion earthquake of 1995 (m_s = 6.2) that killed 26 people; and the most recent Athens earthquake of 1999 (m_s = 5.9) that killed 143 people, and caused many collapses and widespread damage. From the data of table 7.1, it follows that the average annual number of deaths in Greece due to earthquakes in the second half of this century is about 19.

In terms of economic consequences, it is estimated that in recent years, the direct cost of earthquake damage in Greece on an average annual basis, not including costs from loss or interruption of economic activities, is on the order of 150 to 200 million U.S. dollars.

7.1.2   Activities for seismic risk mitigation and current Greek experience

Since earthquake occurrence is beyond human control, and will remain so for the foreseeable future, the only means to reduce seismic risk is by taking measures to minimize the effects of earthquakes. Such measures can be grouped into two lines of action: one aimed at increasing the seismic safety of construction, future and existing, and another aimed at pre-disaster planning and preparedness for coping with the emergency created by a catastrophic event. Research in seismology, earthquake engineering and earthquake prediction contributes ultimately towards these two goals.

Activities for increasing seismic safety of construction

The most effective means for mitigating earthquake risk is by building earthquake-resistant structures that will withstand the strongest anticipated earthquakes for the region. In engineering terminology these are called design earthquakes, they are specified in the codes and their characteristics are determined from seismological research that utilizes historical, instrumental and geotectonic data. Progress in the direction of producing safer structures requires:

(1) The existence of adequate and up-to-date regulations for earthquake-resistant design of structures, for which it is important to have activity in earthquake engineering and seismological research and a mechanism in the governmental bureaucracy for code revision and implementation.

	Name	Date dd.mm.yy	m_s	Depth (km)	MMI	P.G.A. (g's)	N_D	N_B/N_M
1	Argostoli	12.8.53	7.2	-	X+	-	476	27800
2	Acarnania	21.10.53	6.3	-	VIII	-	-	-
3	Sofades	30.4.54	7.0	-	IX+	-	25	6600
4	Magnesia	19.4.55	6.2	-	VIII+	-	1	450
5	Samos	16.7.55	6.9	-	VIII	-	-	40
6	Amorgos	9.7.56	7.5	-	IX	-	53	530
7	Magnesia	8.3.57	6.8	-	IX+	-	2	6930
8	Rhodes	25.4.57	7.2	-	VIII	-	-	16
9	Heraklio	14.5.59	6.3	-	VIII+	-	-	156
10	Zakynthos	15.11.59	6.8	-	VII	-	-	-
11	Corinthos	28.8.62	6.8	95	VIII+	-	1	400
12	Alonnisos	9.3.65	6.1	-	IX+	-	2	71940
13	Agrinio	31.3.65	6.8	78	VIII+	-	6	1460
14	Arcadia	4.1.65	6.1	-	X	-	18	8470
15	Phokida	6.7.65	6.3	-	VIII+	-	1	575
16	Kremasta	5.2.66	6.2	-	IX	-	1	2770
17	Megalopoli	1.9.66	6.0	-	VIII	-	-	240
18	Katouna	29.10.66	6.0	-	VIII	-	1	25
19	Arta	1.5.67	6.4	-	IX	-	9	3500
20	A.Efstratios	19.2.68	7.1	-	IX	-	20	175
21	Korinthos	13.9.72	6.3	75	VIII	-	-	52
22	Thessaloniki	20.6.78	6.5	8	VIII+	0.15/0.14	48	9480/(1)
23	Almyros	9.7.80	6.5	-	VIII+	0.13/	-	520
24	Alkyonides	24.2.81	6.7	8	IX	0.29/0.10	20	22550/(9)
25	N.Aegean	19.12.81	7.2	6	VIII	-	-	9
26	N.Aegean	18.1.82	7.0	10	VI	-	-	-
27	Cephalonia	17.1.83	7.0	8	VI	0.18/0.08	-	-
28	Kalamata	13.9.86	6.0	8	IX	0.30/0.37	20	2000/(7)
29	Kyllini	16.10.88	6.0	8	VIII	0.16/0.13	1	~400
30	Pyrgos	26.3.93	5.2	20	VII	0.45/0.13	-	~610
31	Patra	14.7.93	5.4	21	VII	0.42/0.15	-	~600
32	Kozani	13.5.95	6.6	9	VIII+	0.21/0.08	-	~300
33	Aigio	15.6.95	6.2	26	VIII+	0.53/0.19	26	~750/(3)
34	Konitsa	6.8.96	5.6	6.6	VII+	0.39/0.19	-	~80
35	Zakynthos	18.11.97	6.6	10	VII	0.25/0.08	-	-
36	Athens	1.9.99	5.9	15	IX	0.30/0.15	143	6000/(30)

Table 7.1: Major Greek earthquakes for the period 1950–1999 with $m_s > 6.5$ or MMI>VIII or P.G.A.>0.15 g. N_D is the number of deaths, N_B is the approximate number of totally damaged buildings/number of major collapses. In the last column, numbers in bold show collapsed multistory RC buildings (N_M). Earthquake # 6 generated tsunami waves $\sim$ 25 m high and # 20 $\sim$ 1.2 m high. Data 1 to 28 are from *Papazachos and Papazachou* (1989).

(2) Well trained engineers and construction technicians who will be aware of the seismic risk and of the consequences of poor design and construction practices. This requires academic programs in engineering schools, where earthquake engineering and earthquake resistant design are taught and continuing education programs for older engineers and construction technicians, especially after new codes are introduced.

(3) Quality control procedures that will include design review, as well as inspections at the construction site to verify compliance with the construction drawings and specifications.

Since in every populated area threatened by earthquakes many buildings will have been built either before earthquake resistant design codes were introduced or in periods when the applicable regulations were inadequate in comparison to current standards, effective reduction of seismic risk requires intervention into the existing building stock. This is a difficult and expensive task that has rarely been attempted. When it was tried, it was done selectively for: (a) structures of high importance (e.g. hospitals, buildings where large numbers of people are assembled, important bridges, monumental structures etc.); and (b) specific types of building construction exhibiting the highest seismic risk (e.g. the most vulnerable buildings in areas of highest seismicity).

Given that structures of high importance belong typically to the public sector, the cost of intervention in category (a) will largely be covered by public funds. On the other hand, buildings in category (b) are in most cases privately owned and thus the burden for their seismic upgrading will fall on individuals. However, considering that the owners of such buildings will usually be from the poorest classes of society, seismic upgrading of their houses will generally rank quite low in their priorities. Obviously, any large scale seismic upgrading program of buildings in category (b), before an earthquake strikes, will require a great deal of incentives and even then the chances of successful implementation are quite low. Thus, it is not surprising that such programs have rarely, if ever, been applied and completed anywhere in the world.

In Greece, earthquake resistant design regulations were introduced for the whole country in 1959 and reflected the state of knowledge at that time. Those regulations were primarily for RC (Reinforced Concrete) structures. After the Thessaloniki earthquake of 1978 and the Alkionides earthquake of 1981, which were the first strong earthquakes to hit and damage modern Greek cities, Greek engineers became painfully aware of the inadequacies in their design and construction practices and also of the shortcomings of the applicable regulations. The old code was therefore hastily revised in 1984, mainly through the introduction of additional provisions aimed primarily at construction details of RC structures. The new code, together with the Greek engineers' new awareness of the seismic risk, which was epitomized by the shocking pictures of some collapsed multi-storied buildings in the aforementioned two earthquakes and in the subsequent one of Kalamata in 1986, mark a new period of greatly improved earthquake resistant construction in Greece.

In 1992, a modern new code for earthquake resistant design and another for RC structures were introduced and after a 3-year parallel application with the old codes they became effective in 1995, two days after the Aigion earthquake, as the only applicable standards. The new code is similar to the American UBC and to EUROCODE-8 for earthquake resistant design. Based on a new seismic hazard map, the new code divides Greece into four seismic zones characterized by effective peak ground accelerations of 0.12 g, 0.16 g, 0.24 g and 0.36 g. While these developments are welcome, there is still a need in Greece for a quality control mechanism, which at present is the sole responsibility of the structural engineer who designs the structure and supervises its construction. There is no independent review of the design and the building officials make only superficial checks in most cases. As far as risk reduction of existing structures is concerned, very little has been done so far and only for a few monuments.

To monitor seismic activity in Greece and to support seismological and earthquake engineering research, a telemetered network of 31 permanent seismological stations is operated by the Geodynamics Institute of the Athens National Observatory and by the Geophysics laboratory of the University of Thessaloniki. In addition, a network of about 90 strong motion recorders is also in operation under the responsibility of the Institute for Engineering Seismology and Earthquake Engineering in Thessaloniki and of the Geodynamics Institute in Athens.

Activities for preparedness and pre-disaster planning

The activities listed in the preceding section will have no effect on the existing building stock, except for those few cases of structures specifically targeted for rehabilitation and strengthening. Therefore, earthquakes will continue to be catastrophic in the foreseeable future and thus the basic means for mitigating their consequences will be through effective preparedness and pre-disaster planning in order to optimize emergency response, relief operations and subsequent rehabilitation. This will require the creation at national and local levels of pre- and post-earthquake planning and response units charged with: (a) assessing the consequences of catastrophic earthquakes in high-risk areas, on the basis of risk assessment techniques and scenario studies; (b) preparing detailed response plans, keeping them up to date and testing them with occasional drill exercises at proper time intervals; (c) establishing formal procedures for dealing with possible earthquake predictions; (d) planning in detail large scale operations for post-earthquake

emergency assessment of building safety, including the development of well defined criteria for such assessments and the creation of an appropriate legal framework covering all emergency operations (see e.g. *Anagnostopoulos*, 1996); (e)
educating the general public about appropriate pre- and post-earthquake actions
and maintaining an acceptable level of awareness about earthquake risks; (f) planning the rehabilitation phase which includes developing repair and strengthening
procedures for damaged buildings, as well as cost assessment guidelines for the
pertinent works and criteria by which the amount of financial assistance to be
provided by the State for such works in determined, and developing a financial
plan for securing the funds necessary for repair, strengthening or reconstruction
and more generally for financial assistance to those who suffer losses in the earthquake.

In Greece, the need for pre-disaster planning and preparedness became
painfully apparent after the strong earthquakes of 1978 and 1981. After those
two events, Greek authorities took certain steps aimed first at assisting the rehabilitation of the damaged buildings and second at providing the means that would
reduce the seismic risk in the future. These steps were (1) the creation of a branch
in the Ministry of Public Works that would oversee and support rehabilitation activities, (2) the creation of a governmental agency for 'Earthquake Planning and
Protection', charged with formulating earthquake risk reduction policies and with
coordinating the implementation of such policies, and (3) the creation of an Institute for Engineering Seismology and Earthquake Engineering that would carry
out applied research.

For rehabilitation of damaged buildings, the Greek government has heretofore
provided financial support to their owners, paying for 1/3 of the assessed repair
cost and also the interest of a 15-year loan for the remaining 2/3 of this cost.
Moreover, a three-year grace period is allowed before the owner starts paying
back his zero-interest loan. This policy, however, has created many problems,
e.g. pertaining to verifications of the cost estimates (as many people are tempted
to include renovations in the required repair works) or to loan repayments (even
under these extremely favorable terms) because authorities in the past did not take
any measures against people who did not make their payments and thus defaulting
on the loans became widespread. In addition, when the economy is in recession,
the funds for such loans are not readily available. Because of these factors, it
has been suggested to make insurance against earthquake risks mandatory as an
effective means to spread the rehabilitation costs over many years and thus reduce
the impact on the national budget in years with catastrophic earthquakes. To date,
however, this has not been done.

7.1.3 Risk mitigation policies

The mitigation of seismic risk requires policies for supporting the broad range of activities listed above. These policies should be formulated to address the problem on two time scales: short term and long term.

Short term measures

In the short term, activities should be focused on the following:

(1) Emergency response, preparedness and rehabilitation planning, i.e. on items (b) through (f) on page 257. These plans must be tailored to local needs in high risk areas, ideally on the basis of risk assessment techniques using scenario type studies. Regional (prefecture) and local (city) authorities should be the basic players in the implementation of such activities.

(2) Capacity and operational capability assessment of critical facilities (hospitals, fire stations, communication centers, etc.) under a worst case of earthquake scenario and seismic upgrading where needed.

(3) Establishing a dense national strong motion data recording network that will cover adequately the whole country and all major cities. This is basic infrastructure for most risk mitigation activities.

(4) Requiring seismic upgrading of high risk buildings, i.e. vulnerable buildings, housing many people in areas of high seismic hazard (e.g. multistory RC buildings with weak ground stories), by recommending practical, easy and relatively inexpensive solutions, while providing strong economic incentives to the owners. The time horizon for such interventions could be anywhere from 2 to 5 years. The best timing for implementing such measures is immediately after a catastrophic earthquake when people are shocked by TV pictures of earthquake catastrophes.

(5) Passing the necessary legislation for private earthquake insurance, making it compulsory for all new construction and create a state fund for reinsurance purposes. The latter could grow by allocating a portion of building permit fees and of other building related taxes, at no additional cost to the tax-payer. The Californian experience with private and state earthquake insurance could prove quite useful.

Long term measures

In the long term, activities should be focused on: (1) establishing a mechanism for code revisions at appropriate intervals; (2) introducing earthquake engineering courses in academic programs of engineering schools; (3) supporting special training programs for engineers and construction technicians when new codes are introduced; (4) funding earthquake engineering and seismological research; (5) establishing quality control procedures in the design and construction of buildings, including the production of building materials (steel, concrete, etc.); (6) identifying and mapping the active faults of the country; and (7) encouraging the strengthening or removal of old buildings by providing economic incentives and technical know how.

7.1.4   Contribution of research to seismic risk mitigation

Earthquake related research may be classified into three broad categories:

(1) Seismological research, which deals with the earthquake as a natural phenomenon, studying the causes of earthquakes, their expected locations (fault mapping), generation mechanisms, transmission through the earth, and the ground motions they are expected to generate at any given site.

(2) Earthquake engineering research aimed at improved structural performance under earthquake excitation. Its ultimate goal is to generate the know-how for earthquake resistant structures whose performance in the event of earthquakes will ensure that human lives are protected, damage is limited and structures important for civil protection remain operational. These objectives, which are also the objectives of modern codes (e.g. *EC8*, 1994), must be met with an acceptable cost of construction.

(3) Earthquake prediction research aimed at predicting location, magnitude and time of occurrence of earthquakes with error margins that make the predictions useful for practical applications.

It is not difficult now to see where and how each category of earthquake related research can contribute to seismic risk mitigation.

Seismological research will lead to better seismic hazard mapping and improved predictions of future ground motions. This translates into better predictions of earthquake loadings for structures, i.e. better input for structural engineers. In addition, the level and extent of pre-disaster planning and preparedness will depend on the seismic hazard mapping and its temporal variations. Recent experience in Greece has shown once more that earthquakes may happen along unknown faults, near cities and in areas indicated as low seismic hazard areas. Thus, fault mapping and identification of active faults must remain high in the research priorities for seismic risk reduction.

Earthquake engineering research obviously contributes to safer structures and to limiting damage to economically acceptable levels. It contributes also to pre-disaster planning and preparedness, as it is necessary for reliable assessment of the safety of damaged buildings and successful completion of the post-earthquake in-spection operation, after which usage of the safe buildings will be allowed and the number of homeless people will be reduced. Finally, it contributes to the rehabil-itation phase through improved repair, strengthening or reconstruction methods.

Earthquake prediction research has not yet reached a level of maturity that justifies practical applications. This means that predictions of earthquakes cannot yet be made with the minimum level of reliability and accuracy necessary to take action, e.g. issuing an earthquake warning to the public as is done for other natural phenomena (e.g. hurricanes, tornadoes, floods, etc.).

Results from earthquake prediction research must be handled with the high-est possible level of responsibility and professionalism, because earthquake pre-diction, whether successful or not, is a sensitive issue due to the many and far reaching consequences it can have (see e.g., *National Academy of Sciences*, 1975; *California Earthquake Prediction Evaluation Council*, 1977; *Council of Europe*, 1991). Some experience on this issue has accumulated in Greece in the past 15 years. This experience includes at least three incidents of widespread anxiety in Greek cities due to rumors of an impending catastrophic earthquake and also heated debates in the mass media after each damaging earthquake on whether the earthquake had been predicted, whether the authorities knew about it, and why they did not utilize this information.

7.1.5   Concluding remarks

Looking back on the past 20 years of intense seismic activity in Greece, as well as in other parts of the world, the following conclusions may be drawn.

(1) Effective reduction of seismic risk requires a broad range of activities aimed at safer structures and at preparedness to cope with a catastrophic event.

(2) Earthquake resistant, safe, structures can be produced only if an adequate and up-to-date code exists, along with a mechanism to ensure that this code is properly applied in design and construction.

(3) A basic requirement for success in all seismic risk mitigation activities and especially in the production of earthquake resistant structures is the continu-ous awareness of the earthquake threat, first among engineers and public of-

ficials and second among the general public. Engineers in particular should never forget that their structures will be tested to the limit of their capacity without warning or mercy and that their negligence may cause the death of many people.

(4) Unknown seismic faults near populated areas can generate strong unexpected earthquakes. Therefore, research for identifying such faults must be high in the priority list of policy makers.

(5) For seismic risk mitigation purposes, the basic goal of seismological research should be towards improved prediction of the expected strong ground motion at any given location, as this is the input for designing earthquake-resistant structures or upgrading existing ones. Current trends in earthquake resistant design have created a need for more accurate estimates of ground displacements and this must be given proper attention by seismologists.

(6) Earthquake prediction research and its potential results must be dealt with the utmost level of responsibility and professionalism, otherwise some very negative developments may ensue.

(7) Most victims in earthquakes are not killed or injured by the earthquake directly but rather by unsafe structures. In other words: "Earthquakes do not usually kill people. Unsafe structures do".

7.2   Istanbul case study

[By M. Erdik, E. Durukal, B. Siyahi, Y. Fahjan, K. Sesetyan, M.D̃emircioglu, and H.Ãkman]

7.2.1   Background and general considerations

Turkey ranks high among the countries which have suffered losses of life and property due to earthquakes. In the $20th$ century earthquakes in Turkey caused a total of over 110,000 casualties, about 250,000 requiring hospitalization and almost 600,000 destroyed housing units. The City of Istanbul, situated astride the Bosphorus in both Europe and Asia, is the largest city in Turkey, with a history that extends over twenty-six centuries. Between 1950 and 1990 the population increased from about one million to eight million. Today, Istanbul houses about one-eighth of the total population and one-half of the industrial potential of

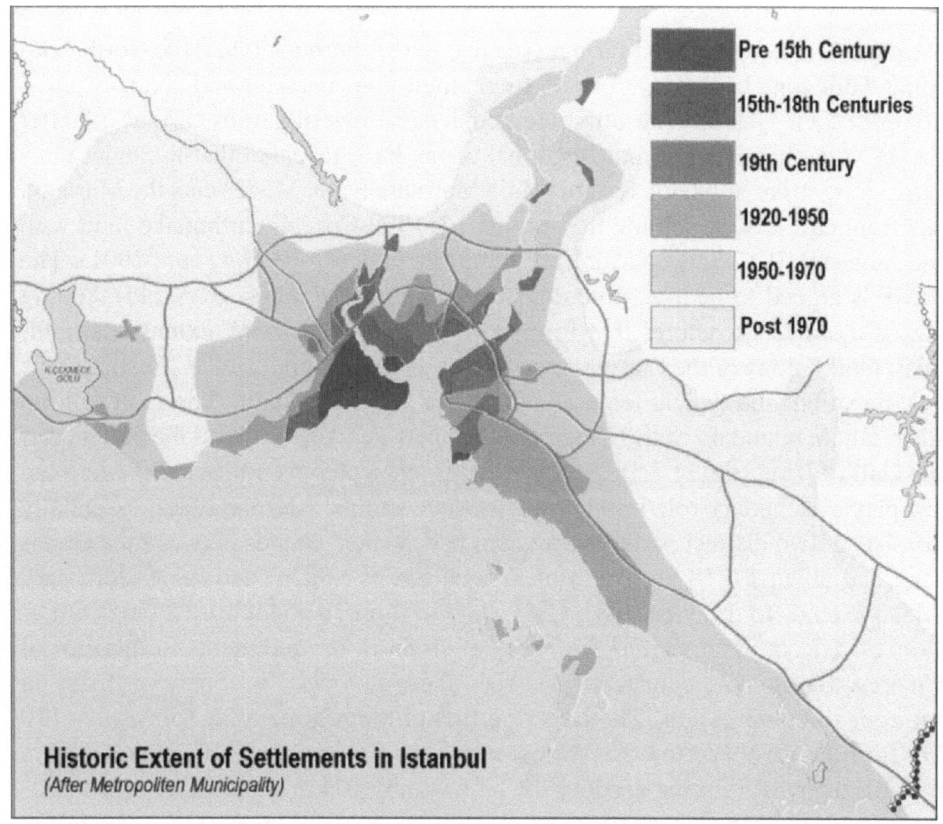

Historic Extent of Settlements in Istanbul
(After Metropoliten Municipality)

Legend:
Pre 15th Century
15th-18th Centuries
19th Century
1920-1950
1950-1970
Post 1970

Figure 7.4: Historic growth of Istanbul.

Turkey. Figure 7.4 illustrates the historic growth of the city. Istanbul has experienced numerous earthquakes. In the recent decades earthquake risk has increased due to overcrowding, improper land-use, inadequate infrasctructure and environmental degradation.

In urban centers the impact of a large earthquake is best portrayed by damage scenarios. The first ingredient of the latter is seismic hazard, the second is the vulnerability of structures and of the socio-economic system. Earthquake scenarios are the basis of risk mitigation and emergency response planning. In this section we present the elements of a damage scenario for a major earthquake in the Istanbul area.

7.2.2   Active tectonics and seismicity

West of 31.5°E toward the Marmara sea region (Mudurnu/Akyazı) the North Anatolian fault zone begins to loose its single fault line character and changes into a
complex fault system. The off-shore geophysical investigations carried out after
the 1999 earthquakes by multi-national teams have revealed that a single, thoroughgoing strike-slip fault system (Main Marmara Fault, MMF) cuts the Marmara
sea from east to west joining the August 17, 1999 Kocaeli earthquake fault with
the August 9, 1912 Sarkoy-Murefte earthquake fault (*Le Pichon et al.*, 2001). The
MMF is argued to be a very young structure (about 200,000 years old), cutting
across the older structures that formed the present NNE-SSW extensional pull-
apart morphology of the Marmara sea. Between 28.8°E and 27.4°E (Yesilkoy) the
MMF exhibits the typical features of a major strike-slip fault. The fault follows
the northern boundary of the Çınarcık Basin between Yeşilköy and the entrance of
the Gulf of Izmit (figure 7.5). All other faults and tectonic entities in the Marmara
sea play a secondary role in releasing tectonic strains. Marmara fault essentially
consists of two distinct parts: the western part, which extends between the Ganos
fault and about 17–18 km south of Kumburgaz at N80°E, and the eastern part,
which extends to the West from Gulf of Izmit through about 10–12 km south of
Princes' Islands with visible evidences of strain in the sediments to the East of
Çınarcık Basin. The connection between these two parts is somewhat fuzzy in
the zone of compression. The larger portion of the Marmara fault measures 110
km from B. Çekmece towards Saros, while the shorter branch measures 65 km
towards the gulf. In some sections the fault is only 10 km from residential areas.

   Ambraseys and Finkel (1991) have shown that in the Marmara region during
the period of 1AD to 1899 there have been thirty-eight earthquakes with magnitude 7 or greater and other twenty-three events with $m_s \geq 5.9$ since 1900 (five
of which, in 1912, 1953, 1957, 1967 and 1970, had magnitude $m_s \geq 7$). This
stands for one large earthquake on average every 45 years, although occurrences
are strongly clustered. Two intervals of clustered activity were the 2nd century
and the period between 355 and 557 AD in which four and nine large earthquakes
are reported. Three large earthquakes occurred between 555 and 557 AD, while
in the following 800 years until 1357 AD there were only four such events, in
740, 989, 1063, and 1344. From 1344 until 1509 AD there was a period of clustered activity during which six large earthquakes occurred, culminating with the
great earthquake of 1509 that destroyed much of Istanbul. Earthquake records
spanning two millennia indicate that, on average, at least one medium intensity
(I_0=VII–VIII) earthquake has affected the city every 50 years (*Ambraseys and
Finkel*, 1991). The epicentral distribution of the major historical earthquakes is
indicated in figure 7.6.

   The average return period for high intensity (I_0=VIII–IX) events has been 300

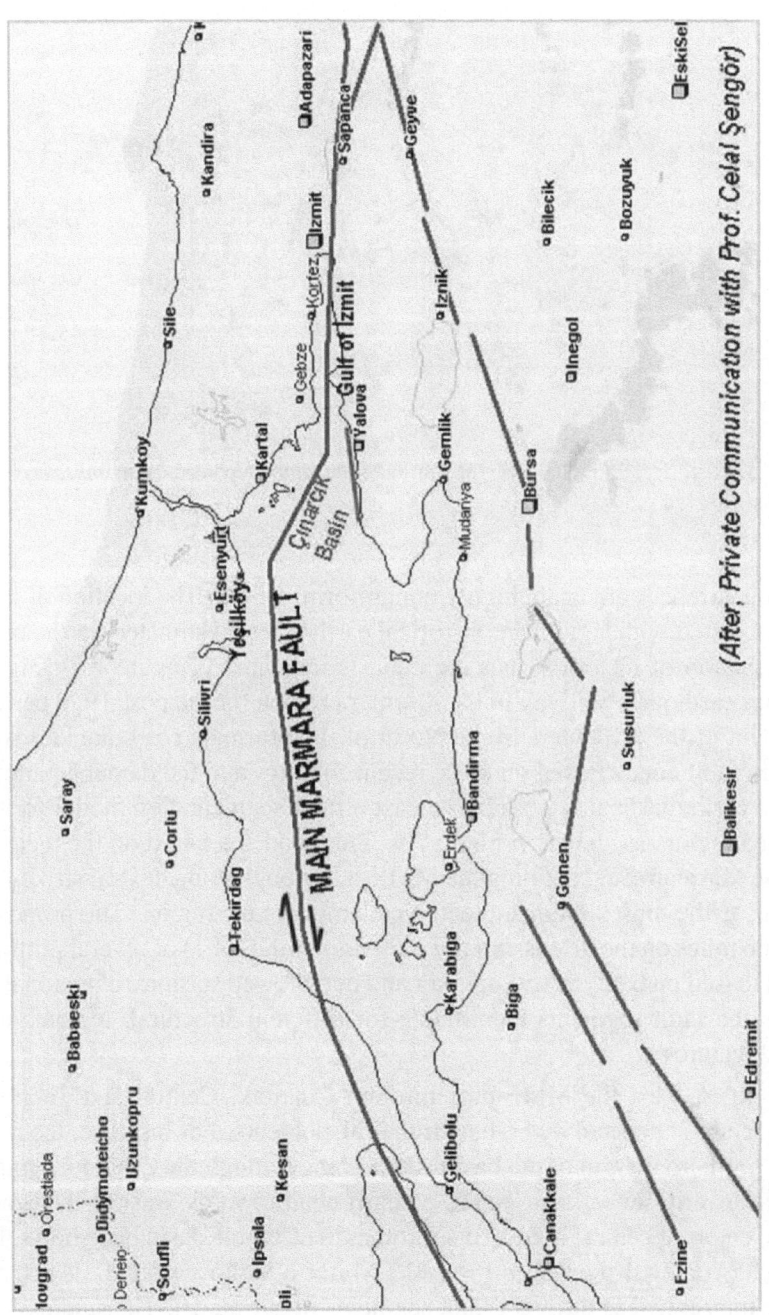

Figure 7.5: Major active tectonic elements in the Marmara sea.

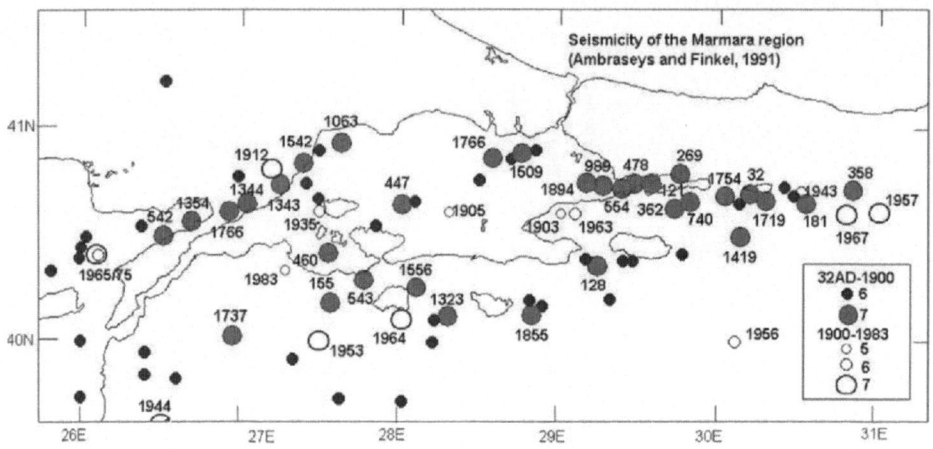

Figure 7.6: Historical Earthquakes in the Marmara sea region, updated from *Ambraseys and Finkel* (1991).

years, and occurrences are again highly nonuniform in time. The location of fault ruptures associated with significant historical earthquakes, estimated on the basis of the distribution of historical damage data, is indicated in figure 7.7. Figure 7.8 illustrates earthquake activity in the Marmara region for the post-1990 period. The alignment of the epicenters in the North of the Marmara sea coincides with the Marmara fault zone. Based on these recent findings and the damage patterns of historical earthquakes it is possible to cast a fault segmentation model for the Marmara Sea region, as shown in figure 7.9. This model is based on the tectonic model of the Marmara Sea, defining the MMF, a thoroughgoing dextral strike-slip fault system, as the most significant tectonic element in the region. The proposed segmentation relies on the discussion of *Le Pichon et al.* (2001) of several portions of the MMF based on bathymetric, sparker and deep-towed seismic reflection data and reflects the fault segments identifiable for different structural, tectonic and geometrical features.

From East to West the MMF cuts through Çınarcık, Central and Tekirdag basins, which are connected by higher structural elements. For instance, the fault follows the northern margin of the basin when going through the Çınarcık Trough in the northwesterly sense, then makes a sharp bend towards West to the South of Yesilkoy, entering central highs, cuts through the Central Basin and shows this alternate behavior until it reaches the 1912 Murefte-Sarkoy rupture. All these features are interpreted as different fault segments in our model. The studies conducted after the 1999 Kocaeli (m_w=7.4) and Duzce (m_w=7.2) earthquakes indicate (assuming that the stress regime in the Marmara Sea remains unchanged) a 65 per

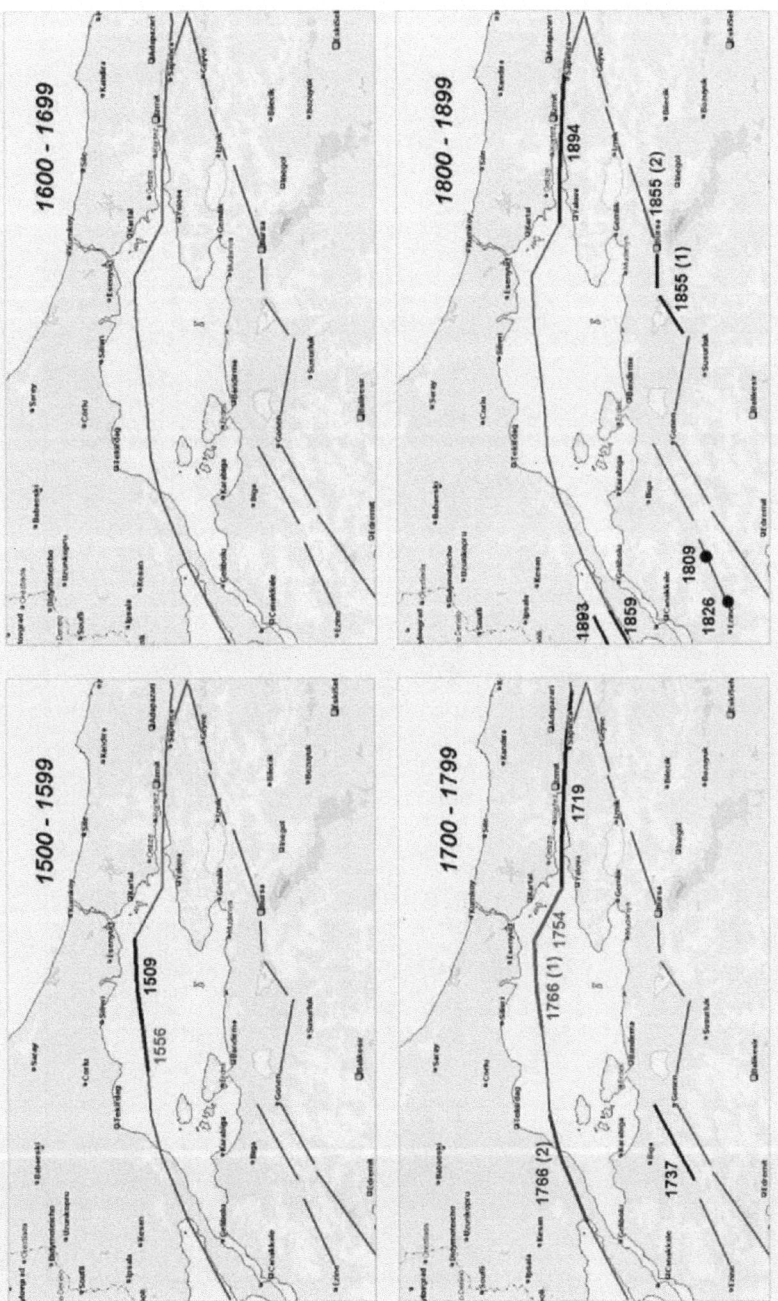

Figure 7.7: Significant earthquakes and associated fault ruptures effecting Istanbul and its vicinity in the period 1500–1900.

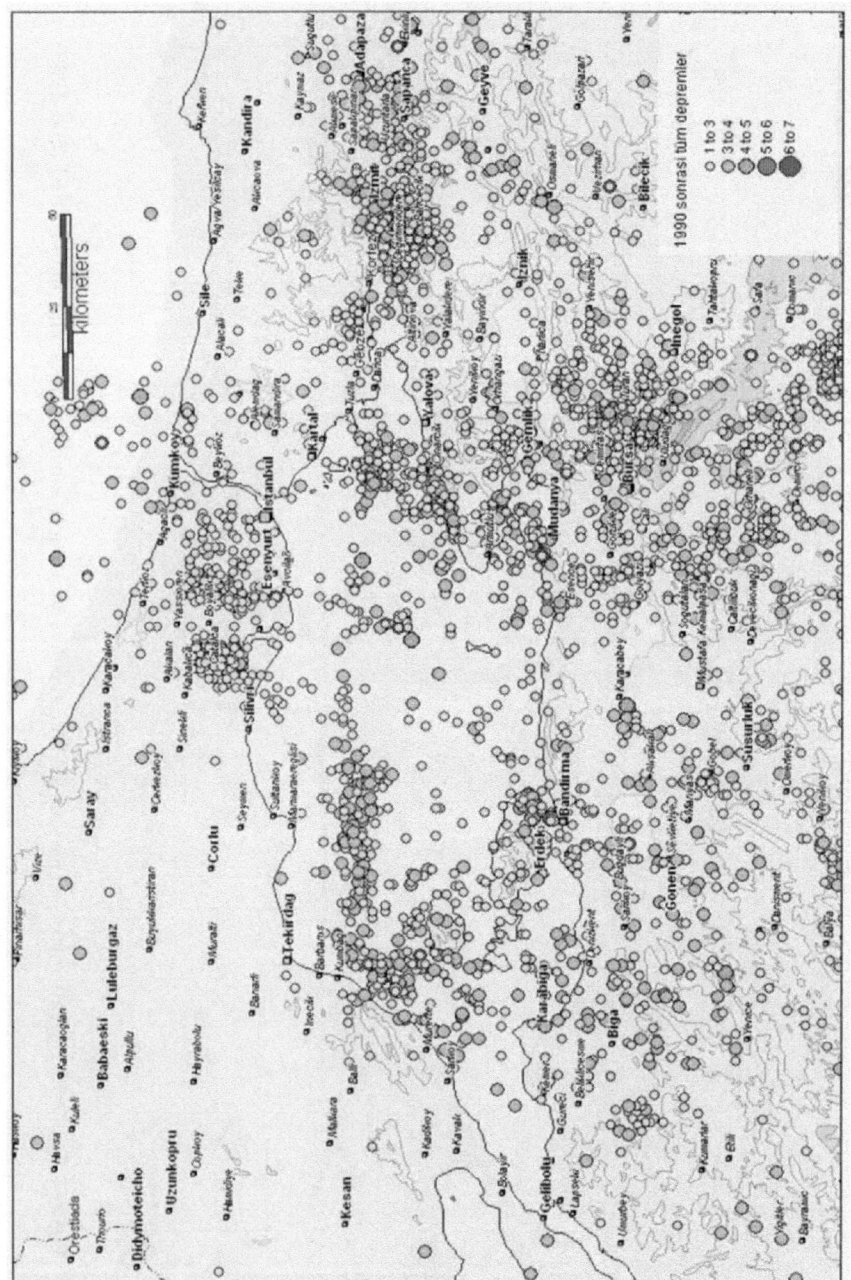

Figure 7.8: Seismic activity of the Marmara region during 1990–2000 (Kandilli Observatory).

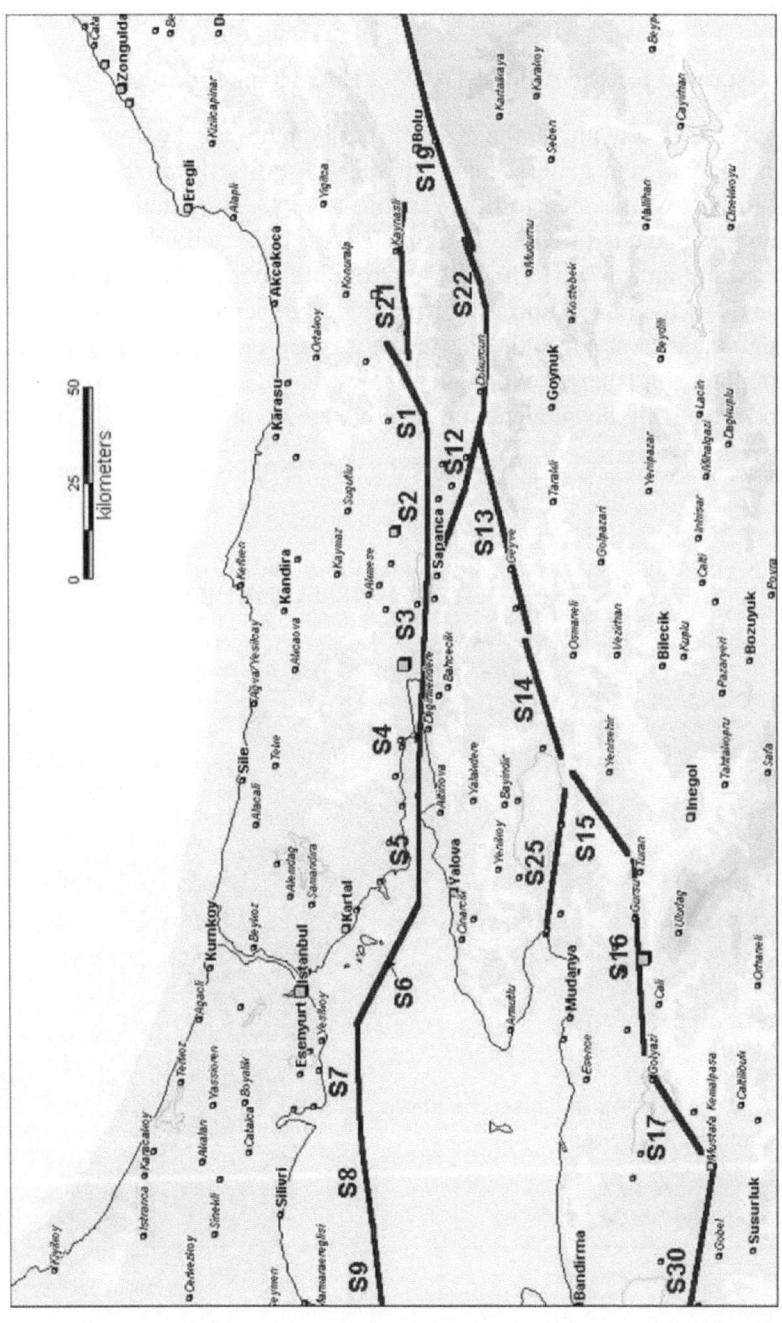

Figure 7.9: Fault Segmentation Model for Northeastern Turkey.

cent probability for the occurrence of a $m_w \geq 7.0$ magnitude earthquake effecting Istanbul (*Parsons et al.*, 2000).

Damage in Istanbul for the August 17, 1999 Kocaeli earthquake

A comprehensive treatment of the Kocaeli earthquake can be found in *Erdik* (2001)[1]. In Istanbul, the general intensity was VI with a limited region of intensity VII in the Avcilar area to the West of the city. The relative distribution of the building damage is provided in figure 7.10. The heavily damaged Avcilar area is indicated in figure 7.11. The 1997 metropolitan population of Istanbul was 8,500,000 and the estimated number of reinforced concrete (from now on RC) buildings amounted to about 800,000. In the Kocaeli earthquake about 1–2 per cent of all buildings in Istanbul experienced some damage causing 454 deaths and 3,600 injuries requiring hospitalization. There were about 450 heavily damaged

---

[1] Available at http://www.iiasa.ac.at/Research/RMS/july2000/Papers/erdik.pdf

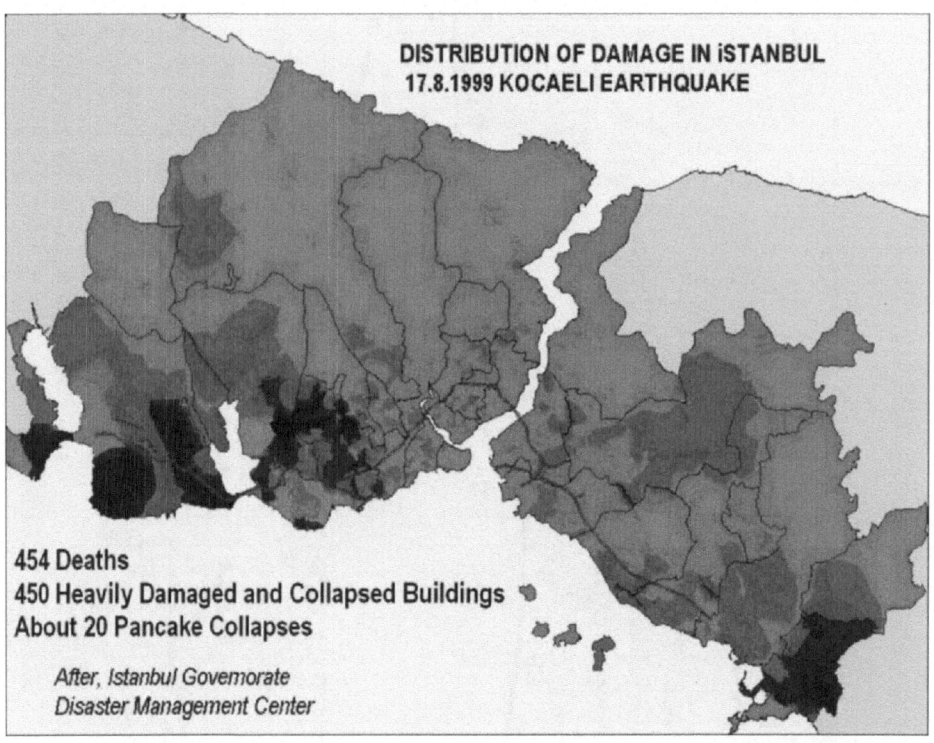

Figure 7.10: General distribution of building damage in Istanbul in August 17, 1999 Kocaeli earthquake (after, Governorate of Istanbul).

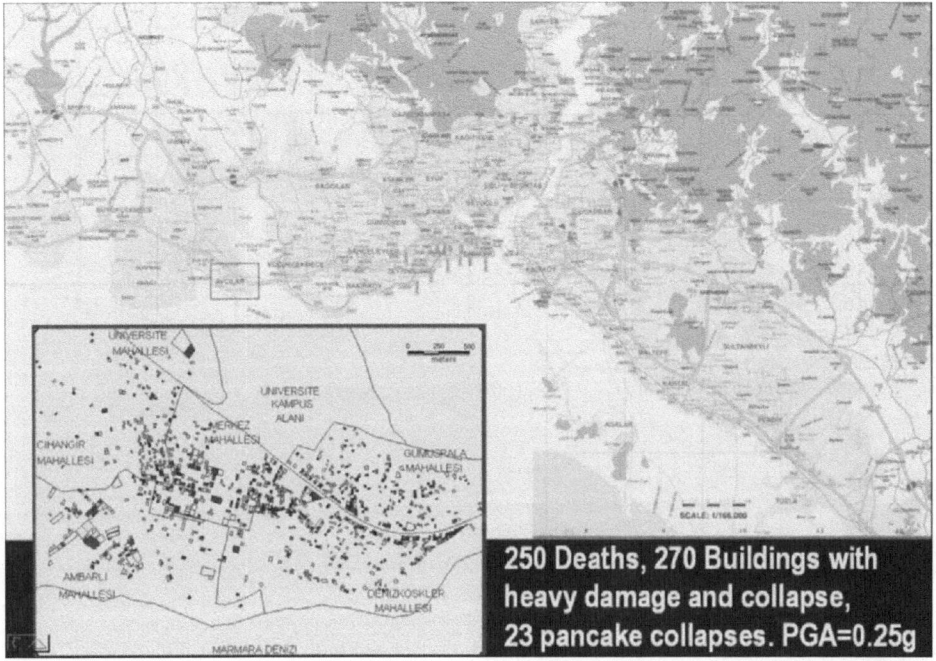

Figure 7.11: Damage in the Avcilar area due to the August 17, 1999 Kocaeli earthquake (After, Earthquake and Soils Directorate of Istanbul Metropolitan Municipality).

or collapsed buildings, approximately half of them located in the Avcilar district. About 3 per cent of the schools (3200) were closed for damage and about 65 hospitals experienced some damage. It should be noted that the buildings in Istanbul were exposed to only a fraction of the ground motion that were specified in the earthquake resistant design code.

7.2.3   Earthquake hazard assessments

Historical seismicity, seismo-tectonic structures, the current tectonic stress regime and the GPS measurements suggest enhanced probabilities of occurrence for a major earthquake in the Marmara sea. Considering the major historical earthquakes, characteristic earthquake properties associated with the North Anatolian Fault and the maximum potential earthquake that the faults can generate, it can be stated with justified prudence that the 'scenario earthquake' will have a magnitude $m_w = 7.5$ and will be located at the closest point of the fault system from Istanbul (figure 7.12). This conclusion agrees with the study of *Erdik* (1995). It should be noted that the 'benefit of the doubt' in this specification of the 'scenario

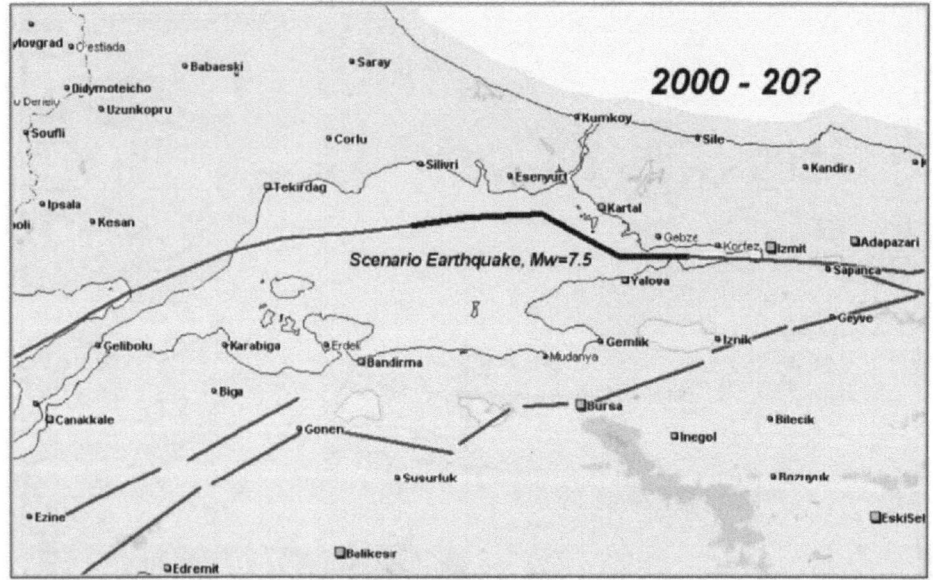

Figure 7.12: Fault rupture and segmentation associated with the scenario earthquake
($m_W = 7.5$).

earthquake' is used for Istanbul. The hazard assessments is based on deterministic
techniques using the regional intensity-based attenuation relationships developed
by *Erdik et al.* (1983, 1985). The latter yield synthetic isoseismal shapes associated with magnitude 6.5, 7.0 and 7.5 earthquakes and their comparison with the
isoseismal map of the magnitude 7.4 Kocaeli earthquake is illustrated in figure
7.13. The seismic induced ground motion is empirically correlated with the soil
conditions. A preliminary *NEHRP* (1997) soil classification map for Istanbul was
prepared using the 1:50000 scale geological map by Istanbul Metropolitan Municipality and site-specific soil data obtained from several sources. Figure 7.14
indicates the synthetic site-dependent isoseismal map to be expected in case of
this scenario earthquake.

7.2.4   Vulnerability analysis

The 1998 European Macroseismic Scale (*EMS*, 1998), an updated version of the
MSK scale, differentiates the structural vulnerabilities into six classes (A to F).
RC buildings with low levels of earthquake resistant design are assigned an average vulnerability class of C. Due to deficiencies in design, concrete quality and
construction practices, the bulk of the RC building population in Istanbul may be

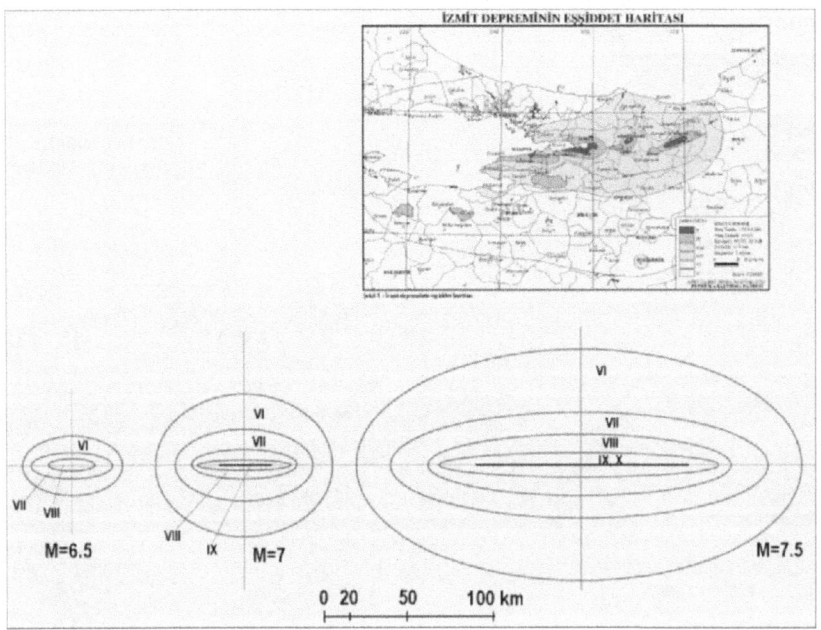

Figure 7.13: Synthetic isoseismal shapes associated with magnitude 6.5, 7.0 and 7.5 earthquakes and their comparison with the isoseismal map of the magnitude 7.4 Kocaeli earthquake.

considered in this vulnerability class. Damage to RC buildings is classified as: D1, negligible to slight damage; D2, moderate damage; D3, substantial to heavy damage; D4, very heavy damage; D5, destruction. For the vulnerability class C *EMS* (1998) provides the following definitions of intensity, where 'few' describes less than 20 per cent and 'many' describes between 20 per cent and 60 per cent:

VI: A few buildings of class C suffer damage of grade 1.

VII: A few buildings of class C suffer damage of grade 2.

VIII: Many buildings of class C suffer damage of grade 2; a few of grade 3.

IX: Many buildings of class C suffer damage of grade 3; a few of grade 4.

X: Many buildings of class C suffer damage of grade 4; a few of grade 5.

Building types and physical vulnerability

The following types of buildings dominate the building population in Istanbul. Only a brief summary of vulnerabilities is given here. For a detailed treatment the reader should refer to *Erdik and Aydinoglu (2002)*.

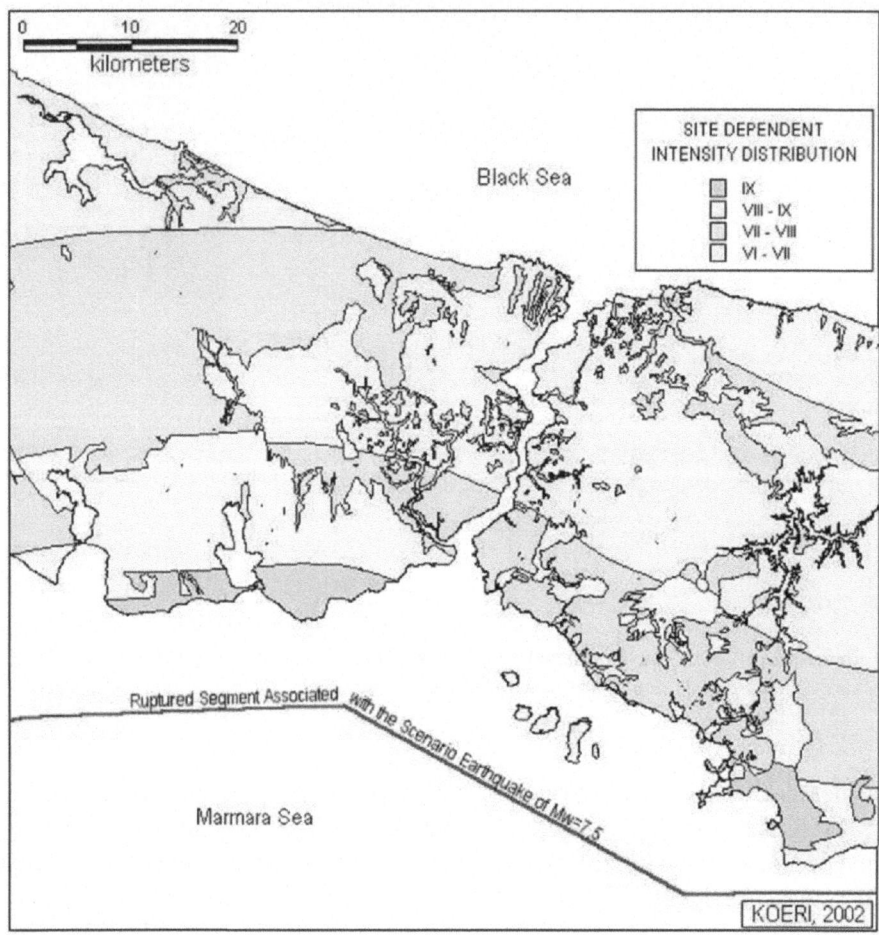

Figure 7.14: Regional distribution of intensities in Istanbul from a m_w=7.5 scenario earthquake.

Unreinforced brick masonry

Usually low rise (up to three storeys). The structure consists of load-bearing fired brick in a cement or lime mortar. Horizontal structure is commonly timber beams, or RC slabs. The use of timber or RC lintels and ring-beam is more common in the better-built houses. Roofs with timber trusses covered by tiles or flat RC slabs are common. Many of the older buildings are of this kind, often ornate and sometimes with stone masonry quoins or stone masonry facades. Some monumental buildings (mosques, old public buildings, etc.) are in load-bearing brick, but their massive construction would be expected to have a different vulnerability function to the residential structures described here.

Reinforced concrete frame with unreinforced masonry infill

This is the most common building type in Istanbul. The most common type of RC structure is the *cast-in-situ RC frame with masonry infill walls*. The height of most of these buildings is 1 to 8 storeys. Ground floors are often left open for shops. Buildings with irregular plan shape are common due to irregular land lots and urban congestion. For infill walls 0.20–0.30 m thick horizontally perforated burned clay bricks or concrete blocks are used with no reinforcement. Buildings in this class showed a substantial increase of damage with the number of stories in 1998 Adana-Ceyhan and 1999 Kocaeli earthquakes.

Dual RC frame and RC shear wall system

Such structures are relatively uncommon but many of the high-rise concrete buildings (10 to 20 storeys) in Istanbul use this system for increased structural strength. On the basis of the empirical earthquake performance of the RC buildings, three vulnerability groups can be identified: Low-rise (1–3 storeys), Mid-rise (4–7 storeys) and High-rise (8 storeys and above).

The vulnerability of the Turkish building population is much higher than in most developed countries. There are several reasons for this. First, poor construction resulted from a high rate of inflation, with consequently very limited mortgage and insurance, impediment to large scale development and industrialization of the construction sector market, high rate of urbanization (which created the demand for inexpensive housing), ineffective control of design and construction, regulations with limited enforcement and government acting as a free insurer of earthquake risk. Modern buildings can suffer major functional and economic loss by damage to the equipment and furniture they house, even though the structures experience little damage. Especially in research laboratories, hospitals and offices, unbolted equipment is highly vulnerable to earthquake damage. The same also applies to exhibited pieces in museums and in art galleries. These losses would constitute a substantial portion of the physical losses even in a small earthquake in Istanbul.

Life loss vulnerability

The ratio of the number of people killed to the number of buildings collapsed is called the 'Lethality Ratio'. This ratio depends on the average number of occupants per building, time of earthquake, number of occupants trapped, and capability of rescue first-aid services. *Ambraseys and Jackson* (1981), using data from Turkey and Greece, give the following statistics regarding the number of casualties per 100 houses destroyed by earthquakes of $m \geq 5$: Rubble masonry houses = 17; Adobe houses = 11; Masonry and reinforced adobe houses = 2; Timber and Brick houses = 1; RC frame houses $\leq$ 1. Data from recent urban earthquakes in Turkey further indicate that there will be one death and four hospitalized injuries per heavily damaged or collapsed RC building.

7.2.5   Earthquake risk to building population

Risk, in the context of disaster management, can be defined as the loss that can result from the occurrence of a hazard. In urban areas the demographic structure, building population, lifeline, infrastructure, major and critical facilities and socio-economic activities constitute the 'Elements at Risk'. Joint consideration of the earthquake hazard, vulnerability and the building inventory would yield the expected damage to the building population in case of the occurrence of the 'scenario earthquake'. The integration of these results indicated that about 40,000 buildings (5 per cent of the total building population) will be damaged beyond repair (i.e. complete damage). A spatial distribution of buildings that will experience damage beyond repair is given in figure 7.15 for a 400 m × 600 m grid.

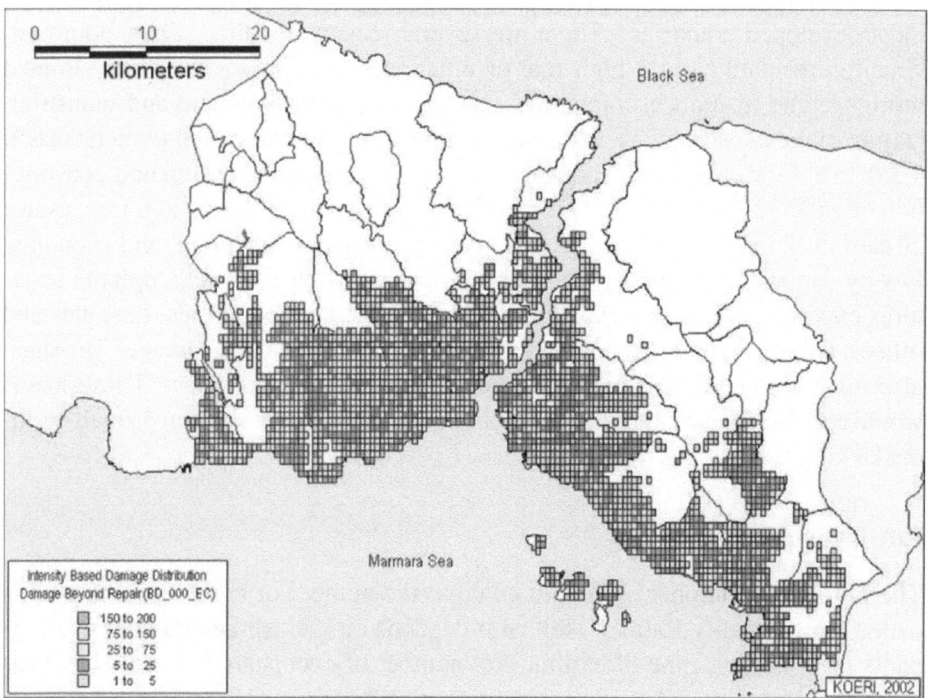

Figure 7.15: Distribution of buildings damaged beyond repair. From *Earthquake Risk Assessment for the Istanbul Metropolitan Area*, Report prepared by Bogazici University Kandilli Observatory and Earthquake Research Institute, Department of Earthquake Engineering.

Most casualties are expected in this damage group, especially in a subset of this group where the collapse will be of the worst 'pancake' form. In pancaked buildings the floors pile up on top of each other making search and rescue very difficult. Estimated collapses for pancaked buildings are 5,000 to 6,000. Furthermore, about 70,000 buildings will receive extensive damage and about 200,000 buildings will be moderately damaged. The expected number of casualties varies between 30,000 and 40,000. A total of about 500,000 households were assessed to be in need of shelters following the 'scenario earthquake'. The total monetary losses due to building damages caused by the 'scenario earthquake' are estimated in the range of about 11 billion US$. These values, especially the casualty figures, entail large uncertainties.

7.2.6   Risk mitigation

General mitigation options

The basic tenets of mitigation are: (1) not to increase the existing risk (i.e. build properly); (2) to decrease the existing risk by retrofit and (3) to transfer the risk costs by insurance. The reduction of the structural vulnerability, the siting and land-use regulations, the design and construction regulations, the relocation of communities, and public education and awareness programs are all viable measures for the mitigation of earthquake risk. The functional characteristics of the settlement can be improved through land-use planning. In Istanbul, the most important issue in mitigating earthquake casualties is the retrofit of existing buildings. The new buildings in Istanbul are in general being built much better than the existing building population. The reasons for this improvement are: (1) the application of the new (1998) earthquake resistant design code; (2) the increased public awareness and demand for earthquake safety; (3) the various training and education programs for engineers; (4) the better zoning regulations and enforcement by municipalities; (5) control by private construction supervision firms. New legislation (Revision Of the Law On Engineering And Architecture, Building Design and Construction Control and Standard Development Regulations for Municipalities) enacted after the 1999 earthquakes was of great help in this respect.

Retrofit of the existing building population

While it is clear that the greatest impact to the reduction of human casualties can be achieved by a retrofit of the existing building population, and in spite of the in fact that several assessment and retrofit applications are under way for both public and commercial buildings, serious initiatives have yet to be taken for residential buildings. A comprehensive retrofit campaign will be a formidable task that would involve assessing the vulnerability of about 800,000 buildings, the development of retrofit alternatives, and the creation of public and private incentives.

The full retrofit (i.e. in compliance with latest code requirements) of a residential building costs about 40 per cent of a new construction and the building has to be vacated for several months. In addition to this high cost and to the inconvenience of moving out, for residential buildings there are strong impediments for retrofitting. The retrofit decision is difficult to take since these buildings are multi-family owned (rented) apartments with tenants having different expectations, budgets and constraints. Due to the high residential mobility (desire for better housing in better neighborhoods) there does not seem to be any sense in spending money for something that is expected to be changed. Furthermore, there is evidence that retrofitting will not increase the sales value or rental fee of the property since it is viewed as an investment with no financial return. Under these conditions, no conceivable reduction in insurance premium or property tax would effectively create an incentive for retrofitting.

Even neglecting the social and legal constraints of retrofit actions to be taken in apartment houses and the highly distressed real estate market in Istanbul, the structural retrofit is not cost effective. For a mid-rise RC Moment Resisting Frame (from now on MRF) building in Istanbul the average loss (Mean Damage Ratio, from now on MDR) in intensity IX region will be 62 per cent. For a mid-rise RC frame building in Istanbul the MDR in intensity VIII region will be 40 per cent. If these buildings are retrofitted to meet the current earthquake resistant design code to its full extent, the MDR will be respectively 16 per cent and 11 per cent in intensity IX and VIII regions. Thus, retrofit actions will save 46 per cent and 29 per cent of the cost of construction of the building, but would require an investment of 40 per cent of the same cost.

For estimated earthquake recurrences, even with average return periods as low as 50 year (as in Istanbul), it is almost impossible to be cost effective in full-scale (meeting code criteria) retrofit applications in intensity IX regions. Only very efficient time-dependent hazard estimates or forecasts would bring this towards the break even point. In intensity VIII (or less) regions retrofit can never be cost effective. Thus, if loss of life and the other social costs associated with an earthquake are disregarded, the expected losses are small in comparison with the immediate cost of retrofit, which translates to economic infeasibility.

Possible incentives for retrofit are being discussed all over the world. These include: (1) earthquake retrofit grants by government or compulsory insurance agencies; (2) below-market interest rates on loans for earthquake retrofit; (3) insurance premium discounts to policy holders upon completion of the retrofit; (4) financial incentives to property owners of properties, such as waiving of building permit fees, city property tax and easing of siting and geometric regulations. It should be noted that the direct use of the Turkish Catastrophe Insurance Pool

(TCIP) in earthquake risk mitigation, such as funding of retrofit applications, does not seem to be realistic. The premiums are far short of the actual risks (about 0.2 per cent per annum, 2 per cent deductible for a flat in Istanbul). These rates should be compared with 0.5 per cent premium rate and 10–15 per cent deductible in San Francisco. In Istanbul, with such low rates, no risk-based adjustment of this premium can serve as an incentive to structurally upgrade private property. The market require that for retrofit actions and campaigns to be successful the insurance premiums should be realistic, such that property owners can see that it would be profitable to retrofit instead of waiting for damage and reimbursement. Although building owners will find the future property losses small compared to the cost of full retrofit and will hardly appreciate any convenience at the individual level, society will greatly benefit from a retrofit campaign through the reduction of physical and social losses, that will eventually be covered by the state. The loss imposed on the public finance by 1999 Kocaeli earthquake was about 6.2 billion US$ (*Erdik*, 2001). About 3.5 billion US$ of this amount was utilized for post-earthquake housing construction. The special earthquake taxes introduced by the government have roused about 3 billion US$ in one year after the earthquakes. International finance agencies contributed another 2.5 billion US$. Although steps are taken, such as TCIP, public funds will continue to be used for rehabilitation after earthquake disasters in Turkey. As such, the use of public funds can be justified for retrofit purposes under a strategy that is designed to maximize benefits with well defined priorities and distributed minimum expenditures. Such a strategy can lead us to the concept of minimum retrofit.

Priorities in the retrofit of residential buildings

Under the circumstances discussed above it seems rational is to give the highest priority to the most vulnerable buildings Istanbul. The objective of the retrofit will be to avoid total collapses which are responsible for most of the deaths (fatality ratio about 10 per cent). The earthquake performance criteria will be the prevention of total collapse at minimum cost. These are the buildings that are expected to collapse in a 'Pancake' manner. Somewhat crude screening criteria for the identification of these most vulnerable residential buildings in Istanbul can be set as follows: buildings higher than 4 storeys, built before 1970 (pre-1975 code, poor concrete quality and corrosion issues), located in zones with PGA (Peak Ground Acceleration) $\geq$0.25 g or SA (Spectral Amplitude) (0.2) $\geq$ 0.60 g or EMS-$I \geq$ VIII. Buildings with added floors (with no engineering services) and buildings that have received structural damage in 1999 Kocaeli Earthquake but not retrofitted properly can also be considered. Initial assessments indicate that about 5,000 buildings fall in this first priority group as illustrated in figure 7.16.

The cost of minimum retrofit is estimated to be on the average 40,000 US$ per building with a total cost of about 200 million US$. Upon proper implementation of this retrofit scheme it is expected that about 20,000 lives can be saved. On the technical side, intelligent retrofit schemes that are suitable for general campaign-type applications needs to be developed.

A possible mode of operation for this minimum retrofit campaign can be the following. Following an identification process, the Municipality declares the buildings as 'Hazardous' and legally enforces that, unless 'Minimum Retrofitting' is effected within a prescribed time, the occupancy permit (if any) will be canceled, the tenants evicted and building sold on behalf of the owners. For financing, each household will have access to a credit of about 5,000 US$. Such a credit scheme is not new in Turkey. After the Kocaeli Earthquake the Turkish government granted about 5,000 US$ low interest credit with long payment terms to eligible owners of medium damaged housing units. The second priority can be the retrofit of about 40,000 buildings that were assessed to be damaged beyond repair. These are mid-rise RC frame buildings located in zones with PGA ≥ 0.2 g or

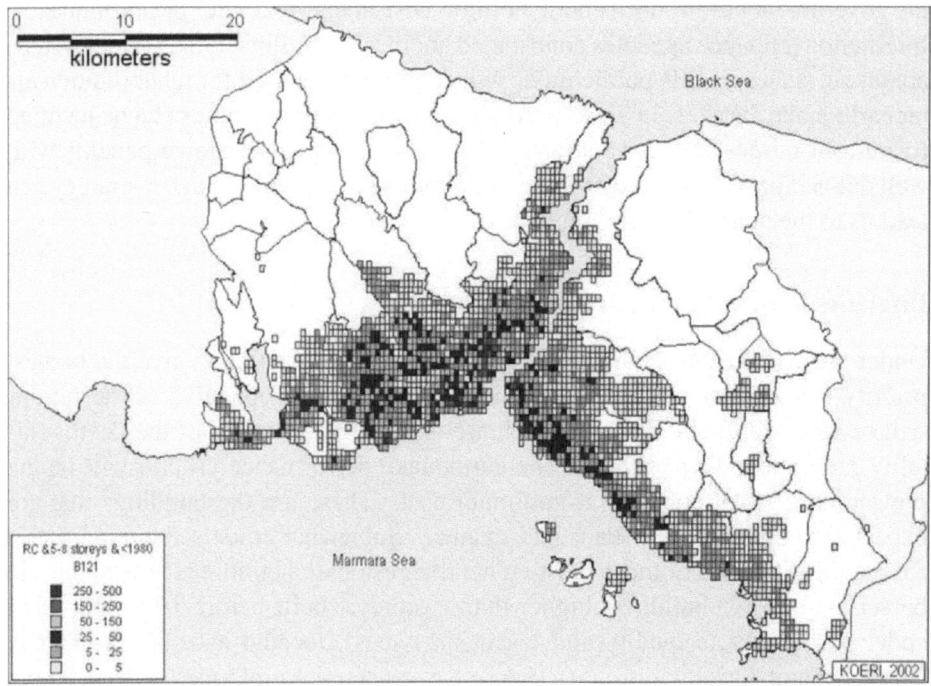

Figure 7.16: RC buildings with 5-8 storeys built before 1980. From *Earthquake Risk Assessment for the Istanbul Metropolitan Area*, Report prepared by Bogazici University Kandilli Observatory and Earthquake Research Institute, Deptartment of Earthquake Engineering.

SA(0.2) $\geq$ 0.50 g or in zones with EMS-I $\geq$ VIII. The retrofit performance criteria will be life safety. The approximate cost of this retrofit will be 1.6 billion US$.

7.3   References

Ambraseys, N. N., and Finkel, C. F., 1991. Long-term seismicity of Istanbul and of the Marmara Sea region, *Terra Nova*, **3**, 527–539.

Ambraseys, N. N., and Jackson, J. A., 1981. Earthquake hazard and vulnerability in the Northeastern Mediterranean: the Corinth earthquake sequence of February-March, *Disasters*, **5**, 355–368.

Anagnostopoulos, S. A., 1996. Large scale operations for post-earthquake emergency assessment of building safety, Paper No. 967, *Proceedings, 11th World Conference on Earthquake Engineering*, Acapulco, Mexico.

California Earthquake Prediction Evaluation Council, 1977. Earthquake prediction evaluation guidelines, *California Geology*, 158–160 (reprinted 1983, *Bull. Seism. Soc. Am.*, **73**, 1955–1956.

Council of Europe, 1991. Conclusions of the International Conference on: Earthquake Prediction: State-of-the-Art, Strasbourg, France, Oct. 1991.

EC8, 1994. Design provisions for earthquake resistance of structures, *European Pre-standard*, CEN, Doc. CEN/TC250/SC8/N.

Erdik, M., 1995. Istanbul earthquake scenario and master plan, in: *Informal Settlements, Environmental Degradation, and Disaster Vulnerability: the Turkey Case Study*, WORLD BANK, Washington D.C.

Erdik, M., 2001. Report on 1999 Kocaeli and Duzce (Turkey) earthquakes, in *Structural Control for Civil and Infrastructure Engineering*, Casciati, F., Magonette, G. (eds.), World Scientific.

Erdik, M., and Aydinoglu, N., 2002. Earthquake risk to buildings in Istanbul and a proposal towards its mitigation, *Proceedings of the Second Annual IIASA-DPRI Meeting: Integrated Disaster Risk Management: Megacity Vulnerability and Resilience*, IIASA, Laxenburg, Austria, 29 - 31 July 2002.

Erdik, M., Doyuran, V., Akkas, N., and Gulkan, P., 1985. A probabilistic assessment of the seismic hazard in Turkey, *Tectonophysics*, **117**, 295–344.

Erdik, M., Doyuran, V., Gulkan, P., and Akcora, G., 1983. Probabilistic assessment of the seismic intensity in Turkey for the siting of nuclear power plants, *Proc., 2nd CSNI Specialist Meeting on Probabilistic Methods in Seismic Risk Assessment for Nuclear Power Plants*, Lawrence Livermore National Laboratory, Livermore, CA, USA.

EMS, 1998. *European Macroseismic Scale 1998*, European Seismological Commission, Luxembourg.

Le Pichon, X., Seng¨or, A. M. C., Demirbag, E., Rangin, C., Imren, C., Armijo, R., G¨or¨ur, N., Ç agatay, N., Mercier de Lepinay, B., Meyer, B., Saatç ilar, R., and Tok, B., 2001. The active main Marmara fault, *Earth Planet. Sci. Lett.*, **192**, 595–616.

National Academy of Sciences, 1975. *Earthquake Prediction and Public Policy*, National Academy of Sciences publication, Washington, D.C.

NEHRP, 1997. Recommended Provisions For Seismic Regulations For New Buildings And Other Structures (FEMA 303), Prepared by the Building Seismic Safety Council for the Federal Emergency Management Agency.

Papazachos, B. C., and Papazachou, C., 1989. *The Earthquakes of Greece*, (in Greek with English summaries), Zitis Publications, Thessaloniki.

Parsons, T., Toda, S., Stein, R. S., Barka, A., and Dieterich, J. H., 2000. Heightened odds of large earthquakes near Istanbul: an interaction-based probability calculation, *Science*, **288**, 661–665.

UNDRO, 1980. *Natural Disasters and Vulnerability Analysis*, Report of Expert Group Meeting, Office of the United Nations Disaster Relief Co-ordinator, Geneva.

Chapter 8

Earthquake prediction and public policy

[by R. B. Olshansky and R. J. Geller]

8.1 Introduction

As discussed in earlier chapters, estimates of future seismicity are subject to considerable uncertainty. Nevertheless, as earthquakes are a real and present danger to society, governments, companies, and individuals must adopt specific and concrete counter-measures. This is inherently a political process, in that it requires a tradeoff between cost and risk, taking into account the various uncertainties.

In this chapter we discuss the public policy aspects of seismological research. We argue that the types of seismological information that could be useful for society are more narrowly circumscribed than many scientists recognize, and that science, for the foreseeable future, can deliver at best only a small part of what would be useful for society. We emphasize issues related to short-term earthquake prediction. As long-term forecasts are discussed above in chapters 4 and 5 we cover this topic only in passing.

As discussed above (see section 1.4 and 7.1), there are no immediate prospects for reliable and accurate short term earthquake predictions. However, as this topic fascinates both the public and journalists (see page vii), it seems inevitable that it will continue to be an issue in the future. The most sensitive way this issue can arise is the public announcement of a short term prediction by an (apparently) credentialed scientist, particularly if it is supported by one or more other (apparently) credentialed scientists. A prompt and forceful response by government authori-

284

ties and responsible scientists is essential to damp down the media frenzy and keep public panic (or near-panic) from spreading.

As evidenced by the extensive attention paid to earthquake prediction by the media (see *Gori*, 1993; *Spence et al.*, 1993; *Geller*, 1997a), this topic is as much or more a public policy issue than a purely scientific matter. Many scientists (e.g. *Allen*, 1976; *Jones*, 1996; *Snieder and van Eck*, 1997) have previously made the same point. As noted by *Lambright* (1985, p. 165), earthquake prediction is a public technology and must therefore "be governed as it develops and as it is used".

Except for a brief period of optimism in the 1970s (see review by *Geller*, 1997b), most seismologists have long been pessimistic about prospects for (short-term) earthquake prediction (e.g., *Wood and Gutenberg*, 1935; *Macelwane*, 1946; *Snieder and van Eck*, 1997; *Jordan*, 1997; *Geller et al.*, 1997a, b; *Jones*, 1996; *Kanamori et al.*, 1997). Even during the 1970s cautious or pessimistic views on prediction were held by a significant segment of the seismological community, but with the notable exception of *Brune* (1974), were generally not published (also see *Brune*, 1979). In contrast, prospects for predicting, or even controlling, earthquakes, have long captured the imagination of the public and the media. The peak of optimism occurred in the mid-1970s when several highly optimistic studies, were published. These studies was prominently reported in the media (e.g., cover story, 'Forecast: earthquake', *Time* magazine, Sept. 1, 1975). The transient of extreme optimism in the 1970s is epitomized by a paper by *Scholz et al.* (1973) that was published as the lead article in *Science*:

> "Earthquake prediction, an old and elusive goal of seismologists and astrologers alike, appears to be on the verge of practical reality as a result of recent advances in the earth and materials sciences".

and by a leader (editorial) in *Nature* (*Anonymous*, 1973):

> "Up to 1970 no one had yet predicted an earthquake and there was no prospect of success [...]. The change has been remarkable. Not only have a whole class of premonitory symptoms been identified for certain earthquakes, but there are the makings of a physical theory [...] It is now urgently necessary to think seriously about the social issues that the new knowledge raises. The situation is in some ways similar to that in 1939 when nuclear fission suddenly became a reality".

The Chief of the Office of Earthquake Research and Crustal Studies of the U.S. Geological Survey (USGS), *Hamilton* (1974), said:

> "The ability to predict the time, location and magnitude of seismic events presents earth scientists with an opportunity to help alleviate

the ravages of earthquakes. The prediction capability is developing much more rapidly than all but a few scientists believed possible. Consequently, serious consideration of how prediction can be used most effectively began only recently. It is important to anticipate public response and to evaluate the various possible actions. Earth scientists, engineers, sociologists, economists, and public officials all have a role to play".

8.1.1   Why should we care now?

The above highly optimistic statements were made about 30 years ago. As almost all scientists are now much much more pessimistic about the prospects for deterministic prediction of imminent large earthquakes (see *Geller* 1997b and the other references cited above) one might question the need to pay heed to the policy implications of this topic. However, there are several important reasons why this topic cannot and should not be neglected.

First, as illustrated below using examples from the US and Japan, many legal and administrative provisions that were adopted during the period of optimism in the 1970s still remain 'on the books' today. These laws and regulations were adopted at a time when short-term deterministic prediction was thought to be on the verge of becoming a reality. Since this is no longer the case, these laws and regulations detract from public safety, by diverting attention and resources away from practical measures to reduce seismic risk. They should therefore be repealed or discontinued, to reflect the scientific realities. In many cases it may not be widely known that these measures remain formally in effect.

Second, the topic of short-term deterministic prediction continues to be raised by some scientists who propose new prediction research programs. In many cases the possible benefits to society are cited in support of the proposed research. However, the summary of research by social scientists presented in this chapter shows that a hypothetical highly reliable and accurate short-term prediction capability would be of benefit to the public, but that less accurate or reliable short-term predictions would at best be of only marginal benefit because of the cost of false alarms and missed predictions. Scientists should be aware of these considerations, but in many cases may not be, because most of the social science research was carried out several decades ago, when there was more optimism than now. The literature review presented here will make this material accessible to current researchers.

8.1.2 Ethical considerations

There continues to be strong latent public support for work on earthquake prediction. Some scientists argue that the earthquake science community should exploit this to obtain funding. A recent opinion article (*Ebel*, 2003), titled 'Seismologists must begin forecasting earthquakes', makes the following arguments.

"[...] I think that earthquake forecasting and prediction must become a major direction of research and public education for earthquake seismologists. The public expects us to predict earthquakes, as is known by anyone who is frequently interviewed about earthquakes. [...] As Chip Groat, Director of the U.S. Geological Survey, indicated at the 2002 annual [Seismological Society of America] meeting, earthquake forecasting and prediction must be important goals if we are to improve future funding for seismological science in the U.S.

"I believe that the seismological community needs to be prominently promoting earthquake forecasting and prediction research research. In particular, we need to convince the public that much of our current research is helping us to forecast future earthquake activity. [...]

"[...] with the public thirst for earthquake forecasting and prediction, fueled in part by the increasingly accurate weather forecast capabilities that are displayed prominently in our media, society wants seismologists to make progress toward predicting earthquakes".

The above views raise questions. It seems unwise for seismologists to treat 'forecasting' and 'prediction' as related entities when they present their proposals for future research. The material in earlier chapters of this book make clear that while there are many problems and caveats, statistically based forecasts of future seismic activity have a sound scientific basis. On the other hand, the material in chapters 2 and 3 suggests that deterministic prediction, which is what the public really wants, is likely to be an inherently unrealizable goal and that there certainly are no immediate prospects. If seismologists present their work to the public as a way to 'make progress toward predicting earthquakes', this is likely to mislead the public, and might be viewed by some observers as an untruth. Nothwithstanding the strong public desire for prediction, it is extremely important for scientists not to oversell their work to the public.

8.1.3 Definitions of earthquake prediction

Many researchers (e.g. *Wood and Gutenberg*, 1935; *Macelwane*, 1946; *Allen*, 1976; *Jones*, 1996) make a common-sense distinction between prediction and generalized forecasting similar to that between predicting weather and climate. *Wallace et al.*, (1984) distinguish between predictions, where "the earthquake can occur at any time from the present through the period of the time window", and statements regarding "long-term earthquake potential", which have no specific time window. The former specify a magnitude range, a geographic area and a time interval 'shorter than a few decades' within which a specific earthquake is predicted. Long-term predictions cover a few years to a few decades, intermediate-term predictions cover a few weeks to a few years, and short-term predictions cover up to a few weeks. The last is further subdivided into alerts (3 days to a few weeks), and imminent alerts (up to 3 days). The term alert carries a sense of urgency, and the 3-day period reflects the maximum time for which disaster response agencies could maintain their highest level of readiness. The USGS defines a 'prediction' as a statement of probability and a 'warning' as a formal statement for which a public response is expected.

Allen (1976) and other authors point out that predictions must not only state the predicted hypocentral parameters in quantitative terms, but also must provide quantitative statements of the 'windows' (the acceptable ranges of the parameters). Predictions must also provide data on the chances of the prediction succeeding by random chance, and must be in writing and publicly accessible. Unless these criteria are met, subsequent objective evaluation of the predictions is almost impossible (e.g. *Kagan and Jackson*, 1996; *Evans*, 1997).

Some researchers (e.g. *Wood and Gutenberg*, 1935; *Macelwane*, 1946; *Allen*, 1976; *Geller*, 1997b) regard the term 'long term prediction' as inappropriate, because the public, government, and media tend to assume that all 'predictions' are alarms of imminent earthquakes, regardless of efforts by scientists to attach qualifying adjectives. Such researchers think that some other term, e.g. *Wood and Gutenberg's* (1935) 'generalized forecast', should be used in preference to 'long-term prediction'.

8.1.4   Proposals for earthquake prediction research

Starting in the 1960s, there have been several calls for the establishment of large-scale earthquake prediction research programs. The proponents commonly cited the potential benefits to society as one of their arguments (e.g., *Press and Brace*, 1966; *Scholz et al.*, 1973; *Davies*, 1974; *Press*, 1975: *Lindh*, 1990; *Tullis*, 1994; *Wyss*, 1997; *Aceves and Park*, 1997; *Silver*, 1998; *Gusev*, 1998). However, a detailed and convincing analysis of the costs and benefits to society has not yet been made.

Some scientists and scientific administrators briefly discussed the public policy aspects of prediction (e.g. *Kissingler*, 1974; *Carlson*, 1976; *Hamilton*, 1976; *McKelvey*, 1976; *Steinbrugge*, 1976), but primarily this question was left for social scientists. *Allen* (1976, 1982), was cautious regarding the unique 'fishbowl science' aspects of earthquake prediction, but reflected the optimism of the times that prediction would gradually become a reality, and that this would be in the public interest. This optimism continued in the literature of the 1980s, although it was tempered somewhat as the scientific difficulties of prediction became more apparent (*Kerr*, 1978, 1979). *Wesson and Wallace* (1985) stated that the success-ful prediction of the eruption of Mt. St. Helens in 1980 ushered in "the era of real-time geology", and that "the geologic event most in need of successful pre-diction – the one whose destructiveness could most be averted – is a catastrophic earthquake".

8.2 Views of social scientists

8.2.1 Report of NAS Panel in 1975

In light of the apparent breakthroughs in scientific progress towards earthquake prediction (see section 8.1, above) in the mid-1970s, the National Academy of Sciences (NAS) of the U.S. established a panel on Public Policy Implications of Earthquake Prediction in April 1974. The resulting report, *Earthquake Prediction and Public Policy*, is hereafter cited as *Panel* (1975). The panel noted several risks of earthquake prediction, but was generally optimistic. The panel's chairman (*Turner*, 1976a, p. 179) said:

> "By 1973, accumulating empirical evidence and a promising the-oretical break-through (*Scholz et al.*, 1973) had made the imminent prospect of earthquake prediction credible. Concerned scientists and public officials quickly began asking what should be done if the pre-monitory signs of a potentially destructive earthquake were detected for a heavily populated area. When and how should the prediction be released? Was there a real danger that releasing the prediction would provoke mass panic, public disorder, and economic disaster? What steps might be taken in order that the community would benefit from a period of advance warning?"

The Panel noted that several aspects of earthquake prediction make it unique among forecasts of natural hazards. For example, because of the long time in-tervals between destructive earthquakes, "every prediction of a serious quake is likely to be the first such experience for most of the inhabitants of the affected area and for most of the personnel charged with preparing for the event" (*Turner*, 1976a, p. 180). Perhaps most significantly, "Unlike floods, hurricanes, and torna-does, earthquakes are preceded by no external signs through which the public can make their own informal confirmations of the prediction or identify the moment of occurrence. Hence earthquakes will be distinctive among natural disasters in the extent to which public response will depend exclusively on the faith people place in scientific predictions" (*Turner*, 1976a, p. 180). It follows that the public (and scientists, for that matter) will be unable to recognize a near-miss, such as when a predicted hurricane suddenly changes directions. A missed prediction would hurt scientific credibility in a way that could not happen with an unsuccessful weather forecast, as weather is a visible physical phenomenon.

Panel (1975) said that issuing warnings should be "the responsibility of pub-lic officials acting in the interests of the people they represent". Public officials would face dilemmas regarding how seriously to take predictions and which ac-tions to take. Evacuation plans, in particular, would likely be politically unaccept-able, unless the danger is imminent and obvious. Thus, prediction poses political dilemmas, because, in the end, governmental bodies must decide how they will use the information provided by scientists (see *Roberts*, 1984). And, as noted by the former Prime Minister of New Zealand (*Marshall*, 1978), "The less accurate the prediction, the more difficult are the political decisions". Both issuing pre-dictious and responding to them (or ignoring them) have legal implications for government agencies (*Huffman*, 1984; *Masood*, 1995).

Panel (1975) made a number of recommendations for further research, be-cause so little was known about how the public would react to earthquake pre-dictions. Proposed research included: the social and economic reactions to pre-dictions, potential legal issues, how people process information regarding low-probability disasters, determination of factors that influence the credibility of pre-dictions and warnings, how the economy would respond to a prediction, the role of insurance, refining loss estimation methods, and comparing the costs and benefits of alternative hazard-reduction measures. A more detailed research agenda was published by the *Committee on Socioeconomic Effects of Earthquake Predictions* (1978) of the NAS. Several studies in this area are summarized below.

8.2.2   Social science research

Thiel (1976) and *Morton* (1980) considered appropriate responses by society to predictions on four different time frames, roughly corresponding to those of *Wal-lace et al.* (1984): long term, intermediate term, short-term, and imminent. Of these four types of predictions, imminent predictions have the greatest potential for saving lives and preventing injuries. As noted by *Turner* (1976b, p. 17), "If people are protected from fire and collapsing structures, and evacuated short dis-tances from low-lying coastal areas, very few lives need be lost in even a strong earthquake".

An assessment of the social impact of earthquake prediction (*Weisbecker et al.*,

1977) noted that earthquake prediction is not science as usual, but is inextricably linked with social concerns. The laboratory for earthquake prediction is the real world inhabited by real people; there is little separation between the scientist and the public. Also, what is hypothesis testing to scientists has actual consequences for the public. The public must know how to respond to probabilistic statements, and incorrect responses could lead to deaths or injuries.

In 1975, the University of Colorado's (UC) Institute of Behavioral Sciences began a major study on the social consequences of earthquake prediction (*Anonymous*, 1977a, b; *Haas and Mileti*, 1976, 1977a, b; *Mileti et al.*, 1981, 1984). The prediction scenarios considered in the UC study were based on a model like *Scholz et al.*'s (1973) dilatancy-diffusion model in which the logarithm of the time duration of the 'precursor' interval was linearly related to the magnitude (*Anonymous*, 1977b). In particular, it was assumed that an $m = 6$ earthquake would show 'anomalies' for several months before the earthquake, an $m = 7$ for 1–3 years, and an $m = 8$ for as long as a decade. Note that the dilatancy-diffusion model is now generally regarded as no longer viable (*Geller*, 1997b).

The UC team conducted interviews to gauge the reactions of news organizations, state and federal agencies, large business firms, local organizations, and families. They concluded that an accurate prediction would significantly reduce deaths, injuries, and property damage, and that it was likely that the prediction would produce neither panic nor apathy. However, if the prediction has an extended lead time, especially a year or longer, the target community would "suffer significant social disruption and decline in the local economy" (*Haas and Mileti*, 1977a, p. S286). The UC studies found that a reliable prediction issued by reputable scientists would be taken seriously by most people. However, they said that to be effective predictions should reflect agreement among scientists.

Mileti et al. (1981) identified three important elements that influence public perception of prediction credibility: (1) the reputation of the person or organization making the prediction, (2) confirmation of the information from other sources, and (3) how certain the predictor is that the earthquake will occur. However, the study of Mileti *et al.* was primarily theoretical. A survey of actual experience (section 8.5.2, below) suggests that the most important factor governing public reaction to predictions is the degree to which a prediction is specific, rather than the credibility of the predictor.

Despite the potential problems, *Mileti et al.* (1981, p. 126) concluded that "Earthquake prediction holds great promise of societal utility, [and] is well worth the cost of development, only if work is begun now to research, draft and implement the political and administrative policies necessary for effective use".

Some social scientists appear to be overly optimistic about the state of the art in prediction research. For example, *Mileti and Fitzpatrick* (1993, p. 60) comment as follows:

"Obviously, the history of earthquake prediction over the past several decades is peppered with successes and failures. But the science of quake prediction is being advanced, societal research is keeping pace, and applications to enhance societal readiness are too numerous to mention. 'As the ability of scientists to predict earthquakes improves, such forecasting capability will likely be put to even increased use' (*Governor's Board of Inquiry*, 1990, p. 91). The belief by many scientists that there will be a highly destructive great earthquake in a densely populated area of the country in the near future leaves us wishing well to those who would develop successful prediction capabilities".

In contrast to the above view that "the science of quake prediction is being advanced", many scientists (e.g. *Evans*, 1997; *Geller*, 1997b; *Geller et al.* 1997a, b; *Kagan*, 1997a) argue that large-scale applied science projects on prediction are unwarranted, in view of the absence of notable progress over the past 30 years or more. Even proponents of prediction research (e.g., *Wyss*, 1997; *Aceves and Park*, 1997; *Silver*, 1998; *Gusev*, 1998) agree that immediate prospects are scant but argue that the problem is of such great potential importance that further intensive research is warranted. Thus a broad consensus seems to exist that reliable and accurate (short-term) predictions that could be useful for society are not possible at present, and will not become feasible in the foreseeable future (for at least the next 10–20 years). Social scientists should be aware of the above scientific opinions when they discuss the public policy aspects of prediction.

8.2.3   Costs and benefits of short-term earthquake prediction

Questions of technical feasibility notwithstanding, several studies have considered the costs and benefits to society of a hypothetical short-term earthquake prediction capability. One difficulty in such studies is the need to place a monetary value on human life, but they can nonetheless shed light on the trade-offs involved in policy decisions.

Ellson et al. (1984) analyzed the effects of a hypothetical earthquake prediction for the Charleston, South Carolina, metropolitan region. Their model explicitly excluded consideration of deaths and injuries, and several of its assumptions are subject to question, but the findings are still illuminating. The total regional economic losses were estimated as follows: 1 billion US$ for an unpredicted earthquake, 900 million US$ for a successfully earthquake, and 294 million US$ for a false alarm. Significantly, the potential gains of a successful prediction are less than the losses from a false alarm by a factor of about three. Thus, a prediction method would have to be very successful, with few false alarms, to be justified on the basis of economic costs (*Matthews*, 1997, also reached a similar conclusion).

Schulze et al. (1990) developed a model of the expected costs and benefits of a short-term (2 day time window) prediction of a great earthquake on the southern San Andreas fault near Los Angeles. The model parameters were chosen to match the stated goals of the Parkfield, California, earthquake prediction experiment (*Bakun and Lindh*, 1985; *Roeloffs and Langbein*, 1993): a 40 per cent rate of false predictions, and successfully predicting 20 to 50 per cent of all large earthquakes. The costs of the prediction program are assumed to be US$ 60 million in capital outlay for the seismic monitoring network, with US$ 2.5 million per year in operating costs (1980 dollars). Because of the short time frame, it is assumed that people will respond positively, analogously to hurricane warnings. Economic costs include the costs of the program, the costs of slowing the economy for two days, and the costs of false predictions. The benefits are lives saved with an assumed value of US$ 1 million per life. Based on published earthquake loss studies, the authors estimate a time-of-day weighted average of 7,800 deaths without prediction, and 3,400 deaths with prediction. They concluded that the net benefits of the earthquake prediction program fall, in the main, within the area of economic feasibility. This conclusion, however, is very sensitive to the false alarm rate.

Furthermore, one wonders whether the public would respond appropriately to a low-probability prediction. For example, the New Zealand Civil Defense Director stated, in the case of a hypothetical prediction "with a probability of 25 per cent or less, it is doubtful than any elaborate protective measures beyond those already taken in the face of the existing general possibility of earthquakes could be justified" (*Holloway*, 1984, p. 910). This suggests that low-probability predictions should not be issued, unless the value placed on hypothetical lives saved is much greater than assumed in the above analysis.

Paté and Shah (1979), instead of initially assuming a monetary value for a human life, estimated the expected costs per life saved of an earthquake prediction, in order to compare the effectiveness of earthquake prediction to other risk reduction measures. They tested their model for the San Francisco Bay Area, with the caveat that the results "are only orders of magnitude" because of the numerous assumptions. They estimated a total net cost of US$ 4.6 billion and 2,943 lives saved, with an expected cost per life saved of US$ 1.56 million. They compared this to a previous analysis (*Paté*, 1978) which showed that the seismic requirements of the Uniform Building Code correspond to an expected cost per life saved of US$ 6.3 million. Thus, they concluded that "earthquake prediction should indeed be considered as a way of saving lives within acceptable economic limits", and that "earthquake prediction is economically competitive with the present requirements of building codes for the purpose of saving lives" (p. 1546). Their key assumption is that the prediction system is and accurate and reliable, without the problem of false alarms.

Stiros (1997) considered the costs and benefits of hypothetical earthquake predictions in Greece. He concluded that even if such predictions could be reliably
issued, the cost per life saved would be considerably higher than that of other life-
saving measures, such as those designed to improve traffic safety. He noted that
almost all lives lost in recent Greek earthquakes were due to the failure of multistory buildings. He therefore argued that reinforcing such buildings is preferable to attempting to develop a hypothetical earthquake prediction capability as a
means for reducing earthquake casualties in Greece.

One significant cost not measured by the above studies is whether the existence
of a hypothetical earthquake prediction capability would lull communities into a
false sense of security at times or in places for which predictions had not been
issued. It is also possible that long-term mitigation actions would suffer, which
would increase the costs of the overall system (*Cochrane*, 1984). Note that a
conservative system, with few false alarms being issued, would also probably
have many unpredicted earthquakes. Under such a system, long-term mitigation
actions necessarily would still be the primary means of reducing risk.

From the standpoint of science, a prediction may be regarded as successful if
the hypocenter, origin time and magnitude fall within the allowable ranges specified in the prediction announcement (see *Kagan and Jackson*, 1996; *Evans*, 1997;
and references cited in the above works). However, as pointed out by *Burton*
(1996) and *Stiros* (1997), to minimize earthquake damage we must know the region in which ground motion will be high enough to cause significant damage,
rather than just the epicenter. The damage caused by the 1989 Loma Prieta, California, earthquake shows the importance of this distinction. Some of the most
severe property damage and loss of life (e.g. the collapse of the elevated section
of the Route 880 freeway overpass in Oakland or the liquefaction in the Marina
District of San Francisco) occurred at distances of about 100 km from the epicenter (*Hanks and Krawinkler*, 1991). The contribution of a hypothetical successful
short-term prediction of the Loma Prieta event to risk reduction might therefore
have been limited if precautionary measures had been concentrated near the epicenter. On the other hand, taking precautionary measures in a large area centered
on the predicted epicenter would have required needless expenses in the vast proportion of the region in which no significant damage occurred.

Stiros (1997) points out that in many cases earthquake damage in Greece was
due to sequences of events of roughly comparable size (over weeks or months)
rather than to a single large shock. This is true for some other regions as well.
Predictions, to be useful for society, must provide information on the source parameters of such sequences rather than just those of a single forthcoming event.

The above discussion shows that short-term earthquake prediction, if highly
reliable and accurate, could cost-effectively save lives, with little long-term dam-

age to the regional economy. Conversely, if, as appears to be the case at present, reliable and accurate methods for short-term prediction do not exist, issuing predictions could be very costly, while failing to benefit society. As noted above, intermediate term predictions seem to offer the worst of all possible worlds, in that even if they were scientifically sound they would probably cause more disruption to society than any risk mitigation that might be achieved.

8.3   U.S. earthquake prediction program

The U.S. earthquake prediction program has been authorized by law since the mid-1970s. However, the fraction of the total seismic hazard mitigation effort devoted to prediction, particularly short-term prediction, has been declining.

The Disaster Relief Act of 1974 (Public Law 93-288, 88 Stat. 145) charged the President with ensuring "that all appropriate Federal agencies are prepared to issue warnings of disasters to State and local officials" (Sec. 202(a)). The warning authority "with respect to disaster warnings for an earthquake, volcanic eruption, landslide, mudslide, or other geological catastrophe [...]" was delegated to the Director of the USGS. In April, 1977, the USGS Director announced and defined a sequence of informational announcements regarding geologic hazards: notice of potential hazard, hazard watch, and hazard warning (*McKelvey*, 1977).

Earthquake prediction played a prominent role in the Congressional deliberations in the 1970s on the bills that eventually became the Earthquake Hazards Reduction Act (EHRA) of 1977 (*Stallings*, 1995). The Act itself (PL 95-124, 91 Stat. 1098, October 7, 1977) contains several references to earthquake prediction. It declared that "A well-funded seismological research program in earthquake prediction could provide data adequate for the design, of an operational system that could predict accurately the time, place, magnitude, and physical effects of earthquakes in selected areas of the United States" (section 2, Findings, subsection (4)). One of the Act's seven objectives is earthquake prediction (section 5(c)). The Act called for the 'development of methods to predict the time, place, and magnitude of future earthquakes' (section 5(e)(2)), and it called for a plan that would provide for, among other things, "evaluation of prediction techniques and actual predictions of earthquakes, [and] warning the residents of an area that an earthquake may occur [...]" (section 5(f)(1)).

The EHRA was amended in 1980, largely to appropriate funds and to recognize the role of the newly-created Federal Emergency Management Agency (FEMA). Section 5(f)(1) was amended to explicitly give the USGS Director "authority [...] to issue an earthquake prediction or other earthquake advisory as he deems necessary" (PL 96-472, 94 Stat. 2258, October 19, 1980).

FEMA (1980, pp. 12-13) considered the problem of responding to a prediction

"The possibility of a credible, scientifically-based prediction of a catastrophic earthquake poses serious challenges to government and our society. The current level of scientific understanding of earthquake prediction and the available resources are such that present instrumentation efforts are directed toward research rather than maintaining extensive monitoring networks for real-time prediction. The transition from research to fully operational capability will require additional scientific understanding as well as resources. Earthquake predictions are possible, perhaps likely, however, from the current research effort. Even with a significant level of uncertainty, any scientifically credible prediction that indicates a catastrophic earthquake is expected within about 1 year or less, will require very difficult and consequential decisions on the part of elected officials at all levels of government. Decisions may include such possibilities as the mobilization of National Guard and U. S. Department of Defense resources prior to the event, the imposition of special procedures or drills at potentially hazardous facilities, such as nuclear reactors or dams, the condemnation or evacuation of particularly unsafe buildings with the subsequent need for temporary housing, and the provisions of special protection of fragile inventories. If the prediction is correct and appropriate actions are taken, thousands of lives can be saved and significant economic losses can be avoided. The costs of responding to a prediction may be substantial, however, and the commitment of resources undoubtedly will have to be made in the face of considerable uncertainty and even reluctance. Indeed, the possibility of an inaccurate prediction must be faced squarely".

Political Pressure in the 1980s

There were several documented instances of strong support for earthquake prediction by politicians in the early 1980s. It is not clear whether these examples are typical or exceptional. On September 26, 1980, E.G. Brown Jr., the Governor of California, wrote to J.E. Carter, the President of the United States, regarding earthquake hazard mitigation. The full text of the letter is given by *FEMA* (1980, pp. 41–42). The portion dealing with earthquake prediction is as follows:

"As you know, significant theoretical and public policy research had already been completed by our Seismic Safety Commission, State Geologist, Earthquake Prediction Evaluation Council and the Office of Emergency Services. Together with the U. S. Geological Survey and the Federal Emergency Management Agency (FEMA), they had

clearly been keeping abreast of the state of the art of earthquake prediction. Indeed, combined state and federal efforts, founded on major
theoretical advances in American, Russian and Chinese seismic and
geological theory since the early 1970s, had shifted the language of
earthquake prediction in California from 'if' to 'when'!

"In light of my personal interest in this subject, I have signed into
law Assemblyman Frank Vicencia's [Assembly Bill Number] 2202, a
jointly funded state-Federal project to design a comprehensive earthquake prediction-response plan. It is the state's intention to prepare
a plan for the greater Los Angeles area as quickly as feasible. In
my view, such a full scale prediction-response program had become
possible only after the research findings of both physical and policy
scientists during the past 5 years. It is my conviction that such a plan
is now timely—neither too early nor too late".

In 1982, the U.S. Senate Committee on Commerce, Science, and Transportation, which oversaw the EHRA, expressed its displeasure that "the United States
is still far from having an earthquake prediction capability comparable to ones
that exist elsewhere in the world". It further requested "a definite commitment
to a prototype prediction system in one or more high risk areas by fiscal year
1988. Such a system could provide short-term warnings of an impending earthquake, save innumerable lives, reduce injuries, and possibly reduce substantially
the property damages from a major earthquake". The Parkfield experiment (see
section 8.3.3, below) was one response to this request.

There apparently was considerable political pressure on the USGS during the
first half of the 1980s. J.R. Filson, Chief of the USGS Office of Earthquakes,
Volcanoes, and Engineering, made the following statement at a meeting of the
National Earthquake Prediction Evaluation Council (NEPEC) on November 16,
1984 (*Shearer*, 1985a, pp. 10):

"The earthquake hazard in Southern California has been a high priority in Washington, DC for several years, and Congress has been push-
ing USGS to become more operational in its approach to earthquake
prediction".

The grounds for the Congressional and gubernatorial optimism in the mid-1980s
are not clear, but perhaps the highly optimistic statements of the mid-1970s (see
section 8.1, above) were still regarded as operative by politicians and their staffers.

Policy in the 1990s

Two national policy documents, from the White House (*National Earthquake
Strategy Working Group*, 1995) and from Congress (*Office of Technology Assess-*

ment, 1995) de-emphasize the importance of prediction, as compared to mitigation. The latter (p. 17) observed that NEHRP was begun at a time of scientific optimism, when it was thought that short-term prediction would soon become reality, but that "Since then, however, prediction has proved more elusive than originally thought". The former emphasizes earthquake-resistant construction and seismic risk assessment tools, but does contain one item (p. 31): "Develop and evaluate methods for short- and intermediate-term forecasts and apply the methodologies to selected regions with high earthquake potential".

The most recent (1998–2002) five year plan (*Page et al.*, 1997) for the U.S. Earthquake Hazards Program (EHP) includes short-term prediction among the elements of the program:

> "For short-term prediction, a critical issue is whether any observable signals, besides foreshocks, are generated before an impending earthquake. Signals preceding the Loma Prieta (electromagnetic and strain) and Kobe (hydrologic and geochemical) earthquakes are possible examples. The decade-long baselines of data already acquired at Parkfield, California, are irreplaceable for identifying intermediate-term changes in strain rates and other variables and assessing their significance".

The following prediction-related tasks are included in the five year plan: (1) "Continue the experiment at Parkfield to monitor possible earthquake precursors under controlled conditions such that their relationship to the earthquake generation process can be established"; (2) "Investigate reports and observations of possible intermediate- and short-term earthquake precursors associated with major earthquakes"; and (3) "Authoritatively evaluate credible methods for earthquake prediction proposed by researchers worldwide through the National Earthquake Prediction Evaluation Council (NEPEC)".

8.3.1   Current Federal and State laws

The EHRA was substantially revised in 1990 by the National Earthquake Hazards Reduction Program (NEHRP) Reauthorization Act (PL 101-614, 104 Stat. 3231, November 16, 1990). The 1990 Act is more detailed, and emphasizes mitigation more than its 1977 predecessor (also see *Page et al.*, 1992). At present (July 2003) earthquake prediction continues to be mentioned by the laws of the United States. A search of the data base for the U.S. Code (http://uscode.house.\-gov/usc.htm) found the following mentions of earthquake prediction. Only the relevant clauses are shown:

42 USC Sec. 7701

The Congress finds and declares the following: [...]

(2) [...] With respect to future earthquakes, such loss, destruction, and disruption can be substantially reduced through the development and implementation of earthquake hazards reduction measures, including [...]
(C) prediction techniques and early-warning systems,
(3) [...]
(4) A well-funded seismological research program in earthquake prediction could provide data adequate for the design, of an operational system that could predict accurately the time, place, magnitude, and physical effects of earthquakes in selected areas of the United States.

42 USC Sec. 7704

(1) [...]
(2) Federal Emergency Management Agency
(A) Program responsibilities
In addition to the lead agency responsibilities described in paragraph (1), the Director of the Agency [FEMA] shall [...]
(vi) provide response recommendations to communities after an earthquake prediction has been made under paragraph (3)(D) [...]
(3) United States Geological Survey
The United States Geological Survey shall conduct research necessary to characterize and identify earthquake hazards, assess earthquake risks, monitor seismic activity, and improve earthquake predictions. In carrying out this paragraph, the Director of the United States Geological Survey shall [...]
(C) develop standard procedures, in consultation with the Agency [FEMA], for issuing earthquake predictions, including aftershock advisories; (D) issue when necessary, and notify the Director of the Agency [FEMA] of, an earthquake prediction or other earthquake advisory, which may be evaluated by the National Earthquake Prediction Evaluation Council. [...]

Earthquake prediction is also mentioned extensively in the laws of the State of California (see http://www.leginfo.ca.gov/calaw.html) but we omit discussion here.

8.3.2   NEPEC

NEPEC is an advisory body to the Director of the USGS. The USGS Earthquake Prediction Council, which was a predecessor of NEPEC, was discussed by the USGS Director (*McKelvey*, 1976, 1977) at a meeting in November 1975 where the USGS plans for earthquake prediction were outlined. These plans, and the role of the Council, were discussed in a news report (*Anonymous*, 1976):

> "Although presently, no operational capability for reliable earthquake prediction exists, it can be expected to be developed soon because of the rapid progress that scientists in the United States, the Soviet Union, Japan and China are making in earthquake research. Therefore it is not premature to present a tentative plan for issuing predictions and warnings. The plan is expected to change as suggestions for improvements and comments are received, more progress is made in prediction research, and experience in issuing predictions is gained".

The news report explained that the USGS plan envisioned a system of predictions ("a statement that an earthquake will occur at a certain time and place, have a certain magnitude [. . .]") and warnings (recommendations for countermeasures such as evacuations). The former were to be issued by the USGS, and the latter by local governments. The plan envisioned that interpretations of raw data by individual USGS scientists would be reviewed by the USGS Earthquake Prediction Council, a body consisting of 5–10 USGS scientists as well as experts from outside USGS. No explicit provision was made for predictions by scientists not funded by USGS, but "it is believed that these scientists would be willing to discuss their data with either the [USGS] Council or state earthquake prediction review groups". The council would review proposed predictions and make a recommendation to USGS headquarters. The news report explained what would then be done:

> "A report of the council would go to USGS Headquarters (Reston, Virginia) where a decision would be made about whether a prediction should be issued and how to issue it. If the case is not sufficiently strong, a decision could be made to issue an advisory notice stating, for example, that possible precursors have been detected in a certain area and that the area is under intensive study. The nature of the action would be tailored to the particular situation. A statement would then be communicated to the governor of the state potentially affected, to federal agencies with responsibilities for disaster preparedness and response [. . .], and to the public. The public would not necessarily be notified simultaneously with the others. It might be judged that the negative impact of a prediction could be lessened if responsible state

and federal officials received prior notice. A strong case can be made
that a warning should be issued with a prediction so that the public
is not left without any recommendation for appropriate action. Upon
receipt of a prediction the governor's office would refer the prediction
to the state office concerned with disaster response and might choose
to call together his own group of experts to evaluate the prediction.
[...] USGS scientists would be available for discussion at any stage
of the process".

The USGS Earthquake Prediction Council was replaced by NEPEC in 1979
(*U.S. Department of the Interior*, 1979) The authority was provided by the Earthquake Hazards Reduction Act of 1977, in furtherance of the objectives of section 202 of the Disaster Relief Act of 1974. NEPEC consists of a Chairman (not a
USGS employee), Vice Chairman (Chief of the USGS Office of Earthquake Studies) and 8–12 members (less than half of whom can be USGS employees) who
are experts in scientific disciplines related to earthquake prediction. The members
are appointed by the USGS Director for staggered 3 yr terms. The Council can
invite non-member scientists to participate in discussions, and can invite other observers. The 1994 version of the NEPEC Charter (*U.S. Department of the Interior*,
1994) does not differ substantively from the 1979 version. NEPEC is mentioned
in the USGS five year plan for 1998-2002 (*Page et al.*, 1997) but has not been
active since 1996 (A. Michael, personal communication). Some key provisions of
the 1979 version of the NEPEC Charter are excerpted below.

> **"Functions/Responsibilities**: The [USGS] Director shall be re-
> sponsible for deciding when and/or whether to issue predictions or
> other information pertinent to the potential for the occurrence of a fu-
> ture significant earthquake (e.g. negative evaluations or advisories).
> A prediction is defined to mean a statement on the time of occurrence,
> location and magnitude of a future significant earthquake, based on
> qualification of the uncertainty of those factors. [...] "The Council
> shall advise the Director on issuing predictions as to the completeness
> and scientific validity of the available data and on related matters as
> assigned by the Director. Specifically, the Council shall be responsi-
> ble for assessing data and issuing reports on its findings in a timely
> manner.
>
> "The Council's duties will involve the evaluation of predictions
> made by other scientists, from within or outside of government, rather
> than issuance of predictions based on data gathered by the Council
> itself.
>
> **"Operating Procedures**: [...] In any case where a member of the
> Council has been personally involved in developing the prediction,

which the Council has been convened to review that member shall not vote in the Council's evaluation of that prediction. [...]

"In evaluating predictions, the Council's objectives are (1) to provide objective and critical review, by a uniform process, of any scientific data or interpretation of scientific data that might warrant issuance of a formal USGS prediction of a specific earthquake, or that might warrant a formal USGS position other than a prediction (e.g. a negative evaluation or advisory); (2) to recommend to the appropriate scientists any actions that might be desirable or required to clarify the basis for a prediction; (3) to maintain an accurate record of predictions evaluated and evidence pertinent to them; and (4) to provide the Director a timely and concisely written review of the evidence relevant to a prediction of any potentially damaging earthquake (usually those of magnitude 5 or greater on the Richter scale) and a written recommendation as to whether the evidence is sufficiently clear that an official prediction by the Director should be issued, or, if not, what if any other official position the Director should take. Where the recommendation is not unanimous the report should include the full range of viewpoints expressed by Council members. [...] The Director bears the ultimate responsibility for a decision as to whether or not a prediction is to be issued".

Role of NEPEC

NEPEC's name makes clear that it is an evaluation council, rather than a body for originating predictions. This is also spelled out by the NEPEC charter (above), which clearly states that NEPEC's mission is the evaluation of predictions made by other scientists, rather than issuance of predictions based on data gathered by NEPEC itself.

Nevertheless, some members of NEPEC favored a more pro-active role. The minutes of the NEPEC meeting of November 16, 1984 (*Shearer*, 1985a, pp. 1–5) gives the arguments made on both sides of the issue. NEPEC Chairman L.R. Sykes said that NEPEC should "take a more active role in earthquake prediction", and Vice-Chairman J.R. Filson suggested that NEPEC can "do more than respond to other people's prediction". NEPEC member R.E. Wallace agreed that NEPEC should be more active. On the other hand, NEPEC member K. Aki felt NEPEC should act as a jury, determining the validity of predictions brought before it. and NEPEC member J. Davies pointed out that if it took an active role in some areas it could compromise its ability to make impartial judgments. NEPEC member T.V. McEvilly also felt that NEPEC should restrict itself to reviewing data brought before it rather than risk conflicts of interest. According to the minutes, the Chairman then called for reaching a consensus, and this was that:

> "The NEPEC will take a more activist role within the confines of the Charter".

Activities conducted by NEPEC

As noted above, the USGS Earthquake Prediction Council (and NEPEC) were established to evaluate and validate work by USGS scientists (as well as work by outside scientists) before the USGS Director formally issued a prediction or took other action. However, although NEPEC has reviewed long-term predictions by USGS-led teams, and has approved the guidelines for the Parkfield prediction experiment, to our knowledge it never has had to review an 'operational' short-term prediction whose primary proponent was a USGS scientist. NEPEC was heavily involved in 1979–80 in reviewing a prediction by an earth scientist at the U.S. Bureau of Mines (see section 8.5.2, below). On the other hand, NEPEC kept a low profile in the case of a public prediction by an amateur in 1990 (see section 8.5.2, below). This is not surprising, as the above discussion shows that debunking predictions by cranks was not originally envisioned as one of NEPEC's responsibilities.

Regional and State evaluation panels

The U.S. has a federal system of government, and several states or regions have also established their own bodies for evaluating predictions. California established an Advisory Group on Earthquake Prediction in 1974. It was subsequently renamed the California Earthquake Prediction Evaluation Council in 1976, and official guidelines were adopted in February 1977 (*California Earthquake Prediction Evaluation Council*, 1977). It was established to advise the Director of California's Office of Emergency Services regarding earthquake predictions. The Council consists of the State Geologist, plus eight members appointed by the OES Director. All prediction statements (even those appearing in the press) are brought to the Chairman for initial screening and possible full Council evaluation. The Council is set up to be a scientific body, and the decision to issue a warning or other statement is up the Director of OES. California also has an earthquake prediction response plan, to guide state agencies in the event of a prediction (*California Office of Emergency Services*, 1990).

The Central U.S. Earthquake Consortium in 1994 established the CUSEC Earthquake Prediction Evaluation Council (*Central U.S. Earthquake Consortium*, 1994). It consists of 6 to 10 members, appointed by the CUSEC State Geologists, with one member also a member of NEPEC. Its tasks are assigned by the Chairman of the CUSEC State Geologists. Any of the State Geologists, with the consent of the majority of the CUSEC State Geologists, may request a meeting of the Evaluation Council. Any of the State Geologists can decide to issue an evaluation or other prediction-related information.

8.3.3 Parkfield earthquake prediction experiment

We present a summary here; for further details see *Geller* (1997b, p. 438–440). It was proposed that characteristic $m = 6$ earthquakes occur on the San Andreas fault at Parkfield, California, at intervals of approximately 22 yr. An $m = 6$ earthquake at Parkfield occurred in 1966. On the basis that there was a 95 per cent probability that "the next characteristic Parkfield earthquake" would occur before 1993, the USGS and cooperating agencies established the 'Parkfield Earthquake Prediction Experiment' (*Bakun and Lindh*, 1985, *Roeloffs and Langbein*, 1994). As of July, 2003, the "next characteristic $m = 6$ Parkfield earthquake" has not yet occurred.

The Parkfield prediction was evaluated and approved by NEPEC and other cognizant agencies (*Shearer*, 1985a, b; *Kerr*, 1985). The minutes of the NEPEC meeting of November 16, 1984 (*Shearer*, 1985a, pp. 8–9) do not record any negative comments, or any doubts that the "characteristic $m = 6$ earthquake" forecast for 1988 ± 4 yr would occur as scheduled. The comments of NEPEC member K. Aki capture the general consensus:

> "Good approach, an earthquake will occur there; short-term precursors are being observed and confirmed in the laboratories. Helps fill the gap between long-term and short-term prediction. Present lack of intermediate precursors".

The minutes of the NEPEC meeting list ten points of consensus regarding the evaluation of the Parkfield prediction, of which the first three were:

1. "The Parkfield Earthquake Prediction Experiment is one of the highest priorities for the U.S. Earthquake Hazards Program and has the highest probability for a successful prediction".

2. "Endorsed the general aspects of the prediction made by Bakun and Lindh that a M 6 earthquake will happen in the Parkfield area by 1988 ± 4 years".

3. "Advise USGS Director to make an information statement regarding the Parkfield earthquake as a long-term prediction. Include the Bakun and Lindh paper in the recommendation".

The following official announcement was issued on April 5, 1985 by the Director of the USGS (*Shearer*, 1985b):

"The forecast that an earthquake of magnitude 5.5 to 6.0 is likely to occur in the Parkfield, Calif., area within the next several years (1985–1993) has been reviewed and accepted by State and Federal evaluation panels according to an announcement today (April 5, 1985) by the U.S. Geological Survey.

"The California Office of Emergency Services has reviewed the evaluation with local officials and will take coordinated action should the extensive monitoring equipment arrayed throughout the Parkfield region indicate that the anticipated earthquake is imminent".

Bakun (1987) summarizes the Parkfield experiment as follows:

"As a prediction experiment, the principal goal of the Parkfield study is a detailed description of the final stages of the earthquake preparation process; observations at Parkfield should aid in the evaluation of the feasibility of intermediate- and short-term earthquake prediction elsewhere. Furthermore, the detailed history of strain accumulation and release over a complete cycle that is being recorded at Parkfield should provide the basis for testing and refining models for earthquake recurrence on plate boundaries.

"A secondary goal of the experiment is the issuance of a short-term warning by the USGS to the Governor's Office of Emergency Services, the agency responsible for public dissemination of any warning. [...] It is likely that the prediction experiment and the plans will serve as prototypes and stepping stones for future earthquake prediction efforts in the United States".

The Parkfield experiment was widely covered by the media. For example, *The Economist* ("Small earthquake somewhere, next year—perhaps", August 1, 1987), reported:

"Parkfield is geophysics's Waterloo. If the earthquake comes there without warnings of any kind, earthquakes are unpredictable and science is defeated. There will be no excuses left, for never has an ambush been more carefully laid for such an event".

At the onset of the Parkfield experiment it was generally taken for granted that the expected $m = 6$ earthquake would occur, and discussion focused on whether or not the monitoring network would detect precursors so that the earthquake could be predicted. However, after the non-occurrence of the expected earthquake within the 95 per cent time window, the basic premises of the experiment were questioned by *Savage* (1993) and *Kagan* (1997b).

Public alarms issued at Parkfield

The Parkfield Earthquake Prediction Experiment included a system for issuing alarms to the public. The highest category is an A-level alert, which means the estimated probability of an $m = 6$ earthquake within the next 72 hr exceeds 37 per cent (*Langbein* 1993). A-level alerts are issued based on potential foreshocks, aseismic creep, or a combination of the two. (Note that *Michael and Jones*, 1998, question many of the assumptions that were originally used to decide the probabilities for the various alert levels at Parkfield.) A-Level alerts were issued on Oct. 20, 1992 (*Kerr*, 1992; *Finn*, 1992; *Langbein*, 1992, 1993) and Nov. 14, 1993 (*Wuethrich*, 1993). Both were false alarms. No panic or significant unrest among the public was reported. However, as the population of the Parkfield area is extremely sparse, the experience from these alarms is not necessarily indicative of what would happen if an alarm was issued for an urban area.

Evaluation of Parkfield experiment

The Parkfield experiment was evaluated by the NEPEC Working Group (*Hager et al.*, 1994):

> "The relatively narrow window that was stated in the original prediction has led to expectations that the experiment would be over relatively quickly, leading to the misconception that the experiment has now somehow 'failed' because the narrow time window has closed.
>
> "The general public now perceives the Experiment primarily as a short-term earthquake forecasting project with an inherent expectation to accurately predict an earthquake, while the scientific community views it not only as a short-term prediction experiment but also as an effort to trap a moderate earthquake within a densely instrumented network. It is important to educate the public that there is great value to this monitoring effort even if the prediction effort is unsuccessful.
>
> "Parkfield remains the best identified locale to trap an earthquake. The consensus is that the annual probability for the expected 'characteristic' event is about 10 per cent per year. At this level, the Working Group concludes that the Experiment should be continued, both for its geophysical and public response benefits".

8.4   Japan's earthquake prediction program

Japan has had an official earthquake prediction program since 1965. This program is sometimes cited by scientists who advocate establishment of a similar program in the United States (e.g. *Press and Brace*, 1966). When the Japanese program was first established (April 1965–March 1969) it was named the Earthquake Prediction Research Program. However, in 1969 the word 'research' was dropped, implying a more operationally oriented program, when the program was renewed for five years. Five year plans have since become the standard, and a numbering system was established when the 2nd Earthquake Prediction Plan began in April 1969.

The Kobe earthquake (January 17, 1995) occurred while the '7th Earthquake Prediction Plan' was underway (April 1994 - March 1999). After the Kobe earthquake part of the earthquake prediction program was renamed the earthquake investigation program, and other tasks (e.g. engineering seismology) were also included in the new program. A detailed discussion of the reorganization is outside the scope of this paper. However, the 5 year plans for earthquake prediction also continue; the current plan is in effect from April 1999 to March 2004. but is officially named 'New Research on Seismological Observations for the Purpose of Earthquake Prediction' rather than '8th Earthquake Prediction Plan'. The JMA's real-time monitoring system for trying to make a short-term prediction of the 'Tokai earthquake' continues to be in effect, with no major changes. Discussions of Japan's earthquake prediction plan are given by *Hagiwara and Rikitake* (1967), *Kanamori* (1970), *Rikitake* (1972), and *Mogi* (1985, 1995), See *Geller* (1997b, pp. 437–438) for additional references.

There has been considerable discussion of whether the Japanese earthquake prediction program should be terminated or overhauled (e.g., *Geller*, 1991a, b; *Hamada*, 1991). A government advisory committee reported that the prediction program has not met its goals and has overstated the chances of developing accurate earthquake forecasts (*Normile*, 1997; *Swinbanks*, 1997). An external panel reviewing the current (1999–2004) five year plan said that the use of the term 'earthquake prediction' in the name of the five year plan was misleading, as no work on operational prediction was being performed. As this could give a misleading impression to the public, the panel said that if a new five year plan is approved from 2004-2009 the term 'earthquake prediction' should be dropped from the plan (*Mainichi Shinbun* newspaper, October 23, 2002). However, as of July 2003, the name of the program has not been changed in proposals for the next five year plan.

8.4.1   Long-term forecast of the 'Tokai earthquake'

Based on arguments assuming quasi-periodicity of earthquake occurrence, *Mogi* (1970) and other authors suggested that a large earthquake might occur in in the Tokai district (off the Pacific Coast of Japan near Shizuoka, between Tokyo and Nagoya). This hypothetical earthquake is commonly called the 'Tokai earthquake' in Japan. *Ishibashi* (1977) argued that a large earthquake in the Tokai district was overdue:

> "There is reason to fear that the 'Suruga Bay earthquake' is imminent. Speaking precisely, the result of long term prediction research is that there is strong reason to fear that precursory phenomena could begin at any time". (authors' translation)

The consensus of Japan's earthquake prediction community was summarized by *Rikitake* (1979):

> "Many Japanese seismologists, earthquake engineers, and national and local officials responsible for disaster prevention are quite convinced nowadays that a great earthquake of magnitude 8 or so will hit the Tokai (literally East Sea) area, an area in central Japan between Tokyo and Nagoya, in the near future. The anticipated epicentral region is off the Pacific coast of Shizuoka Prefecture, including Suruga Bay. Should the feared earthquake occur it would certainly destroy one of the most industrial areas in Japan, through which the world-famous Shinkansen (bullet train) runs and where a few nuclear reactors are operating.
> "The targeted area was often struck by great earthquakes in historical times such as the 1854 and 1707 earthquakes, both of which had magnitudes estimated at 8.4. The mean return period of recurrence of great earthquakes there is estimated as about 120 years. As more than 120 years have already passed since the last shock, there is reason to believe that an earthquake will recur sooner or later".

Ishibashi (1981) comments as follows:

> "The seismic gap for 126 years on the Suruga trough thrust since 1854 and the considerable amount of strain accumulation in this region estimated from geodetic survey data suggest a fairly high probability of a near-future occurrence of this faulting. And, if the event is a little smaller-scale, it may occur earlier".

Ishibashi (1977) said that precursors of the 'Tokai earthquake' "could begin at any time". This phrase is vague and qualitative. However, from the actions taken by the authorities, it is clear that a time scale of at most a few years was initially envisioned. As 25 years have now passed, it seems reasonable to judge that the long-term forecast of the 'Tokai earthquake' was a failure.

Although the earthquake described as 'imminent' by *Ishibashi* (1977) did not occur in the subsequent 25 years, the belief that this scenario earthquake is predetermined to occur in the near future continues to be widespread in Japan, among both Earth scientists and the general public, and this is still the premise of government policy. One rough indication of the degree of public belief in the 'Tokai earthquake' is provided by a Google search of Japanese language material for the phrase 'Tokai earthquake', which showed about 56,400 hits as of October 2002. In contrast a search of English language material found only 512 hits for the 'Parkfield earthquake', and 14,100 and 20,700 hits respectively for the Loma Prieta and Northridge earthquakes.

Effect of naming scenario earthquakes

In both the case of the 'Tokai earthquake' and the 'Parkfield earthquake', assigning a name to a hypothetical scenario earthquake (in the same way that names like 'Loma Prieta', 'Northridge', or 'Kobe' are assigned to significant earthquakes that actually occurred) seems to cause the scenario earthquake to be treated as though it were 'real'. It is over 25 years since the 'Tokai earthquake' scenario was proposed, and over 15 years since the 'Parkfield earthquake' scenario was proposed. Logic would suggest that both scenario earthquakes could be deemed to be failed predictions that do not warrant further consideration, but in both cases a significant fraction of the relevant scientific community and government authorities, as well as the public, continues to treat these 'events' as though they were 'real'.

8.4.2 System for short-term prediction

Ishibashi's warning was reported widely in Japanese news media (*Mogi*, 1985, pp. 270–272). Ishibashi called for establishing observational networks to detect precursors. Powerful politicians took up Ishibashi's call, and legislation was proposed in the Japanese Parliament. While this legislation was pending, the Izu-Oshima-Kinkai earthquake (January 14, 1978, $m = 7.0$) occurred, causing 25 deaths and significant property damage. In the aftermath, the Large-Scale Earthquake Countermeasures Act (LECA), was passed in June 1978, and took effect in Dec. 1978.

Wakita et al. (1980) reported precursory changes in radon concentration before the Izu-Oshima-Kinkai earthquake. The *Mainichi Shinbun* newspaper (February 5, 1978), reported these observations on page one, under the banner headline "Epoch making step towards earthquake prediction", with the subheadline "If the number of observatories is increased prediction is possible". This publicity facilitated passage of the LECA.

Under the LECA if 'anomalous data' are recorded, an 'Earthquake Assessment Committee' (EAC) will be convened within two hours. Within 30 min the EAC must make a black (alarm) or white (no alarm) recommendation. The former would cause the Prime Minister to issue the alarm, which would shut down all expressways, bullet trains, schools, factories, etc., in an area covering seven prefectures. Tokyo would also be effectively shut down. The procedures for rescinding an unsuccessful alarm have never been publicly discussed. The estimated cost of a prediction is US$ 7 B/day; the government is legally immune from demands for compensation (see *Davis and Somerville*, 1982; Mogi 1985, pp. 299-306).

The Japanese EAC, like NEPEC, is an advisory body to the responsible government official, the Director of the Japan Meteorological Agency (JMA) which is responsible for monitoring earthquakes as well as weather. The six members of the EAC are professors of geophysics (or retired professors) who live in the Tokyo area, so that they can easily be summoned on short notice. Unlike NEPEC, the sole task of the EAC is to advise the JMA Director, based on various types of geophysical data, whether or not a short-term prediction of the Tokai earthquake should be issued. Another committee, the Coordinating Committee for Earthquake Prediction, is a forum for the presentation of reports by researchers, but it does not issue evaluations. There is no committee in Japan tasked with the evaluation of predictions in general.

Once a year (on Sept. 1, the anniversary of the 1923 Kanto earthquake, which caused 140,000 deaths) a prediction drill is held. The EAC are driven to JMA headquarters in police cars, arriving at about 7AM. Mock anomalies are analyzed, and a mock prediction ("There is a high risk of a magnitude 8 earthquake [in the Tokai district] within 2 or 3 days") is issued at 9AM on national TV by the mock Prime Minister. This annual performance helps to create the mistaken impression that the 'Tokai earthquake' is an imminent danger and can be predicted.

Notwithstanding this scientific failure, the legal and bureaucratic apparatus for predicting the 'Tokai earthquake' and issuing legally binding public alarms remains in effect. Reports in the news media continue to imply that the likelihood of an earthquake in the Tokai district is much greater than elsewhere on Japan's Pacific Coast.

8.4.3   Public perception

The Science and Technology Agency of Japan periodically conducts 'technology assessment' surveys (*Swinbanks*, 1993). The respondents are professionals from all fields of science, technology, and medicine. The 1971 respondents thought that prediction of earthquakes of magnitude 6 or greater would be possible by 1996. In the 1976 survey the expected date of success receded to 2003, and in 1981 it receded to 2006. The 1986 survey changed the question to cover only prediction of magnitude 7 or greater events a few days in advance; the respondents expected success by 2006. The 1992 respondents expected success by 2010. The 1997 survey results were announced in the *Asahi Shinbun* newspaper (morning edition,

11 July 1997). Prediction is now expected to become possible in 2023.

An opinion poll by Japan's Office of the Prime Minister in September 1995 (Anonymous, 1996) revealed that 34.6 per cent of the public thought the 'Tokai earthquake' could be predicted (about half of these respondents thought all $M \geq$ 7 events could be predicted); 44.5 per cent thought prediction was impossible; 20.9 per cent didn't know or gave other answers. Polls by *Ohta and Abe* (1977), *Hirose* (1985), and *Nishida* (1989) also found substantial public belief that prediction was possible. A comparable poll in the U.S. (*Turner*, 1982) found that 5.4 per cent said earthquakes could be predicted 'quite accurately', and 36.4 per cent said 'somewhat accurately'.

8.5 Public reactions to predictions

'Since my first attachment to seismology, I have had a horror of predictions and of predictors. Journalists and the general public rush to any suggestion of earthquake prediction like hogs toward a full trough. [...] [Prediction] provides a happy hunting ground for amateurs, cranks, and outright publicity-seeking fakers. The vaporings of such people are from time to time seized upon by the news media, who then encroach on the time of men who are occupied in serious research". —*Richter* (1977)

There are no well documented cases of a successful short-term earthquake prediction. (One earthquake in China in 1975 has been claimed as a success, but there are many inconsistencies and ambiguities; see *Geller*, 1997b, pp. 433–434.) In contrast, there are several well documented instances of publicly announced groundless predictions, some of which caused substantial social unrest. Some examples are introduced below. We then discuss common features and differences among these episodes, and then consider possible measures by which the recurrence of similar episodes might be prevented.

8.5.1 Codes of practice for earthquake prediction

Several codes of practice for earthquake predictions have been promulgated. These codes unfortunately cannot be enforced, but it would be desirable for their provisions to be followed by professionals in earthquake science and related fields. It would also be desirable for the media not to publicize predictions that were not made in accordance with such codes.

The IASPEI/UNESCO code (*IUGG*, 1984) outlines a sensible set of procedures. Perhaps the most important is as follows.

"The news media are generally not the appropriate means by which to announce a prediction. The author of a prediction should instead

communicate the prediction confidentially to the government authority designated or best able to deal with such predictions".

Similar codes include that of *California Earthquake Prediction Evaluation
Council* (1977) and *Council of Europe* (1991). Unfortunately, the publically announced predictions discussed below did not follow the procedures recommended
by these codes, and as a result there was considerable needless disruption to society.

8.5.2   Publicly announced predictions

The Jupiter Effect: *Gribben* (1971) suggested that earthquakes were correlated
with sunspots or planetary alignments. Gribben, by then a member of *Nature*'s
editorial staff, published a popular book, *The Jupiter Effect* (*Gribben and Plagemann*, 1974). This book's thesis is (pp. 116):

> "A remarkable chain of evidence, much of it known for decades but
> never before linked together, points to 1982 as the year in which the
> Los Angeles region of the San Andreas fault will be subjected to the
> most massive earthquake known in the populated regions of the earth
> in this century. At the end point of this chain, directly causing this
> disaster, is a rare alignment of the planets in the Solar System".

Insightful book reviews were published by Kaula (1974), Anderson (1974), and
Bolt (1975). Los Angeles was not "subjected to a massive earthquake" in 1982.
Although Gribben's prediction attracted considerable publicity, it did not cause
any panic.

Kawasaki, Japan, 1974: A scare over a possibly anomalous uplift of the crust
took place in Kawasaki (20 km from Tokyo) in 1974 (*Dambara*, 1981). The area
had been subsiding rapidly since 1950 due to pumping out of underground water.
When this pumping ceased, the water level recovered rapidly, causing an apparent
uplift. This was reported extensively as a possible precursor of a $m = 6$ class
earthquake in the news media. The Chairman of the Coordinating Committee for
Earthquake Prediction had to call a press conference to calm public apprehension
(*Geographical Survey Institute*, 1979, pp. 45; *Ohta and Abe*, 1977). Considerable
publicity and public attention was generated, but there was no panic.

North Carolina, 1975-1976: David Stewart, a faculty member of the University of North Carolina, predicted "a major earthquake in the next few decades
or less" in the coastal Wilmington area, near a nuclear reactor (*Stewart*, 1975).
Stewart's data (claims of crustal uplift) were shown to be in error. According
to *Spence et al.* (1993), Stewart then collaborated with a self-proclaimed psychic, Clarissa Bernhardt, who made 13 specific and general earthquake predictions, one of which (*Kerr*, 1991; *Mileti and Fitzpatrick*, 1993, pp. 52) was for an

$m = 8$ earthquake. These predictions were announced by Stewart and Bernhardt in a joint paper, "Some recent psychic earthquake predictions', presented at the Annual Convention of the Southeastern Regional Parapsychological Association. According to the *St. Louis Post-Dispatch* newspaper (October 21, 1990, "He Calls it 'a Fact'—State's Quake Expert Believes in Psychic Phenomena", reproduced by *Spence et al.* 1993, pp. 120) Stewart and Bernhardt publicly predicted an $m = 8.0$ event, most likely within 3 days of January 17, 1976. Widespread public disquiet reportedly occurred; the prediction failed.

Southern California, 1976: *Whitcomb* (1976) reported a seismic velocity anomaly beneath the Transverse Ranges in Southern California. This was reported in *Science* (*Shapley*, 1976; *Hammond*, 1976) as a prediction of an event with $m = 5.5$–6.5. No such event occurred. *Turner et al.* (1986) discuss public reaction; there was considerable interest, but no panic. The prediction was submitted to the California Earthquake Prediction Evaluation Council, which concluded that the evidence was insufficient for an official prediction (*Mileti et al.*, 1981, pp. 33). Whitcomb withdrew the prediction later that year. The apparent anomaly was probably an artifact (see *Geller*, 1997b, pp. 428–429).

Southern California, 1976-c1980: Reports of possible geodetic anomalies that were later found to have been artifacts have generated considerable interest and anxiety. *Richter* (1958, pp. 193) describes one such case. Another is the 1974, Kawasaki, Japan, uplift discussed above. Perhaps the best known case occurred in Southern California. See *Geller* (1997b, pp. 432–433) for a more detailed discussion of technical aspects. As an outgrowth of studies of geodetic deformation before and after the 1971 San Fernando, California, earthquake *Castle et al.* (1976) reported an uplift of about 15-25 cm over an area of about 12,000 km^2 in southern California during (approximately) 1960-1975. This inferred uplift was usually called either the 'Palmdale Bulge' or 'Southern California uplift'. F. Press, then Chair of Earth and Planetary Science at the Massachusetts Institute of Technology, wrote to the Vice President of the United States, N.A. Rockefeller, on January 21, 1976 saying that funding was needed to study the uplift. Excerpts from this letter (see also *Press*, 1976) are given by *Shapley* (1976):

> "The discovery, which will soon be released publicly, is most disturbing because such uplifts in the past have preceded earthquakes of great destructive power. [...]

> "The effect on Los Angeles of an earthquake in the region of the uplift would be quite disastrous. A structural engineer at U.C.L.A., Professor Martin Duke, has estimated that as many as 40,000 buildings would suffer collapse or serious damage.

> "There is no question that the uplift must be taken very seriously even though geophysicists have, as yet, no clear understanding of its

origin or significance. [...]

"The region of the uplift should now be subjected to a most intense scrutiny [...] In Japan a geophysical anomaly of this magnitude
would trigger an intensive study or a public alert. [...]

"Having visited China, I can attest to their technical proficiency
in this field of science, and express my concern that because of insufficient resources a similar achievement may not be possible in this
country".

Shapley (1976) reports that US$ 2.1 million was allocated to the USGS to mon-
itor the apparent uplift. Many reports about the apparent uplift appeared in the
mass media in southern California (*Turner et al.*, 1986). The report of the apparent uplift did not meet the California Earthquake Prediction Evaluation Council's
criteria for a prediction, so it was not officially reviewed as such (*Wesson*, 1984).

There was extensive debate over whether the Palmdale Bulge was an artifact.
Holdahl (1982) concluded:

"Vertical motion in the Palmdale region has now been reassessed.
[...] The new results show 7.5 ± 4.0 cm of apparent upward movement at Palmdale. Part or all of the 7.5 cm might be attributable to
residual systematic error. The great concern associated with Palmdale, if based on motion calculated from leveling data, does not seem
warranted. Castle's original result was strongly influenced by a single
level route to Palmdale. Refraction error accumulates rapidly on this
route because of the long gentle slope and the location of sightings
over railroad ballast where vertical temperature gradients are unusually large. [...] The rigorous analysis of refraction-corrected leveling
data described in this paper supports the approximate computations
of *Strange* (1981) that indicated that what was thought to be uplift at
Palmdale was only the appearance of uplift created by different refraction error accumulation in successive surveys. The primary cause
of this difference was reduced sight lengths starting in 1964".

Los Angeles, 1976: (See *Turner et al.*, 1986, pp. 8, pp. 45-50.) Henry
Minturn, a self-styled Ph.D. geophysicist, claimed in a local TV program on
November 22, 1976 to have predicted many earthquakes successfully in Latin
America; he forecast a large earthquake in Los Angeles on December 20. The
claimed basis of this prediction was "the gravitational pull of the moon on 'weak
arches' in the earth's crust". The prediction received extensive coverage in the
media, but when Minturn's claim to have a Ph.D. was revealed to be false, the
story died out quickly.

Oaxaca, Mexico, 1978: (See *Garza and Lomnitz*, 1979; *Lomnitz*, 1983; *Lomnitz*, 1994, pp. 122–127.) A crank letter to the governor of Oaxaca in January 1978 predicted a large earthquake on April 23, 1978. In an entirely unrelated development, a press conference was called to publicize a long-term forecast by U.S.-based scientists, *Ohtake et al.* (1977). A news story reporting on the press conference, "Texas U Predicts Big Mexico Quake" (reproduced by *Lomnitz*, 1994, pp. 124) was published in the *Mexico City News* newspaper on April 10, 1978. The news story contained phrases such as "a massive earthquake will occur soon in the state of Oaxaca". The long-term prediction by the Univ. of Texas researchers and the short-term prediction by the crank were wrongly taken as identical by the public. Considerable public disquiet and economic loss occurred. No significant earthquake occurred on or near the date of the prediction, but the false prediction caused considerable economic damage in the city of Pinotepa. According to *Garza and Lomnitz* (1979):

> "The Mayor was indignant about the prediction, which he claimed had caused more damage to Pinotepa than the 1968 earthquake. Though he used strong and profane language, he also claimed that the reports of widespread panic were exaggerated. [...] A stroll through the town revealed that perhaps 20 per cent of the homes were shuttered, indicating that the residents were out of town".

Ironically, as pointed out by *Garza and Lomnitz* (1979), *Whiteside and Habermann* (1989) and *Lomnitz* (1994, pp. 122-127), the anomaly reported by *Ohtake et al.* (1977) was an artifact of man-made changes in the seismicity catalogue, rather than a real physical phenomenon.

Peru, 1981: (See *Kerr*, 1981a, b; *Richman*, 1981a, b; *Giesecke*, 1983; *Echevarría et al.*, 1986; *Olson et al.*, 1989.) Brian Brady, a researcher at the U.S. Bureau of Mines, said that events with $m_w = 9.8$ and $m_w = 8.8$ would occur at the subduction zone off the coast of Peru in August, 1981 and May 1982, respectively. A foreshock with $m_w = 7.5$–8 was forecast for June 1981. The scientific rationale for these predictions is given by *Brady* (1976, pp. 1068-1069). NEPEC issued a statement (reproduced in full by *Richman*, 1981a) rejecting the prediction:

> "The council regrets that an earthquake prediction based on such speculative and vague evidence has received widespread credence outside the scientific community. We recommend that the prediction not be given serious consideration by the government of Peru".

When the 'foreshock' failed to occur in June 1981, Brady retracted his prediction. However, an official of the USGS had to travel to Lima in June 1981 to reassure the Peruvian people. *Echevarría et al.* (1986) estimated the economic damages to be approximately US$ 50 million, primarily from loss of tourism, and found that the poorer segments of society bore a disproportionate share of the costs.

Missouri, USA, 1990: (See *Kerr*, 1990, 1991; *Gori*, 1993; *Dalrymple*, 1991, *Spence et al.*, 1993; *Shipman et al.*, 1993; *Showalter et al.*, 1993; *Farley et al.*, 1993.) Iben Browning, a business consultant with a Ph.D. in biology but no background in Earth science, predicted that a $m = 6.5$–7.5 earthquake would strike New Madrid, Missouri (250 km SE of St. Louis) between December 1-5, 1990. Browning's promotional efforts generated a flood of publicity. Due to the panic, public schools were closed on December 3. Browning profited by selling explanatory video tapes at US$ 99/copy. David Stewart (see above), then Director of the Center for Earthquake Studies at Southeast Missouri State University, said that Browning's prediction was worthy of serious consideration. The news media treated this as an authoritative endorsement. Browning claimed to have made a successful prediction of the October 18, 1989 Loma Prieta earthquake. His claim was accepted without investigation by the press, but a transcript of the speech in which Browning claimed he made the prediction (*Spence et al.* 1993, pp. 64) shows that there was no mention of an earthquake occurring in the San Francisco area, or even California. An Ad Hoc Working Group showed that Browning's prediction was without merit (*Spence et al.*, 1993, pp. 45–66). However, despite the convincing arguments of the Ad Hoc Working Group, the print and electronic media continued to cover the Browning story intensively until the predicted earthquake failed to occur in early December 1990. *Spence et al.* (1993) present an extensive collection of newspaper stories that graphically demonstrate the extent of the media frenzy. Also notable are the photographs of the remote transmitting units for live TV coverage on the days of the Browning prediction window in early December. It is remarkable that a bogus prediction by an imposter generated such a media circus.

The Browning prediction aroused great frustration among the leaders of the Earth science community in the U.S. The President of the American Geophysical Union (AGU), (*Dalrymple*, 1991) said in an editorial in AGU's newsletter:

> "December 3, was to have been the date of a major earthquake centered in the New Madrid, [Missouri], seismic zone, or so predicted an individual who believed that a particular alignment of the Sun and Moon with the Earth would trigger movement along the fault zone. Media around the country ran the story and the prediction was taken seriously. The National Guard was placed on alert in some states. Many families packed their belongings and left the region. Local governments canceled vacations for employees and stocked emergency supplies. Schools and businesses were closed to minimize the danger to a frightened public.

"The fact that there was no scientific basis to the prediction got lost in the news reports. Also lost was the historical perspective — a rich history of bogus earthquake predictions based on tidal effects, missing pets, psychic visions, and other zany ideas going back several decades, whose success rate is no better than, as one prominent geophysicist put it, 'throwing darts at a calendar' ".

"A panel of the nation's leading seismologists reviewed the 'methodology' that led to the prediction and concluded that it should not serve as the basis for public policy. The scientists' view was reported subsequently as the counterweight to the bogus prediction, as if the two views were of comparable intellectual standing — just another controversy within the scientific community. Of course there was no major quake in the Midwest. The San Francisco area, the Middle East, and Tokyo, all of which were fingered by the same individual for a December 3 earthquake, also escaped disaster".

Greece, 1991: This controversy involved a prediction by P. Varotsos (a professor of physics at the University of Athens) and his co-workers (commonly known as the 'VAN' group) based on geo-electrical data. For information on VAN see *Geller* (1997b, pp. 436 and the works cited therein). Recent work by *Pham et al.* (2002) suggests it is highly likely that the geo-electrical signals claimed by VAN as earthquake precursors are actually artifacts. The 1991 prediction by the VAN group is outlined by *Drakopoulos and Stavrakakis* (1996), with a rebuttal by *Varotsos et al.*, 1996. A prediction of an $m = 5.5$ to 6.0 event in Thessaloniki was announced on TV and in newspapers by Varotsos and his coworkers. The prediction, which was ultimately a failure, caused considerable unrest.

Central California, 1995: C. G. Sammis predicted an $m = 6.0$–6.5 earthquake in central California during the widow from February 1–July 9, 1995. The prediction (*Los Angeles Times* newspaper, April 29, 1995, "USC Geology Chairman Forecasts Quake") failed (follow-up story, 'Quake Prediction Error Can't Shake 2 Scientists' Core Beliefs', 3 August 1995). The basis for Sammis's prediction is the proposed method of 'time-to failure analysis' (e.g., *Bufe and Varnes*, 1993; *Sornette and Sammis*, 1995). However, a systematic test of time-to-failure models (*Gross and Rundle*, 1998) showed that they were less effective in predicting earthquakes that the pure chance Poisson model. Given that it was based on an as yet unproven (and apparently ineffective) method, Sammis should probably not have released his prediction to the media.

Tokyo, 1995: (See Geller 1996c) A 'former JMA employee' (whose routine duties had no connection with seismology) announced in the spring of 1995 that an $m = 8.0$ quake would hit Tokyo at 12:37 PM on 9 September 1995. The nominal basis was planetary alignment. No panic resulted, but this groundless prediction was extensively publicized. The JMA received many calls from worried citizens.

Athens, 1999: A magnitude 6.0 earthquake (USGS/NEIC) struck Athens on September 7, 1999, causing extensive damage and 140 deaths. After this earthquake, the VAN group (see above) claimed to have observed geo-electrical signals that were precursors of another, even larger, earthquake. VAN's prediction was sent to the government, but also was published in the media. A meeting of a committee (roughly corresponding to the U.S. NEPEC) of the Greek government's Anti-Seismic Planning and Protection Organisation (OASP) was held on September 16, 1999 to evaluate the prediction. Varotsos did not attend, but one of his assistants did. Dimitris Papanikolaou, president of the official earthquake risk assessment committee said after the meeting that the information submitted by VAN was "insufficient for scientific review and of no practical use". Several days later Ota Kulhanek, professor at Uppsala University and chairman of the Swedish National Committee for Geophysics, arrived in Greece and made a statement in Varotsos' favor on Mega television. Matters escalated further as highly specific rumors that VAN had predicted an earthquake of magnitude 6 over the first weekend in October were disseminated by the media, causing widespread unrest and a sharp increase in tranquilizer usage. The predicted earthquake did not occur. The above information is based on the newspaper's web site: http://www.athensnews.gr/

Other Prediction Scares: (1) Unconfirmed information on the internet (but duplicated at two web sites) said that on July 7, 1998 thousands of residents of Tehran had fled the city in response to a prediction by a psychic that a major earthquake would destroy Tehran within a few days. (2) The *Japan Times* newspaper reported on October 1, 1998 (based on an AP story) that a rumor that a large earthquake would strike the city of Baguio, Philippines, at 11AM caused panic among residents. (3) A large earthquake struck Gurjat, India, on January 26, 2001. The USGS/NEIC website says there were 18,602 confirmed deaths. According to several web sites, several days after the earthquake an astrologer's prediction that another large earthquake would strike between February 3 and 15, most likely on February 3, caused panic among many residents.

8.5.3   Common features

The above episodes, taken collectively, show certain common features.

1. In many cases predictions caused social unrest, severe in some cases, but they made no contribution to reducing seismic risk.

2. The provenance (credentialed scientist vs. amateur) of a prediction seems to be much less important than the nominal degree of precision, particularly that of the time window, in governing the degree of media and public attention.

3. Regardless of whether the predictor is a credentialed scientist or an amateur, the media and public seem to pay greater heed if a prediction is endorsed by a credentialed scientist.

4. Efforts by professional scientists to debunk invalid predictions have not been notably successful.

5. Public and media sensitivity to predictions is much greater in the immediate aftermath of a damaging earthquake.

6. The above examples of false predictions by credentialed scientists occurred because random fluctuations or artifacts were misinterpreted as earthquake precursors.

7. Some predictions were reviewed by official or quasi-official bodies, but only after they had already been publicized. No systematic procedures are in place for vetting predictions before they are publicized.

8. Codes of ethics for earthquake predictors have been promulgated, but they are not legally binding, and they were generally not followed in the above episodes.

8.5.4 Countermeasures

As noted above, many of the individual instances of panics or near-panics caused by bogus predictions have been studied by social scientists, but the fact that many such episodes have occurred in many countries does not seem to be widely recognized. However, as shown above, this is indeed a significant problem. Unfortunately, even when highly reputable scientists debunk bogus predictions, the media tend to treat the debunkers and the predictor as having essentially equal credibility (especially if there is at least one other credentialed scientist backing the prediction). In such cases the controversy drags on, or may even escalate, until the end of the prediction window.

Perhaps the only way to stop such panics and near-panics would be to have an authoritative government body which could review and debunk bogus predictions. However, such a body must be established in advance, and must be able to move quickly, in order to damp out the panic before it takes on a life of its own. It is important for the prediction evaluation/debunking body to define its procedures and standards in advance. Unfortunately, to our knowledge, this has not been done anywhere.

Problems can arise if the members of an evaluation panel are themselves active in earthquake prediction research. In such a case the panelists are in a position where they may want to criticize a specific bogus prediction without criticizing prediction research in general. This may tend to lead to rather ambiguous statements that are also hard to explain to the public and to the media.

8.6   Discussion and conclusion

As discussed in the resolution adopted by the ARW, the goal of applied earthquake science should be to reduce seismic risk. If efficient and reliable earthquake prediction were possible, this would make a significant contribution, but unfortunately it is clearly an unrealizable goal for the foreseeable future, and may well be inherently impossible. That being the case, it is clear that earthquake prediction (in the 'journalistic' sense of the term), has been greatly overemphasized over the past thirty or forty years. Specific efforts to make operational predictions (Parkfield in the U.S. or Tokai in Japan) serve to unrealistically build up public hopes, and in the end detract from practical measures that can be taken to reduce seismic risk. The organizations and underlying laws and regulations regarding short-term deterministic prediction should probably be abolished. On the other hand, highly specific but completely bogus predictions have caused social unrest in many countries, and methods for authoritatively debunking them should be set up in advance.

Because of the problematic nature of short-term prediction, both scientific and social, it is increasingly obvious that public policy must be based on long-term probabilistic prediction as the driver of mitigation actions.

Seismologists and other earthquake scientists justifiably seek funding for a variety of valuable research and also for the operation of observational networks. At times it can unfortunately be quite difficult to persuade government authorities of the necessity for such projects, and it is perhaps tempting to claim that these project should be funded because they will lead to 'progress towards predicting earthquakes'. However, the long term loss of credibility in the eyes of the public and government authorities suggests strongly that such misleading tactics should not be used to secure public support for earthquake science.

8.7 References

Aceves, R. L. and Park, S. K., 1997. Cannot earthquakes be predicted?, *Science*, **278**, 488.

Allen, C. R. 1976. Responsibilities in earthquake prediction, *Bull. Seism. Soc. Am.*, **66**, 2069–2074.

Allen, C. R. 1982. Earthquake prediction–1982 overview, *Bull. Seism. Soc. Am.*, **72**, S331–S335.

Anderson, D. L., 1974. The Jupiter effect (book review), *Am. Scientist*, **62**, 721–722.

Anonymous, 1973. Using prediction properly, *Nature*, **245**, 174.

Anonymous, 1976. Earthquake prediction, *Eos, Trans. Am. Geophys. Un.*, **57**, 122–123.

Anonymous, 1977a. Prediction: the negative aspects, *Nature*, **265**, 4.

Anonymous, 1977b. After the prediction, what? *Eos, Trans. Am. Geophys. Un.*, **58**, 1085–1086.

Anonymous, 1996. Money doesn't buy quake prediction faith, *Science*, **271**, 915.

Bakun, W. H., 1987. Future earthquakes, *Rev. Geophys.*, **25**, 1135–1138.

Bakun, W. H., and Lindh, A. G., 1985. The Parkfield, California, earthquake prediction experiment, *Science*, **229**, 619–624.

Bolt, B. A., 1975. Book review of *The Jupiter Effect*, *Physics Today*, **28**(4), 74–75.

Brady, B. T., 1976. Theory of earthquakes —IV. General implications for earthquake prediction, *Pure Appl. Geophys.*, **114**, 1031–1082.

Brune, J. N., 1974. Current status of understanding quasi-permanent fields associated with earthquakes, *Eos, Trans. Am. Geophys. Un.*, **55**, 820–827.

Brune, J. N., 1979. Implications of earthquake triggering and rupture propagation for earthquake prediction based on premonitory phenomena, *J. Geophys. Res.*, **84**, 2195–2198.

Bufe, C. G., and Varnes, D. J., 1993. Predictive modeling of the seismic cycle of the greater San Francisco Bay region, *J. Geophys. Res.*, **98**, 9871–9833.

Burton, P. W., 1996. Dicing with earthquakes, *Geophys. Res. Lett.*, **23**, 1379–1382.

California Earthquake Prediction Evaluation Council, 1977. Earthquake prediction evaluation guidelines, *California Geology*, 158–160 (reprinted 1983, *Bull. Seism. Soc. Am.*, **73**, 1955–1956).

California Office of Emergency Services, 1990. California Short-Term Earthquake Prediction Response Plan, Sacramento, October 1990.

Carlson, J. W., 1976. Earthquake forecasting: an opportunity, in *Earthquake Prediction – Opportunity to Avert Disaster*, U.S. Geological Survey, Geological Survey Circular 729, 2–3.

Castle, R. O., Church, J. P., and Elliott, M. R., 1976. Aseismic uplift in southern California, *Science*, **192**, 251–253.

Central U.S. Earthquake Consortium, 1994. Charter, CUSEC Earthquake Prediction Evaluation Council, Organization of CUSEC State Geologists, Memphis, Tennessee, February 28, 1994.

Cochrane, H. C., 1984. An economic evaluation of earthquake prediction under current and possible future conditions, in *Earthquake Prediction, Proceedings of the International Symposium on Earthquake Prediction*, Unesco Press, Paris; Terra Scientific Publishing Company, Tokyo, 713–735.

Committee on Socioeconomic Effects of Earthquake Predictions, National Research Council, 1978. *A Program of Studies on the Socioeconomic Effects of Earthquake Predictions*, National Academy of Sciences, Washington, D.C.

Council of Europe, 1991. European code of ethics concerning earthquake prediction, promulgated at Intl. Conf. on Earthq. Prediction: State of the Art, Strasbourg, France, Oct. 1991.

Dalrymple, G. B., 1991. Good press for bad science, *Eos, Trans. Am. Geophys. Un.*, **72**, 41.

Dambara, T., 1981. Geodesy and earthquake prediction, in *Current Research in Earthquake Prediction I*, Rikitake, T. (ed.), pp. 167–220, D. Reidel, Dordrecht/CAPJ, Tokyo.

Davies, D., 1974. Quake prediction no bonanza, *Nature* **249**, 7.

Davis, J. F., and Somerville P., 1982. Comparison of earthquake prediction approaches in the Tokai area of Japan and in California, *Bull. Seism. Soc. Am.*, **72**, S367–S392.

Drakopoulos, J., and Stavrakakis, G.N., 1996. A false alarm based on electrical activity recorded at a VAN-station in northern Greece in December 1990, *Geophys. Res. Lett.*, **23**, 1355–1358.

Ebel, J., 2003. Seismologists must begin forecasting earthquakes, *Seism. Res. Lett.*, **74**, 3–5.

Echevarría, J. A., Norton, K. A., and Norton, R. D., 1986. The socio-economic consequences of earthquake prediction: a case study in Peru, *Earthq. Predict. Res.*, **4**, 175–193.

Ellson, R. W., Milliman, J. W. and Roberts, R. B., 1984. Measuring the regional economic effects of earthquakes and earthquake predictions, *Journal of Regional Science*, **24**, 559–579.

Evans, R. 1997. Assessment of schemes for earthquake prediction, *Geophys. J. Int.*, **131**, 413–420.

Farley, J. E., Barlow, H. D. Finkelstein, M. S., and Riley, L., 1993. Earthquake hysteria, before and after: a survey and follow-up on public response to the Browning forecast, *International Journal of Mass Emergencies and Disasters*, **11**, 305–321.

FEMA (Federal Emergency Management Agency), 1980. An Assessment of the Consequences and Preparations for a Catastrophic California Earthquake: Findings and Actions Taken, November, 1980.

Finn, R., 1992. Rumblings grow about Parkfield in wake of first earthquake prediction, *Nature*, **359**, 761.

Garza, T., and Lomnitz, C., 1979. The Oaxaca gap: a case history, *Pure Appl. Geophys.*, **117**, 1187–1194.

Geller, R. J., 1991a. Shake-up for earthquake prediction, *Nature*, **352**, 275–276.

Geller, R. J., 1991b. Unpredictable earthquakes, *Nature*, **353**, 612.

Geller, R. J., 1996. One year after the Kobe earthquake (in Japanese), *Shincho 45*, Feb. 1996 issue, 22–31.

Geller, R. J., 1997a. Predictable publicity, *Astron. & Geophys.*, **38**(1), 16–18, (reprinted 1997, *Seism. Res. Lett.*, **68**, 477–480).

Geller, R. J., 1997b. Earthquake prediction: a critical review, *Geophys. J. Int.*, **131**, 425–450.

Geller, R. J., Jackson, D. D., Kagan, Y. Y., and Mulargia, F., 1997a. Earthquakes cannot be predicted, *Science*, **275**, 1616–1617.

Geller, R. J., Jackson, D. D., Kagan, Y. Y., and Mulargia, F., 1997b. Cannot earthquakes be predicted?, *Science*, **278**, 488–490.

Geographical Survey Institute, 1979. *Summary of the First 10 Years of the Coordinating Committee for Earthquake Prediction* (in Japanese), Tsukuba.

Giesecke, A. A., 1983. Case history of the Peru prediction for 1980-81, in *Proceedings of the Seminar on Earthquake Prediction Case Histories*, Office of the United Nations Disaster Relief Coordinator, Geneva, 51–75.

Gori, P. L., 1993. The social dynamics of a false earthquake prediction and the response by the public sector, *Bull. Seism. Soc. Am.*, **83**, 963–980.

Governor's Board of Inquiry, 1990. *Competing Against Time*, North Highlands, California, State of California, Office of Planning and Research. (A summary of this report was published by Thiel, C. C. Jr., G. W. Housner, L. T. Tobin, 1991, State response to the Loma Prieta earthquake: competing against time, *Bull. Seism. Soc. Am.*, **81**, 2127–2142.)

Gribben, J., 1971. Relation of sunspot and earthquake activity, *Science*, **173**, 558.

Gribben, J. R., and Plagemann, S. H., 1974. *The Jupiter Effect*, MacMillan, London.

Gross, S., and Rundle, J., 1998. A systematic test of time-to-failure analysis, *Geophys. J. Int.*, **133**, 57–64.

Gusev, A., 1998. Earthquake precursors, banished forever?, *Eos, Trans. Am. Geophys. Un.*, **79**, 71–72.

Haas, J. E., and Mileti, D. S., 1976. Socioeconomic Impact of Earthquake Prediction on Government, Business, and Community, Institute of Behavioral Science, University of Colorado, Boulder, 39.

Haas, J. E., and Mileti, D. S. 1977a. Socioeconomic and political consequences of earthquake prediction, *J. Phys. Earth*, **25**, S283–S293.

Haas, J. E., and Mileti, D. S., 1977b. Socioeconomic impact of earthquake prediction on government, business, and community, *California Geology*, **30**, 147–157.

Hager, B. H., (chair), Cornell, C. A., Medigovich, W. M., Mogi, K., Smith, R. M., Tobin, L. T., Stock, J., and Weldon, R., 1994. Earthquake Research at Parkfield, California, for 1993 and Beyond– Report of the NEPEC [National Earthquake Prediction Evaluation Council] Working Group, U.S. Geological Survey Circular 1116.

Hagiwara, T., and Rikitake, T., 1967. Japanese program on earthquake prediction, *Science*, **157**, 761–768.

Hamada, K., 1991. Unpredictable earthquakes, *Nature*, **353**, 611–612.

Hamilton, R. M., 1974. Earthquake prediction and public reaction, *Eos, Trans. Am. Geophys. Un.*, **55**, 739–742.

Hamilton, R. M., 1976. The status of earthquake prediction, in *Earthquake Prediction – Opportunity to Avert Disaster*, U.S. Geological Survey, *Geological Survey Circular*, **729**, 6–9.

Hammond, A. L., 1976. Earthquakes: an evacuation in China, a warning in California, *Science*, **192**, 538–539.

Hanks, T. C., and Krawinkler, H., 1991. The 1989 Loma Prieta earthquake and its effects: introduction to the special issue, *Bull. Seism. Soc. Am.*, **81**, 1415–1423.

Hirose, H., 1985. Earthquake prediction in Japan and the United States, *International Journal of Mass Emergencies and Disasters*, **3**, 51–66.

Holdahl, S. R., 1982. Recomputation of vertical crustal motions near Palmdale, California, 1959–1975, *J. Geophys. Res.*, **87**, 9374–9388.

Holloway, R. H. F., 1984. Earthquake prediction–the role of institutions–a New Zealand view, in Earthquake Prediction, *Proceedings of the International Symposium on Earthquake Prediction*, Unesco Press, Paris; Terra Scientific Publishing Company, Tokyo, 903–912.

Huffman, J. L., 1984. Government liability for the external costs of earthquake prediction, in Earthquake Prediction, *Proceedings of the International Symposium on Earthquake Prediction*, Unesco Press, Paris; Terra Scientific Publishing Company, Tokyo, 857–867.

Ishibashi, K., 1977. Re-examination of a great earthquake expected in the Tokai district, central Japan—possibility of the 'Suruga Bay earthquake" (in Japanese), *Rep. Coord. Comm. Earthq. Pred.*, **17**, 126–132.

Ishibashi, K., 1981. Specification of a soon-to-occur seismic faulting in the Tokai District, Central Japan, based upon seismotectonics, in *Earthquake Prediction: an International Review* Simpson, D. W. and Richards, P. G. (eds.), *Ewing Monograph Series*, **4**, 297–332, Am. Geophys. Un., Washington.

IUGG (1984). IASPEI/UNESCO Code of Practice for Earthquake Prediction, IUGG Chronicle No. 165, pp. 26–29.

Jones, L. M., 1996. Earthquake prediction: the interaction of public policy and science, *Proc. Natl. Acad. Sci. USA*, **93**, 3721–3725.

Jordan, T. H., 1997. Is the study of earthquakes a basic science?, *Seism. Res. Lett.*, **68**, 259–261.

Kagan, Y. Y., 1997a. Are earthquakes predictable?, *Geophys. J. Int.*, **131**, 505–525.

Kagan, Y. Y., 1997b. Statistical aspects of Parkfield earthquake sequence and Parkfield prediction experiment, *Tectonophysics*, **270**, 207–219.

Kagan, Y. Y., and Jackson, D. D., 1996. Statistical tests of VAN earthquake predictions: comments and refections, *Geophys. Res. Lett.*, **23**, 1433–1436.

Kanamori, H., 1970. Recent developments in earthquake prediction research in Japan, *Tectonophysics*, **9**, 291–300.

Kaula, W. M., 1974. The next California earthquake (book review), *Science*, **186**, 728–729.

Kerr, R. A., 1978. Earthquakes: prediction proving elusive, *Science*, **200**, 419–421.

Kerr, R. A., 1979. Prospects for earthquake prediction wane, *Science*, **206**, 542–545.

Kerr, R. A., 1981a. Prediction of huge Peruvian quakes quashed, *Science*, **211**, 808–809.

Kerr, R. A., 1981b. Earthquake prediction retracted, *Science*, **213**, 527.

Kerr, R. A., 1985. Earthquake forecast endorsed, *Science*, **228**, 311.

Kerr, R. A., 1990. Earthquake—or earthquack, *Science*, **250**, 511.

Kerr, R. A., 1991. The lessons of Dr. Browning, *Science*, **253**, 622–623.

Kerr, R. A., 1992. Seismologists issue a no-win earthquake warning, *Science*, **258**, 742–743.

Kisslinger, C., 1974. Earthquake prediction, *Physics Today*, **27**(3), 36–42.

Lambright, W. H., 1985. Policymaking for earthquake prediction in America and Japan, in Proceedings of the Third U.S.-Japan Science Policy Seminar, February 19-22, 1984, Bartocha, B., and Okamura, S. (eds.), 164–179.

Langbein, J., 1992. The October 1992 Parkfield, California, earthquake prediction, *Earthq. Volc.*, **23**, 160–169.

Langbein, J., 1993. Parkfield: first short-term warning, *Eos, Trans. Am. Geophys. Un.*, **74**, 152–153.

Lindh, A. G., 1990. Earthquake prediction comes of age, *Technology Review*, **93**(February/March issue), 43–51.

Lomnitz, C., 1983. Oaxaca, Mexico, 1978, predictions, in *Proceedings of the Seminar on Earthquake Prediction Case Histories*, Office of the United Nations Disaster Relief Coordinator, Geneva, 29–32.

Lomnitz, C., 1994. *Fundamentals of Earthquake Prediction*, Wiley, New York.

Macelwane, J. B., 1946. Forecasting earthquakes, *Bull. Seism. Soc. Am.*, **36**, 1–4 (reprinted in *Geophys. J. Int.*, **131**, 421–422, 1997).

Marshall, J., 1978. The political and legal effects (in Proceedings of the Seminar on the Social and Economic Effects of Earthquake Prediction), *Bulletin of the New Zealand National Society for Earthquake Engineering*, **11**, 5–7.

Masood, E., 1995. Court charges open split in Greek earthquake experts, *Nature*, **377**, 375.

Matthews, R. A. J., 1997. Decision-theoretic limits on earthquake prediction, *Geophys. J. Int.*, **131**, 526–529.

McKelvey, V. E., 1976. Foreword, in *Earthquake Prediction – Opportunity to Avert Disaster*, U.S. Geological Survey, Geological Survey Circular 729, iii–iv.

McKelvey, V. E., 1977. Warning and preparedness for geologic-related hazards, *Federal Register*, **42**, 19,292–19,296.

Michael, A. J., and Jones, L. M., 1998. Seismicity alert probabilities at Parkfield, California, revisited, *Bull. Seism. Soc. Am.*, **88**, 117–130.

Mileti, D. S., and Darlington, J. D., 1997. The role of searching in shaping reactions to earthquake risk information, *Social Problems*, **44**, 89–103.

Mileti, D. S., and Fitzpatrick, C., 1993. *The Great Earthquake Experiment: Risk Communication and Public Action*, Westview Press, Boulder, Colorado.

Mileti, D. S., Hutton, J. R., and Sorensen, J. H., 1981. Earthquake Prediction Response and Options for Public Policy, Program on Technology, Environment and Man, Monograph #31, Institute of Behavioral Science, University of Colorado, Boulder.

Mileti, D. S., Hutton, J. R., and Sorensen, J. H., 1984. Social factors affecting the response of groups to earthquake prediction: implications for public policy, in *Earthquake Prediction, Proceedings of the International Symposium on Earthquake Prediction*, Unesco Press, Paris; Terra Scientific Publishing Company, Tokyo, 649–657.

Mogi, K., 1970. Recent horizontal deformation of the Earth's crust and tectonic activity in Japan (1), *Bull. Earthq. Res. Inst. Tokyo Univ.*, **48**, 413–430.

Mogi, K., 1985. *Earthquake Prediction*, Academic, Orlando.

Mogi, K., 1995. Earthquake prediction research in Japan, *J. Phys. Earth*, **43**, 533–561.

Morton, D., 1980. Actions to be taken in response to earthquake predictions, (unpublished) Natural Hazards Research and Applications Information Center, University of Colorado, Boulder.

Mulargia, F., 1997. Retrospective validation of time association, *Geophys. J. Int.*, **131**, 500–504.

National Earthquake Strategy Working Group, 1995. Strategy for National Earthquake Loss Reduction, prepared for National Science and Technology Council, and White House Office of Science and Technology Policy, October 1995.

National Science Foundation, and U.S. Geological Survey, 1976. Earthquake Prediction and Hazard Mitigation Options for USGS and NSF Programs, U.S. Government Printing Office, Washington, D.C., September 15, 1976.

Nishida, N., 1989. Public opinion regarding earthquake prediction (in Japanese), *Seism. Soc. Jpn. Newsletter*, **1**(1), 58–59.

Normile, D., 1997. Earthquake prediction: report slams Japanese program, *Science*, **275**, 1870.

Office of Technology Assessment, 1995. Reducing Earthquake Losses, U.S. Congress, Office of Technology Assessment, OTA-ETI-623, U.S. Government Printing Office, Washington, D.C.

Ohta, H., and Abe, K., 1977. Responses to earthquake prediction in Kawasaki City, Japan in 1974, *J. Phys. Earth*, **25**, S273–S282.

Ohtake, M., Matumoto, T., and Latham, G. V., 1977. Seismicity gap near Oaxaca, southern Mexico as a probable precursor to a large earthquake, *Pure Appl. Geophys.*, **115**, 375–385.

Olson, R. S., Podesta, B., and Nigg, J. M., 1989. *The Politics of Earthquake Prediction*, Princeton University Press, Princeton, New Jersey.

Page, R. A., Boore, D. M., Bucknam, R. C., and Thatcher, W. R., 1992. Goals, Opportunities, and Priorities for the USGS Earthquake Hazards Reduction Program, U.S. Geological Survey Circular 1079.

Page, R. A., Mori, J., Roeloffs, E. A., and Schweig, E. S., 1997. Earthquake Hazards Program Five-Year Plan 1998-2002, U. S. Geological Survey Open-File Report 98-143. http://earthquake.usgs.gov/docs/5yrplan.html

Panel on the Public Policy Implications of Earthquake Prediction, 1975. *Earthquake Prediction and Public Policy*, National Academy of Sciences, Washington, D.C.

Paté, M.-E., 1978. *Public Policy in Earthquake Effects Mitigation: Earthquake Engineering and Earthquake Prediction*, Doctoral Thesis, Tech. Rept. No. 30, The John A. Blume Earthquake Engineering Center, Department of Civil Engineering, Stanford University, Stanford, CA.

Paté, M.-E., and Shah, H. C., 1979. Public policy issues: earthquake prediction, *Bull. Seism. Soc. Am.*, **69**, 1533–1547.

Pham, V. N., Boyer, D., Chouliaras, G., Savvaidis, A., Stavrakakis. G., and Le Mou'el, J. J., 2002. Sources of anomalous transient electric signals (ATESs) in the ULF band in the Lamia region (central Greece): electrochemical mechanisms for their generation, *Phys. Earth Planet. Int.*, **130**, 209-233.

Press, F., 1975. Earthquake prediction, *Scientifi c American*, **232**(5), 14–23.

Press, F., 1976. Haicheng and Los Angeles: a tale of two cities, *Eos, Trans. Am. Geophys. Un.*, **57**, 435–436.

Press, F., and Brace, W. F., 1966. Earthquake prediction, *Science*, **152**, 1575–1584.

Richter, C. F., 1958. *Elementary Seismology*, W. H. Freeman, San Francisco.

Richter, C. F., 1977. Acceptance of the Medal of the Seismological Society of America, *Bull. Seism. Soc. Am.*, **67**, 1244–1247.

Rikitake, T., 1972. Problems of predicting earthquakes, *Nature*, **240**, 202–204.

Rikitake, T., 1979. The large-scale earthquake countermeasures act and the earthquake prediction council in Japan, *Eos, Trans. Am. Geophys. Un.*, **60**, 553–555.

Roberts, J. L., 1984. The political and administrative consequences of prediction, in *Earthquake Prediction, Proceedings of the International Symposium on Earthquake Prediction*, Unesco Press, Paris; Terra Scientific Publishing Company, Tokyo, 885–895.

Roelffs, E., and Langbein, J., 1994. The earthquake prediction experiment at Parkfield, California, *Rev. Geophys.*, **32**, 315–336.

Savage, J. C., 1993. The Parkfield prediction fallacy, *Bull. Seism. Soc. Amer.*, **83**, 1–6.

Scholz, C. H., Sykes, L. R., and Aggarwal, Y. P., 1973. Earthquake prediction: a physical basis, *Science*, **181**, 803–810.

Schulze, W. D., Brookshire, D. S., Hageman, R. K., and Ben-David, S., 1990. Should we try to predict the next great U.S. earthquake?, *Journal of Environmental Economics and Management*, **18**, 247–262.

Shapley, D., 1976. Earthquakes: Los Angeles prediction suggests faults in federal policy, *Science*, **192**, 535–537.

Shearer, C. F., 1985a. Minutes of the National Earthquake Prediction Evaluation Council, November 16-17, 1984, USGS Open File Report 85–201.

Shearer, C. F., 1985b. Minutes of the National Earthquake Prediction Evaluation Council, March 29-30, 1985, USGS Open File Report 85–507.

Shipman, M., Fowler, G., and Shain, R., 1993. Whose fault was it? An analysis of newspaper coverage of Iben Browning's New Madrid Fault earthquake prediction, *International Journal of Mass Emergencies and Disasters*, **11**, 379–389.

Showalter, P. S., 1993. Prognostication of doom: an earthquake prediction's effect on four small communities, *International Journal of Mass Emergencies and Disasters*, **11**, 279–292.

Silver, P., 1998. Why is earthquake prediction so difficult?, *Seism. Res. Lett.*, **69**, 111–113.

Snieder, R., and van Eck, T., 1997. Earthquake prediction: a political problem?, *Geol. Rundsch.*, **86**, 446–463.

Sornette, D., and Sammis, C. G., 1995. Complex critical exponents from re-normalization group theory of earthquakes: implications for earthquake predictions, *J. Phys. I France*, **5**, 6-7–619.

Spence, W., Herrmann, R. B., Johnston, A. C., and Reagor, G., 1993. Responses to Iben Browning's Prediction of a 1990 New Madrid, Missouri, Earthquake, U.S. Geological Survey Circular 1083, U.S. Government Printing Office, Washington, D.C.

Stallings, R. A., 1995. *Promoting Risk: Constructing the Earthquake Threat*, Aldine De Gruyter, New York.

Steinbrugge, K. V., 1976. Earthquake prediction is the beginning of problems – not the answer, in *Earthquake Prediction – Opportunity to Avert Disaster*, U.S. Geological Survey, Geological Survey Circular, **729**, 20–21.

Stewart, D. M., 1975. Is a major earthquake imminent in North Carolina centered near Southport? (abstract), Geol. Soc. Am. Southeastern Sect. Mtg., April 9–12, 1975, Memphis, Tennessee, *Geol. Soc. Am. Abstracts with Programs*, **7**(4), 540.

Stiros, S. C., 1997. Costs and benefits of earthquake prediction studies in Greece, *Geophys. J. Int.*, **131**, 478–484.

Strange, W. E., 1981. The impact of refraction correction on leveling interpretations in southern California, *J. Geophys. Res.*, **86**, 2809–2824.

Swinbanks, D., 1993. Japan disappoints seekers of foresight, *Nature*, **366**, 4.

Swinbanks, D., 1997. Quake panel admits prediction is 'difficult', *Nature*, **388**, 4.

Thiel, C. C., 1976. Possible loss-reduction actions following an earthquake prediction, in *Earthquake Prediction – Opportunity to Avert Disaster*, U.S. Geological Survey, Geological Survey Circular, **729**, 13–16.

Tullis, T. E., 1994. Predicting earthquakes and the mechanics of fault slip, *Geotimes*, **39**(7), 19–22.

Turner, R. H., 1976a. Earthquake prediction and public policy: distillations from a National Academy of Sciences report, *Mass Emergencies*, **1**, 179–202.

Turner, R. H., 1976b. Social, economic, and political implications of earthquake prediction, in *Earthquake Prediction – Opportunity to Avert Disaster*, U.S. Geological Survey, Geological Survey Circular, **729**, 17–19.

Turner, R. H., 1982. Media in crisis: blowing hot and cold, *Bull. Seism. Soc. Am.*, **72**, S19–S28.

Turner, R. H., Nigg, J. M., and Paz, D. H., 1986. *Waiting for Disaster: Earthquake Watch in California*, University of California Press, Berkeley.

U.S. Department of the Interior, Office of the Secretary, 1979. Charter, National Earthquake Prediction Evaluation Council, August 9, 1979.

U.S. Department of the Interior, Office of the Secretary, 1994. Charter, National Earthquake Prediction Evaluation Council, April 11, 1994.

U.S. Geological Survey, 1976. *Earthquake Prediction – Opportunity to Avert Disaster*, Geological Survey Circular 729.

U.S. Senate, Committee on Commerce, Science, and Transportation, 1982. Earthquake Hazards Reduction Act of 1977 Authorization, Report to Accompany S. 2273, 97th Congress, Report No. 97-336, April 20, 1982.

Varotsos, P., Eftaxias, K., and Lazaridou, M., 1996. Reply to 'A false alarm based on electrical activity recorded at a VAN-station in northern Greece in December 1990," by J. Drakopoulos and G. Stavrakakis, *Geophys. Res. Lett.*, **23**, 1359–1362.

Wakita, H., Nakamura, Y., Notsu, K., Noguchi, M., and Asada, T., 1980. Radon anomaly: a possible precursor of the 1978 Izu-Oshima-Kinkai earthquake, *Science*, **207**, 882–883.

Wallace, R. E., Davis, J. F., and McNally, K. C., 1984. Terms for expressing earthquake potential, prediction, and probability, *Bull. Seism. Soc. Am.*, **74**, 1819–1825.

Weisbecker, L. W., Stoneman, W. C., Arnold, R. K., Halton, P. M., Ivy, S. C., Kautz, W. H., Kroll, C. A., Levy, S., Mickley, R. B., Miller, P. D., Rainey, C. T., VanZandt, J. E., and Ackerman, S. E., 1977. Earthquake Prediction in Society, Stanford Research Institute, Center for Resource and Environmental Systems Studies, February 1977.

Wesson, R. L., 1984. Procedures for the evaluation and communication of earthquake predictions within the United States, in *Earthquake Prediction, Proceedings of the International Symposium on Earthquake Prediction*, Unesco Press, Paris; Terra Scientific Publishing Company, Tokyo, 967–981.

Wesson, R. L., and Wallace, R. E., 1985. Predicting the next great earthquake in California, *Scientifi c American*, **252**(2), 35–43.

Whitcomb, J. H., 1976. Time-dependent V_P and V_P/V_S in an area of the Transverse Ranges of southern California (abstract), *Eos, Trans. Am. Geophys. Un.*, **57**, 288.

Whiteside, L., and Haberman, R. E., 1989. The seismic quiescence prior to the 1978 Oaxaca, Mexico, earthquake IS NOT a precursor to the earthquake (abstract), *Eos, Trans. Am. Geophys. Un.*, **70**, 1232.

Wood, H. O., and Gutenberg, B., 1935. Earthquake prediction, *Science*, **82**, 219–220.

Working Group on California Earthquake Probabilities, 1990. Probabilities of Large Earthquakes in the San Francisco Bay Region, California, U.S. Geological Survey Circular 1053.

Wuethrich, B., 1993. Waiting for Parkfield to quake, *Eos, Trans. Am. Geophys. Un.*, **74**, 553–554.

Wyss, M., 1997. Cannot earthquakes be predicted?, *Science*, **278**, 487–488.

Acknowledgments

The Editors wish to thank NATO for supporting the Advanced Research Workshop on 'State of Scientific Knowledge Regarding Earthquake Occurrence and Implications for Public Policy" held in Arbus, Sardinia, Italy, October 14–19, 2000, which sowed the seeds for this book.

The authors of section 4.2 and 5.1 would like to acknowledge their many colleagues both at the USGS and at other institutions for the work on which these sections are based. They also thank Michael Blanpied and Paul Reasenberg for helpful reviews.

For section 5.2, Yan Kagan, Yufang Rong and David Jackson appreciate partial support from the Southern California Earthquake Center (SCEC). SCEC is funded by NSF Cooperative Agreement EAR-0106924 and USGS Cooperative Agreement 02HQAG0008. Publication 731, SCEC.

ERS-1 and ERS-2 SAR data for section 6.1 were provided by ESA-ESRIN under contract no. 13557/99/I-DC. The continuous support of ESA and namely of L. Marelli, M. Doherty, B. Rosich, and F. M. Seifert is gratefully acknowledged. The authors are also grateful to the Southern California Integrated GPS Network (in particular M. Heflin of JPL) and its sponsors, the W.M. Keck Foundation, NASA, NSF, USGS, SCEC, for providing data.

The data for section 7.2 come from several sources including Istanbul Metropolitan Municipality, Governorate of Istanbul, State Statistical Institute and Turkish Telekom. The study has been partially funded by American Red Cross Field office in Turkey, Munich-Re and Bogazici University Research Fund.

Addresses of Principal Contributors

Dario Albarello
Dip, di Scienze della Terra
Universit`a di Siena
Via Laterina 8, 53100 Siena, Italia
albarello@unisi.it

Stavros A. Anagnostopoulos
Dept. of Civil Engineering
University of Patras
26500 Patras, Greece
saa@upatras.gr

Silvia Castellaro
Dip. di Fisica, Settore di Geofi sica
Universit`a di Bologna
V.le Berti Pichat 8, 40127 Bologna, Italia
silvia@ibogeo.df.unibo.it

Matteo Ciccotti
Dip. di Fisica, Settore di Geofi sica
Universit`a di Bologna
V.le Berti Pichat 8, 40127 Bologna, Italia
matteo@ibogeo.df.unibo.it

Mustafa Erdik
Kandilli Observatory and Earthquake Res. Inst.
Bogazici University
81220 Cengelkoy, Istanbul
T'urkiye
erdik@boun.edu.tr

Robert J. Geller
Dept. of Earth and Planetary Science
Graduate School of Science, Tokyo University
Hongo 7-3-1, Bunkyo-ku, Tokyo 113-0033, Japan
bob@eps.s.u-tokyo.ac.jp

David D. Jackson
Dept. Earth & Space Sciences, UCLA
595 Charles Young Drive East
3806 Geology Building, Box 951567
Los Angeles, CA 90095-1567, USA
djackson@ucla.edu

Janos Kert´esz
Institute of Physics
Budapest University of Technology
Mûegyetem rkp. 3-9. Budapest 8, H-1111 Hungary
kertesz@phy.bme.hu

Yan Y. Kagan
Dept. Earth & Space Sciences, UCLA
595 Charles Young Drive East
3806 Geology Building, Box 951567
Los Angeles, CA 90095-1567, USA
ykagan@ucla.edu

Ian Main
Dept. of Geology & Geophysics,
University of Edinburgh
West Mains Road, Edinburgh EH9 3JW, UK
ian.main@ed.ac.uk

Andrew Michael
U.S. Geological Survey
Mail Stop 977 - 345 Middlefi eld Rd, MS 977
Menlo Park, CA 94025, USA
michael@usgs.gov

Marco Mucciarelli
Dip. Strutture, Geotecnica, Geologia
Universitella Baslicata)
Campus Macchia Romana, 85100 Potenza, Italy
mucciarelli@unibas.it

Francesco Mulargia
Dip. di Fisica, Settore di Geofi sica
Universiti Bologna
V.le Berti Pichat 8, 40127 Bologna, Italia
francesco.mulargia@unibo.it

Robert B. Olshansky
Dept. of Urban and Regional Planning
University of Illinois at Urbana-Champaign
611 East Lorado Taft Drive, Champaign
IL 61820, USA
robo@uiuc.edu

Paolo Antonio Pirazzoli
CNRS, Laboratoire de G´eographie Physique
1 Place Aristide Briand
92190 Meudon-Bellevue, France
pirazzol@cnrs-bellevue.fr

Fabio Rocca
Dip. di Elettronica e Informazione
Politecnico di Milano
Ponzio 34, 20133 Milano
Italia
rocca@elet.polimi.it

Philip Stark
Dept. of Statistics
Code 3860, University of California Berkeley
Berkeley, CA 94720-3860 - USA
Stark@stat.berkeley.edu

Index

334